Studies of Sonoran Geology

Edited by
Efrén Pérez-Segura
and
César Jacques-Ayala

Instituto de Geología
Universidad Nacional Autonoma
Apartado Postal 1039
Hermosillo, Sonora
México

© 1991 The Geological Society of America, Inc.
All rights reserved.

All materials subject to this copyright and included
in this volume may be photocopied for the noncommercial
purpose of scientific or educational advancement.

Copyright is not claimed on any material prepared
by government employees within the scope of their
employment.

Published by The Geological Society of America, Inc.
3300 Penrose Place, P.O. Box 9140, Boulder, Colorado 80301

Printed in U.S.A.

GSA Books Science Editor Richard A. Hoppin

Library of Congress Cataloging-in-Publication Data

Studies of Sonoran geology / edited by Efrén Pérez-Segura, César
    Jacques-Ayala.
        p.   cm. — (Special paper / Geological Society of America ;
    254)
    Includes bibliographical references.
    ISBN 0-8137-2254-3
    1. Geology—Mexico—Sonora (State) I. Pérez Segura, Efrén.
II. Jacques-Ayala, César, 1946–   . III. Series: Special papers
(Geological Society of America) ; 254.
    QE203.S65S78   1991
    557.2'17—dc20   90-25952
                                                       CIP

Cover Photo: Late Miocene to late Pleistocene clastic fill in the
Moctezuma River basin. The basin is bounded by northwest-
southeast normal faults of the Basin and Range Province in east-
central Sonora. Locally, these deposits are covered by Quaternary
basalts. In the skyline, the mountain is overlain by early Tertiary
acidic volcanic rocks. Photograph by Jaime Roldán-Quintana.

10 9 8 7 6 5 4 3 2

Contents

Introduction 1
    Efrén Pérez-Segura and César Jacques-Ayala

The relation between the Paleozoic strata on opposite sides
   of the Gulf of California 7
    Gordon Gastil, Richard Miller, Paul Anderson, James Crocker,
    Michael Campbell, Philip Buch, Carl Lothringer, Paula Leier
    Englehardt, Mark DeLattre, John Hobbs, and Jaime Roldán-Quintana

Upper Triassic Barranca Group; Nonmarine and shallow-marine
   rift-basin deposits of northwestern Mexico 19
    John H. Stewart and Jaime Roldán-Quintana

Depositional environment, petrology, and provenance of the
   Santa Clara Formation, Upper Triassic Barranca Group, eastern
   Sonora, Mexico 37
    Isabelle Cojan and Paul Potter

Paleontology and biostratigraphy of Cretaceous rocks,
   Lampazos area, Sonora, Mexico 51
    Robert W. Scott and Carlos Gonzalez-Leon

Geology and chemical composition of the Jaralito and Aconchi
   batholiths in east-central Sonora, Mexico 69
    Jaime Roldán-Quintana

Geology of the Yecora area, northern Sierra Madre Occidental, Mexico 81
    Jean-Jacques Cochemé and Alain Demant

Quaternary shorelines along the northeastern Gulf of California;
   Geochronological data and neotectonic implications 95
    Luc Ortlieb

Coal in Sonora, Mexico 121
    Luis Obregon-Andria and Francisco Arriaga-Arredondo

Introduction

Efrén Pérez-Segura
Departamento de Geología, Universidad de Sonora, 83000 Hermosillo, Sonora, México
César Jacques-Ayala
Instituto de Geología, UNAM, Apartado. Postal 1039, 83000 Hermosillo, Sonora, México

INTRODUCTION

The purpose of this volume is to bring together important contributions to the geology of Sonora, northwestern Mexico, made mainly in the last 10 years. The topics cover very different aspects of the geology of this region. But, considering that not much of the research being done at present is published, it is not easy to collect papers on only one topic.

The geology of Sonora began to be of interest in the 18th century when Ignatius Pfefferkorn described "apparently petrifications of birds, animals or different objects" in the Cerro Las Conchas, near Arivechi, west-central Sonora. This description probably induced Remond (1866) to visit the locality to collect some of those "different objects," which were later identified by Gabb (1869). In spite of this century-long interest—and in spite of the mineral wealth of the area—the geology of Sonora still remains largely unknown. It has been mainly in the last 15 years that researchers have realized the great importance that this region has to the understanding of the tectonic evolution of the southwestern margin of the North American continent.

We must pay homage to those who ventured in the first half of this century, and before, into the inhospitable desert, or the unsettled central part of the state and the Sierra Madre Occidental to make the first geological reports. The pioneering works of Dumble (1900a, b, 1901), Burckhardt (1930), Keller (1928), Jaworski (1929), Flores (1929), Taliaferro (1933), King (1934, 1939), Imlay (1939), Arellano (1946), and Cooper and Arellano (1946) are cornerstones in the geological knowledge of Sonora. No wonder they are still present in most reference lists.

During the last 15 to 20 years, a great amount of work has been done. Researchers from Mexico, the United States, and France have made important contributions. Students doing their thesis work (B.Sc., M.Sc., and Ph.D. theses) have also made important contributions in new information and interpretations. Listing the many contributors would be too long, and inevitably would leave some out. So, instead of just presenting what has been done in the geology of Sonora, we prefer to speculate a little on what problems can be addressed.

In spite of all the work done, there is still no systematic geological cartography available. The Instituto Nacional de Estadística, Geografía e Informática (INEGI, formerly DETENAL) has published geological maps at a 1:250,000 scale, but for all practical purposes a scale of 1:50,000 would be highly desirable. Most of the state is unmapped, or many of the maps remain as confidential data. Undertaking this task requires the participation of several institutions, as well as an ensured budget from federal, state, and private sources.

In Sonora, rocks as old as Middle Proterozoic are exposed, and the stratigraphic column is almost completely represented. This makes a complex picture which, at present, is not fully understood.

Much isotopic work has been done in the igneous and metamorphic Proterozoic rocks south of Caborca (Silver and Anderson, 1974; Anderson and Silver, 1979; and unpublished work by them), but these have not been completely mapped or studied petrographically or geochemically. Besides, several types of "basement" have been identified, but at present we do not know if any of these are autochthonous. In the Mesozoic sequences there are several intercalations of quartz-pebble conglomerates, most probably derived from a Proterozoic/Paleozoic basement. Were these derived from the Proterozoic/Paleozoic rocks now present in the area? If not, where is the source now? The Triassic appears to be derived from a source to the north (Potter and Cojan, 1985; Stewart and Roldán, this volume), but block rotations cannot be ruled out at present.

The sedimentary Late Proterozoic and Paleozoic rocks, which have been the main focus of interest for American geologists since the works of G. A. Cooper and A.R.V. Arellano, are now becoming better understood. The works of Peiffer-Rangin (1979), Stewart and others (1984), Poole and Madrid (1986), Noll (1981), Radelli and others (1987), and Bartolini (1988) have shed some light on the sedimentation and structural position of these rocks. Regionally, the paleogeography and tectonic evolution during the Late Proterozoic–Paleozoic of Sonora appears to be clearer now; the regions are central Sonora, Caborca, and Cabullona, in the northeastern part of the state. Curiously

---

Pérez-Segura, E., and Jacques-Ayala, C., 1991, Introduction, *in* Pérez-Segura, E., and Jacques-Ayala, C., eds., Studies of Sonoran geology: Geological Society of America Special Paper 254.

enough, sedimentation in the three regions corresponds mainly to a stable platform, but we still do not know how these three areas relate to each other. Were they part of only one basin? Were they deposited in far-away basins brought together by megashears? Gastil and others (this volume) contribute an integration of the Paleozoic of Sonora and Baja California, providing a different insight into this controversial issue. There are many isolated outcrops of probable Paleozoic rocks in the northern half of Sonora, many of which have been metamorphosed by Laramide intrusives. Could these be the missing link between the three different regions mentioned before?

The Mesozoic is less well understood than the Paleozoic, mainly because in many areas it displays a much stronger deformation. The Triassic of central Sonora has been studied with interest (Wilson and Rocha, 1949; Alencaster, 1961; Potter and Cojan, 1985; Weber, 1985; Cojan and Potter, this volume; Stewart and Roldán, this volume). A stratigraphic and sedimentologic model has been proposed, but all the work has been done in the La Barranca area, and the many reported outcrops in southern Sonora, between La Barranca and Alamos, have been neglected. In the Sierra El Alamo, in the northwest, the Triassic still does not have a sedimentological reconstruction or a sediment provenance study. Such a study would provide the means to compare it with the Barranca Group. Also, between these two areas, exposures could be found, such as in Sierra La Flojera (Roldán and González, 1985), to correlate the two regions and define their geologic relation. The Triassic deserves more stratigraphic and sedimentological study.

The Jurassic is in the same situation as the Triassic. Two kinds of Jurassic sequences have been identified. One is clearly sedimentary, whereas the other is volcanic-volcaniclastic. The sediments have been identified in scattered areas and are well dated as Early Jurassic, such as in the Sierra de Santa Rosa (Hardy, 1973), Sierra El Alamo (González, 1980), Cerro Basura (DeJong, oral communication, 1989), and Sierra de López (Amaya and Rodriguez, 1988). The Late Jurassic has been identified in Cerros Pozo de Serna (Beauvais and Stump, 1976) and in the Cucurpe area (Rangin, 1977, 1982; Rodríguez, 1984). Other areas, such as Cerro Basura, have been assigned a Jurassic age based on degree of deformation (Corona, 1979). Magmatic and volcaniclastic rocks have been reported in the Cucurpe area (Rangin, 1977; Rodríguez, 1984), in the Sierra La Ceniza (Damon, *in* Solano, in preparation), probably in the Pajarito Mountains (Riggs, 1987) and in Planchas de Plata, SW of Nogales (Segerstrom, 1987), as well as in western Sonora (Anderson and Silver, 1978). These different localities have not been clearly related to each other, do not have enough good stratigraphic and sedimentologic studies, and the paleogeographic and paleotectonic setting cannot be defined. How do the two Jurassic sequences relate to each other? Are the sedimentary sequences in Sonora part of a fore-arc, intra-arc, or back-arc setting?

The Cretaceous is probably the best known period within the geology of some regions of Sonora. In the northern and central parts of the state, the Bisbee Group has been well identified, though with different names. The work by Scott and González (this volume) is one of the first biostratigraphic studies of the Lower Cretaceous. As more sequences of this age are described, the geometry of the basin will be better defined. It has been usually considered as a narrow basin, restricted to southeastern Arizona and northeastern Sonora. Now it has been documented as far south as Cerro de Oro (González and Jacques, 1988), 60 km northeast of Hermosillo, and as far west as Cerro El Chanate (Jacques and Potter, 1987; Willard, 1988). It could even extend farther south into Sinaloa (Bonneau, 1972). The volcanic facies have been reported only along the coast of Sonora (Anderson and others, 1969; Gastil and Krummenacher, 1977). What is more important, in the past, all volcanic-sedimentary sequences were assigned to the Jurassic, and all andesitic volcanics were assigned either to the Jurassic or the Tertiary. Now we know that much Cretaceous volcanic activity occurred in Sonora. Recent work has extended the presence of volcanic sequences inland. Jacques and Potter (1987) report the presence of two volcanic units in the Sierra El Chanate. One is in the El Chanate Formation (Albian-Cenomanian?), and the other in the El Charro Formation (Late Cretaceous). A third volcanic unit is present in the Aptian-Albian, Mural Limestone–equivalent, Arroyo Sasabe Formation (Jacques, in preparation). The work by Bojorquez and Rosas (1988) reports the presence of volcanic and sedimentary rocks of Neocomian age north of Huepac, in central Sonora, and Pedro Cruz (oral communication, 1989) described rocks of probable Turonian-Coniacian age with volcaniclastic sediments. These rocks are from the Agua de Obispo Canyon, central Sonora, described by Dumble (1901) as Early Cretaceous. Recent work has documented the presence of widespread Upper Cretaceous strata with intercalated volcanic rocks, as well as intrusive events (Anderson and Silver, 1974). Several questions arise concerning the volcanism of the Cretaceous and, in general, volcanism during the Jurassic and Cretaceous. Was there only one volcanic arc, migrating eastward and back, as proposed by Damon and others (1981, 1983), or were there several arcs superimposed in time and space? Did the Jurassic arc cease its activity during the Nevadan orogeny? How did this arc, or arcs, evolve? Many age dates are available, but the stratigraphy has not been integrated, and no geochemistry has been done. Probably, relating geochemistry and age would help in discriminating between arcs, or between inter-arc settings.

The Cenozoic is widely exposed, and yet poorly known. In an abuse of simplicity, all the sediments assigned to the Tertiary are dumped into the Báucarit Formation, even though several sedimentary events can be recognized, some of which are separated by angular unconformities. Probably because of this broad grouping, no sedimentation has been reported for the Paleocene-Eocene. The fill of large structural basins is mainly of Miocene age, as determined by limited paleontological and isotopic data. Many volcanic rocks, normally considered Tertiary, are now being reassigned to the Cretaceous. Tertiary volcanism is usually thought of as Sierra Madre Occidental volcanism, but its age and geographic distribution are not clearly constrained. How far west does this volcanism extend? Where is the boundary in space and time with the volcanic activity related to the opening of the Gulf of California?

The Quaternary has received more attention in the last 10 years. The publications by Ortlieb (1986, among others, and this volume) and Célis (1979) can be considered the most important contributions. But there are many areas that are still waiting to be studied.

If the stratigraphy is poorly known, the structure and tectonic evolution are even more problematic. Nevertheless, in the last few years, several deformational events have been documented or postulated. Not considering events during the Proterozoic, at the end of the Paleozoic there must have been an important deformation, first documented by King (1939), for the Triassic lies unconformably on the Paleozoic. More recently, evidence has been found for thrusting of basinal lower Paleozoic strata upon the Paleozoic platform sequence, which is as young as Early Permian (Radelli and others, 1987). This appears to explain the distribution of sedimentary facies proposed by Peiffer-Rangin (1979), to which Gastil and others (this volume) add support.

As for the Mesozoic, the Nevadan orogeny has been a matter of great controversy. For some, Jurassic tectonism is related to a Jurassic arc developed along the left-lateral Mohave-Sonora megashear (Anderson and Silver, 1979). Others consider it to be evidenced by a poorly documented unconformity between the Jurassic and the Cretaceous. On the other hand, still others postulate that the boundary between the Jurassic and the Cretaceous is conformable and gradational (Rodríguez, written communication, 1988). Herrera (1989) and Pérez-Segura (1989) state that it is not possible to understand this orogeny without unraveling the problem of the narrow metamorphic belt that extends from Estación Llano to Sonoyta. The metamorphism of this belt has been dated as Laramide in age (Damon and others, 1962; Hayama and others, 1984), and Jacques-Ayala (1986) provides limited data on the age of the protolith as Early to Late Cretaceous. On the other hand, a mid-Cretaceous deformation, poorly recognized, or even ignored, seems to be gaining importance (Pubellier, 1987). Calmus and Radelli (1987) consider that it could be the true Cretaceous orogeny, the Laramide being a minor event, responsible only for the great magmatic event that occurred from 40 to 90 Ma (Damon and others, 1983). The main objection to this hypothesis is that the Upper Cretaceous is just being recognized in other areas besides Cabullona (Taliaferro, 1933), such as in El Chanate (Jacques and Potter, 1987; Jacques and García, 1988), Cerro de Oro (González and Jacques, 1988), Sahuaripa (Pubellier, 1987) and Esqueda (M. Grajales, oral communication, 1989). Here we can ask: is the Upper Cretaceous recording an orogenic event beginning during the Albian or Cenomanian? If so, when did the thrust front arrive in Sonora, and what were its effects? This still remains unsolved.

One of the most cited events in Sonora, but unfortunately the least studied one, is the Tertiary Basin and Range (*s.l.*) extension. The big question is: what effect did the Basin and Range (*s.l.*) have on older structures? Is it valid to speak about vergency, or structural trends, of older deformations? Is there a relation between the Basin and Range and the opening of the Gulf of California? The Gulf of California, a topic in itself, has received much more attention than the mainland. The most recent report on neotectonic studies quantifies the displacement of the Baja California Peninsula with respect to the continent (Lesage and others, 1988).

Finally, we would like to thank the authors for their contributions and patience during the long wait. Working on this volume has been an excellent experience, learning the hard way. The most time consuming task was assisting as much as possible non-English-proficient contributors. We would also like to acknowledge the reviewers, who kindly devoted time to improve the manuscripts. They are:

Eduardo Aguayo
Instituto de Ciencias del Mar y Limnología, UNAM
México, D.F. MEXICO

Gloria Alencaster de Félix
Instituto de Geología, UNAM
México, D.F., MEXICO

Jorge Aranda
Instituto de Geología, UNAM
Guanajuato, Gto., MEXICO

Thierry Calmus
University of Sonora
Hermosillo, Son., MEXICO

Kenneth L. Cameron
University of California,
Santa Cruz, CA, U.S.A.

Ana Luisa Carreño
Instituto de Geología, UNAM
México, D.F., MEXICO

Kenneth F. Clark
University of Texas
El Paso, TX, U.S.A.

Kees A. DeJong
University of Cincinnati
Cincinnati, OH, U.S.A.

William R. Dickinson
University of Arizona
Tucson, AZ, U.S.A.

Ismael Ferrusquía
Instituto de Geología, UNAM
México, D.F., MEXICO

César Jacques
Instituto de Geología, UNAM
Hermosillo, Son., MEXICO

Margaret Klute
University of Arizona
Tucson, AZ, U.S.A.

Fred W. McDowell
University of Texas
Austin, TX, U.S.A.

Gifford H. Miller
University of Colorado
Boulder, CO, U.S.A.

Allison R. Palmer
Geological Society of America
Boulder, CO, U.S.A.

Claude Rangin
Pierre and Marie Curie Univ.
Paris, FRANCE

Rafael Rodríguez
Instituto de Geología, UNAM
Hermosillo, Son., MEXICO

Guillermo A. Salas
University of Sonora
Hermosillo, Son., MEXICO

Joseph F. Schreiber
University of Arizona
Tucson, AZ, U.S.A.

John H. Stewart
U.S. Geological Survey
Menlo Park, CA, U.S.A.

Jean-Marie Vila
Centre Universitaire Antille Guyane
Pointe-à-Pître, GUADELOUPE

REFERENCES CITED

Alencaster, G., 1961, Estratigrafía del Triásico Superior de la parte central del Estado de Sonora: Instituto de Geología, Universidad Nacional Autónoma de México, Paleontología Mexicana 11, part 1, 18 p.

Amaya-M. R., and Rodríguez-C., J. L., 1988, Presencia de rocas volcaniclásticas de edad jurásica en la parte sur de la Sierra de López: Sociedad Geológica Mexicana, IX Convención Anual, Resúmenes.

Anderson, T. H., and Silver, L. T., 1974, Late Cretaceous plutonism in Sonora, Mexico, and its relationship to circumpacifique magmatism: Geological Society of America Abstracts with Programs, v. 65, p. 484.

—— , 1978, Jurassic magmatism in Sonora, Mexico: Geological Society of America Abstracts with Programs, v. 10, p. 359.

—— , 1979, The role of the Mojave Sonora megashear in the tectonic evolution of northern Sonora, in Anderson, T. H., and Roldán-Quintana, J., eds., Geology of northern Sonora; Geological Society of America Annual Meeting Guidebook, Field Trip 27: San Diego, California, San Diego State University Department of Geological Sciences, v. p. 59–68.

Anderson, T. H., Silver, L. T., and Córdoba, D. A., 1969, Mesozoic magmatic events of the northern Sonora coastal region: Geological Society of America Abstracts with Programs, v. 1, part 7, p. 3–4.

Arellano, A.R.V., 1946, Noticias geológicas del Distrito de Altar, Sonora: Geological Society of Mexico Bulletin, v. 21, p. 53–58.

Bartolini, C., 1988, Regional structure and stratigraphy of the Sierra El Aliso, central Sonora, Mexico [M.Sc. thesis]: Tucson, University of Arizona, 189 p.

Beauvais, L., and Stump, T. E., 1976, Corals, molluscs, and paleogeography of Late Jurassic strata of the Cerro Pozo Serna, Sonora, Mexico: Palaeogeography, Palaeoclimatology, Palaeoecology, v. 19, p. 275–301.

Bojorquez-O.J.A. and Rosas-H., J. A., 1988, Geología de la hoja Aconchi (H12-D13), municipio de Aconchi, Sonora, México [B.Sc. thesis]: Hermosillo, Sonora, University of Sonora, 92 p.

Bonneau, M., 1972, Données nouvelles sur les series cretacées de la côte pacifique du Mexique: Société Géologique Francaise Bulletin, v. 7, no. 14, p. 55–65.

Burckhardt, C., 1930, Etude synthétique sur le Mesozoique mexicain: Society Paleontology Suisse Memoir 39, no. 50, 280 p.

Calmus, T., and Radelli, L., 1987, Mid-Cretaceous orogeny and Laramide event of Sonora and Baja California: University of Sonora Department of Geology Bulletin, v. 4, p. 51–56.

Célis, S., 1979, Les foraminifères quaternaire des anciennes lignes de rivage de la côte de Sonora et de Basse Californie, Mexique [Doctorate thesis]: Paris, Université Pierre et Marie Curie, 110 p.

Cooper, G. A., and Arellano, A.R.V., 1946, Stratigraphy near Caborca, northwest Sonora, Mexico: American Association of Petroleum Geologists Bulletin, v. 30, p. 606–611.

Corona, F. F., 1979, Preliminary reconnaissance geology of Sierra La Gloria and Cerro Basura, in Anderson, T. H., and Roldán-Quintana, J., eds., Geology of northern Sonora; Geological Society of America Annual Meeting Guidebook, Field Trip 27: San Diego, California, San Diego State University Department of Geological Sciences, p. 32–48.

Damon, P. E., Livingston, D. E., Mauger, R. L., Giletti, B. J., and Pantoja-Alor, J., 1962, Edad del Precambrico "Anterior" y de otras rocas del zócalo de la región de Caborca-Altar de la parte noroccidental del Estado de Sonora: Universidad Nacional Autónoma de México Instituto de Geología Boletín, no. 64, parte 2, p. 11–44.

Damon, P. E., Shafiqullah, M., and Clark, K. F., 1981, Age trends of igneous activity in relation to metallogenesis in the southern Cordillera: Arizona Geological Society Digest, v. 14, p. 137–154.

—— , 1983a, Geochronology of the porphyry copper deposits and related mineralization of Mexico: Canadian Journal of Earth Sciences, v. 20, p. 1052–1071.

Damon, P. E., Shafiqullah, M., Roldán-Q. J., and Cocheme, J. J., 1983b, El batolito laramide (90 a 40 m.a.) de Sonora: 15th Convencion Nac. de Ing.

Minas, Metal. y Geol. de México, Memoria, p. 63–95.
Dumble, E. T., 1900a, Triassic coal and coke in Sonora: Geological Society of America Bulletin, v. 11, p. 10–14.
——, 1900b, Notes on the geology of Sonora: American Institute of Mining and Metallurgical Engineers Transactions, v. 29, p. 122–152.
——, 1901, Cretaceous of Obispo Canyon, Sonora, Mexico: Texas Academy of Sciences Transactions, v. 4, p. 81.
Flores, T., 1929, Reconocimientos geológicos en la región central del estado de Sonora: Universidad Nacional Autónoma de México, Instituto Geológico Mexicano, Boletín 49.
Gabb, W. M., 1869, Cretaceous and Tertiary fossils: Geological Survey of California, Paleontology, v. 11, sections 1, 11, and 111, 299 p.
Gastil, R. G., and Krummenacher, D., 1977, Reconnaissance geology of coastal Sonora, between Puerto Lobos and Bahia Kino: Geological Society of America Bulletin, v. 88, p. 189–198.
González-Leon, C., 1980, La Formación Antimonio (Triásico Superior-Jurásico Inferior) en la Sierra del Alamo, Estado de Sonora: Universidad Nacional Autónoma de México Instituto de Geología Revista, v. 4, p. 13–18.
González-Leon, C., and Jacques-Ayala, C., 1988, La secuencia del Cretácico Temprano del area de Cerro de Oro, Sonora; Implicaciones paleogeograficas: Resumenes, 2nd Simposio sobre geología y minería de Sonora, Hermosillo, Sonora, p. 23–25.
Hardy, L. R., 1973, The geology of an allochthonous Jurassic sequence in the Sierra de Santa Rosa, northwest Sonora, Mexico [M.S. thesis]: San Diego, California State University, 92 p.
Hayama, Y., Shibata, K., and Takeda, H., 1984, K-Ar ages of the low-grade metamorphic rocks in the Altar Massif, northwest Sonora, Mexico: Journal of the Geological Society of Japan, v. 90, p. 589–596.
Herrera, S., 1989, Geología de la región de Estación Llano, Sonora: Resultados preliminares: Universidad de Sonora, Departamento de Geología, 9 p. (unpublished).
Imlay, R. W., 1939, Paleogeographic studies in northeastern Sonora: Geological Society of America Bulletin, v. 50, p. 1723–1744.
Jacques-Ayala, C., 1986, Las rocas cretácicas del área de Caborca-Altar y sus deformaciones: Simposio Nuevas aportaciones a la geología de Sonora: Centenario Instituto de Geología, Universidad Nacional Autónoma de México, Hermosillo, Sonora, p. 56–68.
Jacques-Ayala, C., and Garcia y B., J. C., 1988, Geology of the Cretaceous rocks of the Sierra El Chanate, northwest Sonora: Resumenes 2nd Simposio sobre Geologia y Mineria de Sonora, Hermosillo, Sonora, p. 31–33.
Jacques-Ayala, C., and Potter, P. E., 1987, Stratigraphy and paleogeography of Lower Cretaceous rocks, Sierra El Chanate, northwest Sonora, Mexico: Arizona Geological Society Digest, v. 18, p. 203–214.
Jaworski, E., 1929, Eine Lias-fauna aus Nordwest Mexico: Schweizerisch palaeontologisch Gesellschaft Abhandlung, v. 48, 12 p.
Keller, W. T., 1928, Stratigraphische Beobachtungen in Sonora (Nordwest Mexico): Eclogae Geologic Helvetiae, v. 21, no. 2, p. 327–335.
King, R. E., 1934, Geological reconnaissance of central Sonora: American Journal of Science, v. 228, no. 164, p. 81–101.
——, 1939, Geological reconnaissance in northern Sierra Madre Occidental of Mexico: Geological Society of America Bulletin, v. 50, p. 1625–1722.
Lesage, Ph., and 11 others, 1988, Mediciones geodésicas de largas distancias; Aplicación al estudio de movimientos de placas en el Golfo de California: Geofísica Internacional, v. 27-3, p. 351–377.
Noll, J. H., 1981, Geology of the Picacho Colorado area, northern Sierra de Cobachi, central Sonora, Mexico [M.S. thesis]: Flagstaff, Northern Arizona University, 69 p.
Ortlieb, L., 1986, Néotectonique et variations du niveau marin au Quaternaire dans la région du Golfe de Californie, Mexique [Ph.D. thesis]: Marseille, Université Aix-Marseille, 1221 p.
Pfeiffer-Rangin, F., 1979, Les zones isopique de Paleözoique inferieur de NW mexicain, témoins de relais entre les Appalaches et la Cordillère ouest americaine: Paris, Comptes Rendus l'Academie des Sciences, v. 286, p. 1517–1519.

Pérez-Segura, E., 1989, Descubrimiento de una paragénesis de Au-Te en el yacimiento de San Francisco, Sonora: 18th Conv. Nac. Asoc. Ing. Minas, Metal. y Geol. de Mexico, Acapulco, Gro., Memoria, p. 63–69.
Poole, F. G., and Madrid, R. J., 1986, Paleozoic rocks in Sonora (Mexico) and their relation to the southwestern continental margin of North America: Geological Society of America Abstracts with Programs, v. 18, p. 720–721.
Potter, P. E., and Cojan, I., 1985, Description and interpretation of the type section of the Barranca Group east of Rancho La Barranca, Municipio de San Javier, Sonora, in Weber, R., ed., Simposio sobre floras del Triásico Tardío, su fitogeografía y paleoecología: 3er Congreso latinoamericano de Paleontología, Universidad Nacional Autónoma de México, Instituto de Geología, Memoria, p. 101–105.
Pubellier, M., 1987, Relations entre domaines cordillerain et mesogéen au nord du Mexique; Etude géologique de la vallée de Sahuaripa, Sonora central [Doctoral thesis]: Paris, France, Université Pierre et Marie Cure, 219 p.
Radelli, L., and 9 others, 1987, Allochthonous paleozoic bodies of central Sonora: Universidad de Sonora Departamento de Geología Boletín, v. 4, no. 1, p. 1–15.
Rangin, C., 1977, Sobre la presencia del Jurasico Superior con amonitas en Sonora Septentrional: Universidad Nacional Autónoma de México Instituto de Geología Revista, v. 1, p. 1–4.
——, 1982, Contribution à l'etude géologique du Système Cordillerain du nord-ouest du Mexique [Doctorate thesis]: Paris, France, Université Pierre et Marie Curie, 588 p.
Remond, A., 1866, Notice of geological explorations in northern Mexico: Proceedings of the California Academy of Science, v. 3, p. 244–257.
Riggs, N., 1987, Stratigraphy, structure, and geochemistry of Mesozoic rocks in the Pajarito Mountains, Santa Cruz County, Arizona: Arizona Geological Society Digest, v. 18, p. 165–176.
Rodriguez-Castañeda, J. L., 1984, Geology of Tuape region, north-central Sonora, Mexico [M.S. thesis]: Pittsburgh, Pennsylvania, University of Pittsburgh, 157 p.
Roldán-G., J., and González-L., C. M., 1985, Notas sobre el Triásico Superior de la Sierra de La Flojera, Sonora: 3er Congreso latinoamericano de Paleontología, México, Memoria, p. 83–85.
Segerstrom, L., 1987, Geology of the Planchas de Plata area, northern Sonora, Mexico: Arizona Geological Society Digest, v. 18, p. 153–164.
Silver, L. T., and Anderson, T. H., 1974, Possible left-lateral early to middle Mesozoic disruption of the southwestern North American craton margin: Geological Society of America Abstracts with Programs, v. 6, p. 955.
Solano, O., 1990, Estudio fotogeológico de las Sierra de la Ceniza, Agua Prieta, Sonora [B.S. thesis]: Hermosillo, Sonora, (in preparation).
Stewart, J. H., McMenamin, M.A.S., and Morales-R., J. M., 1984, Upper Proterozoic and Cambrian rocks in the Caborca region, Sonora, Mexico; Physical stratigraphy, biostratigraphy, paleocurrent studies, and regional relations: U.S. Geological Survey Professional Paper 1309, 36 p.
Taliaferro, N. L., 1933, An occurrence of Upper Cretaceous sediments in northern Sonora, Mexico: Journal of Geology, v. 41, no. 1, p. 12–13.
Weber, R., 1985, Las plantas fósiles de la Formación Santa Clara (Triásico Tardío), Sonora, México, in Weber, R., ed., Simposio sobre floras del Triásico Tardío, su fitogeografía y paleoecología: 3er Congreso latinamericano de Paleontología, Universidad Nacional Autónoma de México, Instituto de Geología, Memoria, p. 107–124.
Willard, J. S., 1988, Geology, sandstone petrography, and provenance of the Jurassic(?)–Cretaceous rocks of the Puerto El Alamo area, northwestern Sonora, Mexico [M.S. thesis]: Cincinnati, Ohio, University of Cincinnati, 250 p.
Wilson, I. F., and Rocha, V. S., 1949, Coal deposits of the Santa Clara District, near Tonichi, Sonora, Mexico: U.S. Geological Survey Bulletin 962, 80 p.

MANUSCRIPT ACCEPTED BY THE SOCIETY APRIL 18, 1990

Printed in U.S.A.

The relation between the Paleozoic strata on opposite sides of the Gulf of California

Gordon Gastil, Richard Miller, Paul Anderson, James Crocker, Michael Campbell, Philip Buch, Carl Lothringer, Paula Leier-Engelhardt, Mark DeLattre, and John Hoobs
Department of Geological Sciences, San Diego State University, San Diego, California 92182

Jaime Roldán-Quintana
Instituto de Geología, Estacion Regional del Noroeste, Hermosillo, Sonora 83000, Mexico

ABSTRACT

Workers in Sonora recognize two distinct provinces of Upper Proterozoic and Paleozoic strata; one of shallow-water origin and one of deeper-water origin. The shallow-water miogeoclinal strata are best known in areas west and southwest of Caborca, and range in age from late Proterozoic to Permian. These correlate with strata in the Sierra Agua Verde–Cerro Cobachi area east of Hermosillo. Strong stratigraphic similarities between these rocks and those of Nevada have fueled hypotheses that they were adjacent in Paleozoic time and have been separated 800 km by left-lateral strike-slip displacement of mid-Jurassic age. Deeper-water strata occur farther to the south and southeast in Sonora where rocks of Ordovician to Permian age contain bedded cherts and carbonate turbidites. These rocks in the Sierra Cobachi and at Barita de Sonora resemble similar-age slope- and basinal-facies rocks in Nevada, and in like fashion appear to be thrust northward or eastward over the miogeoclinal facies.

Both of these facies also occur in the basement strata of peninsular California. Carbonate rock, quartzite, amphibolite, and subordinate argillaceous strata crop out from the San Jacinto Mountains of southern California to the southeastern Sierra San Pedro Martir (Baja California Norte). The only age control on these rocks are the Early Ordovician conodonts found on Coyote Mountain, Imperial County, California.

Deep-water strata with pillow basalt and a cratonal sedimentary contribution are found in the Sierra las Pinta and along the Gulf Coast of Baja California between Puerto Calamujue and Bahia Los Angeles. Ages based on poorly preserved conodonts and corals range from Devonian to Mississippian, but the rocks may not be limited to this age span. Similar strata, dominated by bedded chert, occur on Isla Angel de la Guarda, at the southeastern tip of Isla Tiburón and on the adjacent coast of Sonora, thus effectively connecting across the Gulf of California. Analogous deep-water–facies strata, dominated by bedded chert, occur in northern Sinaloa. Conodonts from San José de Gracia, Sinaloa, indicate a Pennsylvanian age.

If the apparent correlations of both miogeoclinal and deeper-water facies of Paleozoic strata across the Gulf of California are correct, they put constraints on the paleomagnetic reconstructions that place peninsular California in southern Mexico in Late Cretaceous time.

Gastil, G., Miller, R., Anderson, P., Crocker, J., Campbell, M., Buch, P., Lothringer, C., Engelhardt, P. L., and Roldán-Quintana, J., 1991, The relation between the Paleozoic strata on opposite sides of the Gulf of California, *in* Perez-Segura, E., and Jacques-Ayala, C., eds., Studies of Sonoran geology: Geological Society of America Special Paper 254.

INTRODUCTION

For the past 100 years, geologists have recognized the Cordilleran geosyncline extending from Canada southwest through Idaho, Utah, and Nevada to southeastern California, where it becomes lost among the granitic rocks of the Sierra Nevada and the Mojave desert (Stewart and Pool, 1975). It has long been understood that the southeastern facies of carbonate rocks and well-sorted clastics lap onto the adjacent Precambrian basement, with their detrital provenance demonstrably the craton of North America.

Western facies represent a thick succession of fine-grained and less well-sorted clastics, associated with thin-bedded carbonate rocks, bedded chert, and locally, volcanic strata. These rocks are generally believed to be slope or basinal deposits, and their original relation to North America is not established. Earlier workers referred to the western rocks as the "eugeosyncline."

Some formations, such as the Valmy Formation of Nevada, consist of a mixture of rocks, such as bedded chert representing the basinal facies and mature quartz arenites typical of the miogeocline. Such formations have been referred to as transitional facies, implying that there is indeed an in situ transition between the two facies. Coney and others (1980) have suggested that the deeper-water facies, and possibly the so-called transitional facies as well, might be exotic to North America: slices of originally distant terranes sutured onto the edge of this continent in late Paleozoic or early Mesozoic time.

Many have noted that to the southwest these belts appear truncated against rocks of the Mesozoic orogen. However, near Caborca in northern Sonora, Mexico (Cooper and Arellano, 1946; Stewart and others, 1984), is a sequence of Upper Proterozoic to Mississippian strata of miogeoclinal facies that correlates formation for formation with similar-age rocks in the Cordilleran geosyncline of the southern Great Basin. Silver and Anderson (1974) have argued that the Caborca rocks are so similar to those of southern Nevada and southeastern California that they must once have been adjacent, and their present position is best explained by left-lateral displacement on what they term the Mojave-Sonora megashear.

Stewart and others (1984) have suggested that this similarity could also be explained by bending the strand line of the miogeocline eastward around the southern edge of the continent, and they point out that the alternative explanations are not mutually exclusive.

Noll (1981), Poole (1983), Poole and Madrid (1986), and Ketner (1986) have discovered not only that the miogeoclinal sequence of Caborca reappears some 300 km farther southeast in Sonora (25, Fig. 1), but that the age and lithologic analogs of the deeper-facies rocks of the southern Great Basin can be found south of the miogeoclinal rocks, in the Sierra Cobachi and at Barita de Sonora (26, Fig. 1). Thus, it would appear that both of the geoclinal facies are present, and Pfeiffer-Rangin (1979), Stewart (1981), Poole (1983), and Ketner (1986) have suggested extending the correlation around what by this hypothesis would have been the southern edge of North America, to join similar rocks in western Texas and Arkansas.

These earlier hypotheses did not consider the possibility that the facies of the geocline might have persisted southwestward through the plutonic/metamorphic overprint of the Pacific margin orogen to be found in peninsular California. Indeed, until 1981 the only confirmed report of Paleozoic fossils in the Peninsular Ranges was by McEldowney (1970) in the Sierra las Pinta of Baja California. Since 1981 it has been discovered that Paleozoic strata of a variety of ages and facies are widespread throughout the northern half of the peninsula. How these discov-

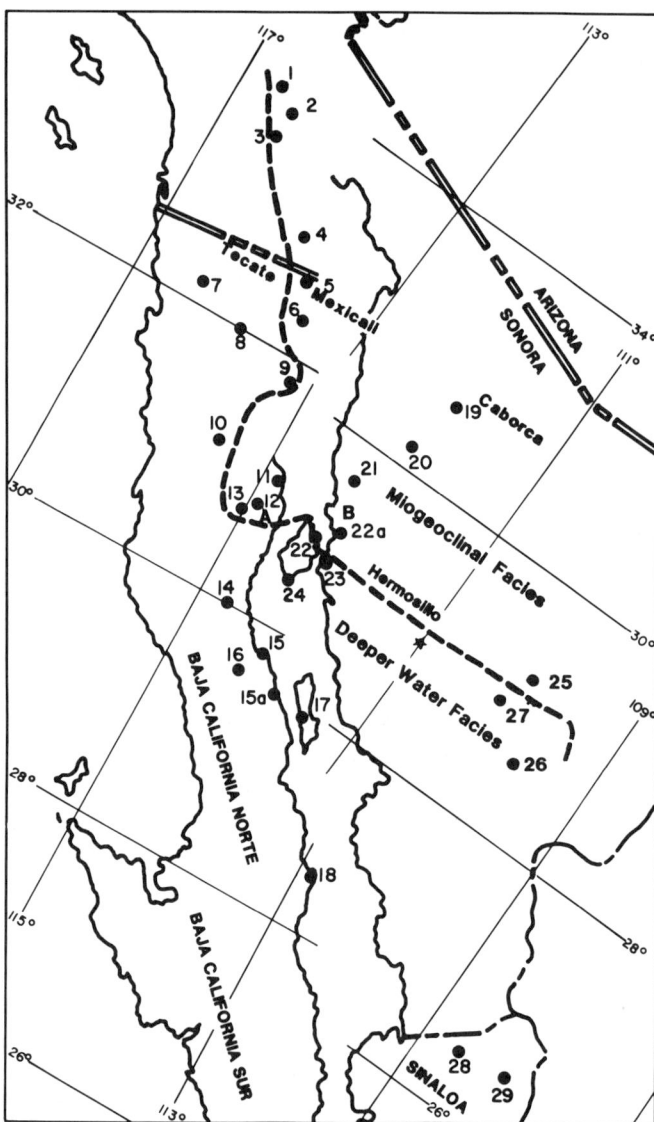

Figure 1. Distribution of miogeoclinal and deeper-water-facies Paleozoic strata in northern Baja California and adjacent Sonora and Sinaloa. The numbered localities are referred to in the text. The peninsula of Baja California has been palinspastically relocated using 300-km right-lateral motion on the San Andreas plate boundary system. The relocation juxtaposes localities A (Santa Rosa Mountains, Baja California) and B (State of Sonora), which possess unique conglomerates believed to have been continuous in early Miocene time (Gastil and others, 1973).

eries relate to the discussion in our introduction is the topic of this chapter.

PALEOZOIC STRATA IN PENINSULAR CALIFORNIA

Analogs are present of miogeoclinal-facies, deeper-water-facies, and in one locality, even transitional-facies rocks. Lower Paleozoic miogeoclinal strata (Fig. 1), characterized by thick sequences of carbonate rock and mature quartz arenite, extend from the San Jacinto block (1), through the Santa Rosa Mountains (2) and Borrego area (3), and south through western Imperial County to the international border. Southward they appear in the Sierra Cucapah (5) and Sierra Mayor (6), the San Felipe area (11), the Sierra Santa Rosa (4), and the southeastern Sierra San Pedro Martir (13). No traces of these lower Paleozoic rocks are found farther south in the peninsula.

Deeper-water facies are probably also present in the eastern peninsular ranges north of the border, but the grade of metamorphism makes it difficult to distinguish them from Mesozoic strata of similar facies. The farthest north distinguishable, deeper-water, Paleozoic rock is found in the Sierra las Pinta (9), about 60 km south of Mexicali; it is Devonian or Mississippian in age. Slope-facies strata of Permian age are found on the 30th parallel (14), and deep-water rocks of Devonian and Mississippian age are found near and along the gulf coast between latitudes 29°00' and 29°40' (15–15A). Similar strata persist somewhat farther south, but trend into the gulf at latitude 28°25' (18) and have not been reported farther south in the peninsula.

THE MIOGEOCLINE IN PENINSULAR CALIFORNIA

The first identification of lower Paleozoic rocks in peninsular California was made near Rancho San Marcos, 45 km southeast of Tecate, Baja California Norte (Gastil and Miller, 1981), and on Coyote Mountain in eastern Imperial County (4, Fig. 1; Dockum 1982; Miller and Dockum, 1983). The section on top of Coyote Mountain is largely magnesium-rich carbonate rocks, bearing some resemblance to the Pogonip Formation of southern Nevada. It structurally overlies ultra-pure quartz arenite and pebbly arkose, which suggest Cambrian and Upper Precambrian formations in southeastern California. From Coyote Mountain it is easy to extend one's lithologic correlation north through similar rocks to the San Jacinto block (1) and south across the international border.

In the San Jacinto Mountains, miogeoclinal rocks have been named the Desert Divide Group (Brown, 1980). To quote from Hill (written communication, 1987):

The desert Divide group consists of at least 4,300 m of quartz-rich schists and gneisses, metacarbonate rock, and quartzite. Although now at upper amphibolite grade the protolith appears to have been a thick sequence of quartz and clay-dominated sediment that contained lenses of limestone near its top, and was capped by at least 800 m of quartz sand. A multi-cycled sediment derived from an old, stable continent is indicated. Strontium isotope ratios suggest that the sequence may be late Proterozoic in age.

In the Palm Canyon and Santa Rosa Mountains (1, 2), Erskine (written communication, 1987) described these rocks as follows:

Compositions of the Santa Rosa Formation range from nearly mono-mineralic quartzite to quartz-rich garnet-sillimanite-cordierite schists, and gneiss, interbedded with marble, and a variety of diopside-tremolite-scapolite-almandine bearing calc-silicate rocks with minor amphibolite. The Palm Canyon Formation is found throughout the length of the Santa Rosa Mountains occupying sheets within a stacked sequence of imbricate low-angle faults associated with the eastern Peninsular Ranges mylonite zone. These rocks are similar to the Santa Rosa Formation except that they are strongly migmatitic and generally lack pure metaquartzite. Although structural and temporal relations are unclear, both the Santa Rosa and Palm Canyon Formations were probably part of a Paleozoic miogeoclinal clastic wedge or prism shed westward from a Precambrian source terrane and deposited on cratonal crust.

Other regional compilations in the area are by Engel and Schultejohn (1984), and Todd and others (1988). Nowhere in this area have miogeoclinal rocks younger than Ordovician been identified, nor have either depositional or structural boundaries between different prebatholithic facies been located.

South of the international border, the miogeoclinal rocks are exposed discontinuously for the length of the Sierra Cucapah (5) and Sierra Mayor (6) (Scott Fenby, personal communication, 1987; Rick Siem, personal communication, 1990).

Metasedimentary rocks of the Sierra Cucapah and Sierra Mayor vary in metamorphic grade from lower- to upper-amphibolite facies. These are predominantly thick carbonate rock units, with calcitic rock predominating over dolomitic rock Near Rio Hondo, carbonate rock is intimately interbedded with pelite. Quartz arenite, amphibolite, and thick pelitic units occur locally.

The next miogeoclinal rocks south from the Sierra Mayor are in the San Felipe area (11). Here, Paul Anderson (1982) reports at least 3,800 m, subdivided into eight map units consisting of orthoquartzite, marble, meta-argillite, and micaceous quartzite. These are believed to represent rapidly deposited blanket sands and sublittoral sheet sands deposited in a shallow-marine environment. Silty limestones and dolomites may have been of cyanobacterial origin.

The composite section displays lithologic and stratigraphic similarities to sequences of late Proterozoic and Early Cambrian age in the Caborca region of central Sonora, Mexico; in particular, unit G resembles the Proveedora Quartzite.

Continuing south from San Felipe, very similar rocks are exposed discontinuously in the Sierra Estrella (east of 12) and the Sierra Santa Rosa (Bryant, 1986; locality A); the southern Sierra San Felipe (Hiner, 1972; Bryant is 12; Hiner is just southwest of 12); and in the southeastern Sierra San Pedro Martir (13). Common to each of these areas are the predominant lithologies of marble, clean quartzite, and amphibolite. Except for the San Felipe area, the grade of metamorphism is such that it is impossible to determine if the amphibolite is derived from intrusive rock or basalt and basalt-derived sedimentary strata. In the Sierra San Pedro Martir (13), the apparently homoclinal sequence begins with a quartz schist (protolith of argillaceous sandstone?). This is overlain by more than 2 km of relatively clean quartzite; it then grades upward through mixed quartz arenite/carbonate rock into locally cherty carbonate rock. The amphibolite portion of this section has only been observed from the air.

South from the Sierra San Pedro Martir, no sequences resembling those described above have been found anywhere in Baja California Norte.

ROCKS POSSIBLY EQUIVALENT TO THE VALMY FORMATION

In the area northeast of Guadalupe Valley (7) are several occurrences of associated quartzite, carbonate rock, and bedded chert. At Rancho San Marcos, Lothringer (1984) reported an exposure measuring about 5.5 km in length and as much as 1 km in width, dated as Early Ordovician by conodonts (Gastil and Miller, 1981; Fig. 2).

The quartzite is 99 percent pure, moderately well sorted, mostly well rounded quartz grains. Original grain diameters range from 0.1 to 5.0 mm, but most are 0.5 to 1 mm. They are silica-cemented and exhibit irregular dissolution boundaries. Most grains are monocrystalline. Heavy minerals include purple zircons and tourmaline (pleochroic clear to dark brown). A clast-supported pebble-to-cobble conglomerate occurs in one locality. The clasts are poorly sorted and moderately well rounded. Chert clasts predominate, with fewer clasts of quartzite and quartz siltstone. The quartzite unit is on the order of 200 m thick, and is in apparent stratigraphic contact with both carbonate rock and bedded chert.

The chert ranges in color from very light gray to brownish gray, to dark gray and black. Some of the lighter chert contains fluorapatite (Coles, written communication, 1986). Individual beds range in thickness from millimeters to centimeters, and are rarely more than 5 cm thick. They are interlayered with varying proportions of argillaceous material, and locally grade into siliceous shale. The darker occurrences contain considerable carbon. There is no evidence of bioturbation. The bedded chert section appears to be about 150 m thick.

The marble is medium to dark gray on fresh surfaces. Distinct bedding is observed only where 1-cm-thick sandstone laminae occur, or where it is interbedded with chert. In thin section the marble shows a mosaic of fine to coarsely crystalline calcite, with less than 5 percent dolomite. Fragments of crinoid columnals, articulate brachiopods, calcareous sponges(?), and other unidentified organic debris are locally abundant, as well as oval cyanobacteria(?) structures and 5-mm concentric ooids. The thickest section is 40 m. Locally, marble is found as lensoidal bodies within the bedded chert.

Most of the conodonts were found in two localities about 1.5 km apart. The original locality (Gastil and Miller, 1981) revealed 10 taxa of the Early Ordovician (Arenigian) *Prioniodusevae/Oipikodus evae* biozone recognized in the Baltic Shield region by Sergeeva (1962, 1963). This biozone is diagnostic of the North Atlantic biogeographic province, and is considered equivalent to North American midcontinent faunas E and 1 of the upper Ibexian or lower Arenigian (Ross, 1977).

Subsequently, Keith Ketner and John Repetski obtained conodonts from a second locality that is also of Arenigian age, but belongs to the Mid-continent biogeographic province (Ketner, written communication, 1985). A similar close association of North Atlantic and Mid-continent faunas is reported in the Vinini Formation of northern Nevada (Harris and others, 1979).

The association of ultramature quartzite with bedded chert, the bimodal character of the quartz grains, the occurrence of dark quartz grains among colorless ones, the heavy-mineral assemblage limited to zircon and tourmaline, and the appearance of phosphate within the bedded chert all bear resemblance to the similar-age Valmy Formation of north-central Nevada (Ketner, written communication, 1984). However, the rocks at San Marcos are coarser grained than anything described in the Valmy, and the Valmy includes very little carbonate rock. The Valmy-equivalent rocks are treated separately because they are considered the westernmost miogeoclinal rocks and possibly correlative to the basinal Vinini Formation. D'Allura and others (1977), for example, show it located on the continental slope, off the edge of the cratonal basement.

These Valmy-like rocks at Rancho San Marcos rest on a melange of similar-composition blocks in a matrix of immature sandstone and shale. Lothringer (1984) and Reed (1985) believe that this 1 × 5 km block of Ordovician-age rocks is an olistolith gravitationally deposited in a submarine fan basin of Triassic-Jurassic age.

DEEPER-WATER FACIES IN THE PALEOZOIC STRATA OF PENINSULAR CALIFORNIA

Sierra las Pinta

The farthest-north exposures of definitely Paleozoic slope- and basinal-facies strata in peninsular California are found in an area located at the northern and southern ends of the Sierra las Pinta (9), some 60 to 75 km south of Mexicali. These rocks were first described by McEldowney (1970) and later by James (1973). McEldowney found corals, crinoids, and brachiopods and suggested a Carboniferous age. These rocks have recently been described in greater detail by Leier-Engelhardt (1986; Fig. 2). We will here paraphrase her descriptions as follows: the rocks in the northern area (520 m in probable upward sequence) consist of calcareous siltstone and calcareous sandy siltstone; siltstone, and calcareous siltstone, interbedded with rare crinoidal grainstone and granule conglomerate; normally graded sandy crinoidal grainstone; massive and normally graded beds of coarse sandstone and clast-supported granule to cobble conglomerate; massive basalt flows and rare pillow breccia and hyalotuff; bedded chert and calcareous argillite. The rocks in the southern

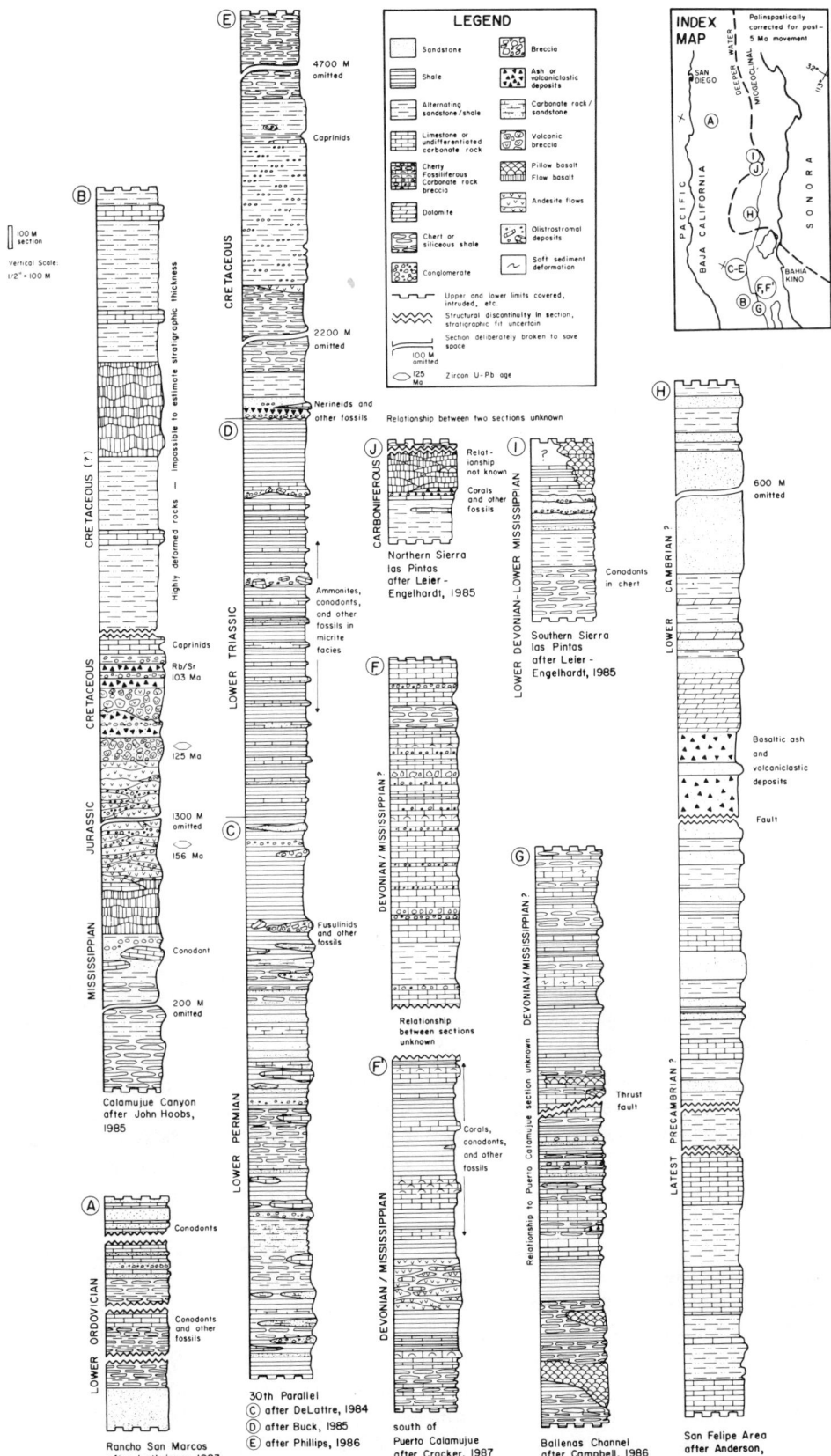

Figure 2. Stratigraphic columns of prebatholithic rocks of northeastern Baja California.

area (720 m) consist in sequence of bedded chert and argillite; fine sandstone, coarse siltstone, and conglomeratic debris flows; limestone and shale; pillow basalt and massive basalt flows.

Lophopyhyllid corals and brachiopods from the northern area suggest a Carboniferous age, and conodont bar fragments recovered from the chert and argillite unit of the southern area indicate an Early Devonian to Early Mississippian age. Detailed thin section analysis suggests that the clastic rocks of both sections were derived from a similar cratonal source and were deposited in a slope to base-of-slope environment with input from nearby carbonate buildups.

The lithic grain population includes quartz arenite chert grains, a small but persistent percentage of volcanic grains, but no metamorphic grains. Minor- and trace-element analyses indicate that the basalts from both areas were erupted during a rifting event, perhaps near a continental margin.

North of the international border there are no recognized deeper-water strata known to be of Paleozoic age, but there are many candidate rocks. Todd and Shaw (1987) have drawn a map showing the extent of candidate strata lying west of the miogeoclinal rocks and east of the presumably Mesozoic flysch rocks ("Julian Schist" and equivalents). These rocks, which they dubbed "transitional," are largely fine-grained and poorly sorted clastic rocks with minor limestone and what is now amphibolite. They differ from the miogeocline succession by lacking quartzite, and from the Julian Schist by containing significant carbonate rock. Hill (1984) describes the Windy Point metamorphic rocks of the San Jacinto Mountains as follows: they contain large thicknesses of plagioclase-rich amphibolites, as well as schists, gneisses, and marbles. The protolith was apparently a laterally variable sequence of immature, possible volcanigenic, sediments with abundant lenses of limestone. Strontium isotope ratios are consistent with late Paleozoic to early Mesozoic age."

West and southwest of the Sierra las Pinta, extensive areas of candidate correlative rock occur in the Sierra Juarez (8) and high Sierra San Pedro Martir (10). In both these ranges the predominant rocks are immature clastics with minor but persistent proportions of carbonate rock, bedded chert, and amphibolite. Unfortunately, both areas have undergone upper amphibolite-grade metamorphism. The less metamorphosed rocks to the west (Reed, 1985; Chadwick, 1987) contain no carbonate rock, bedded chert, or basalt, and are believed to be of Mesozoic age.

On the 30th parallel (14), DeLattre (1984), Buch (1984), and Phillips (1984) have described a well-preserved sequence of rocks at least as old as Early Permian (Fig. 2). Here the slope-facies rocks of Permian age are disconformably overlain by shelf strata of Early Triassic age, which are unconformably overlain by back-arc deposits of mid-Cretaceous age.

Permian rocks on the 30th parallel. DeLattre (1984) and Buch (1984) divided the strata below the Lower Triassic into three informal formations. Paleontologic control is located only at the base of the middle unit.

The lowest of these units consists of a 2,000-m succession of interstratified pelitic and quartzose-carbonate clastic rocks, characterized by thinly interbedded argillite, quartz arenite, sparsely fossiliferous carbonate rock, and bedded chert, with intermittent thicker, laterally discontinuous bodies of calcareous sandstone, impure carbonate rock, and conglomerate.

The thin-bedded rocks account for 75 percent of the unit. They occur in monolithic packets and are rhythmically interbedded in flysch-like succession. Argillites vary from gray to dark gray, platy to block-siliceous, with varying amounts of quartz silt, and are locally calcareous. Normal grading is visible. Chert and chert-like rocks are dark gray to black, and contain abundant carbon. Cherts grade into siliceous argillite. Some chert may result from post-depositional silicification of silt or mudstone. Carbonate rocks are gray to black, and weather to yellowish brown. Concentrations of detrital quartz, argillaceous impurities, and bioclastic debris produce crude laminations and normal grading.

Sandstones are massive to finely laminated, slightly calcareous, and range from arenite to wacke. Grains are monocrystalline quartz, quartzite, and chert. They are moderately to poorly sorted and locally grade into conglomerate.

Conglomerates, in lenses of up to 10 m, contain clasts of quartzite, chert, carbonate rock, and argillite, which are supported in a matrix of sand, silty argillite, and in places carbonate rock. Clasts are pebble to cobble size, poorly sorted, with a random fabric. Fossils include disarticulated crinoid columnals, fragments of bryozoa, and fusulinids(?).

Deposition is interpreted as having occurred by hemipelagic and sediment gravity flows, which accumulated basinward of a carbonate-producing shelf. The tabular bedform, rhythmic appearance, and sequence of sedimentary structures in the flysch-like rocks imply deposition by high-density turbidity currents.

The lenticularity of the conglomerates is attributed by De-Lattre (1984) to deposition within channels. This juxtaposition of channelized, coarse-grained, gravity-flow deposits; hemipelagites; and turbidites suggests deposition proximal to the source, i.e., a base-of-slope wedge or submarine fan.

The overlying middle unit consists of 90 percent argillite with minor sandstone and chaotic argillite–quartzite–carbonate rock conglomerate. The medium gray to black argillite contains finely disseminated pyrite. Locally, grading, flame, and load structures can be observed. Sandstones are generally massive, but some display amalgamated units marked by basal scour and fill, with intraformational rip-ups as much as 0.5 m in length.

The chaotic conglomerates occur in lenses as much as 10 m thick. They range from massive sedimentary breccias to slurries of mud, sand, and fossiliferous carbonate rock. The carbonate-rock clasts measure up to several meters in diameter, and vary from impure unfossiliferous limestone to laminated encrinite with articulated segments as long as 15 cm. There are also brachiopods, fusulinids, and encrusting bryozoa.

Crinoids, with columnals as much as 9 cm in diameter, are the largest ever reported (Webster, 1985). Fusulinids are tentatively identified as *Parafusulina kummeli* Roberts (Calvin Stevens, written communication, 1983), indicating a Leonardian age.

The association of slumps, massive breccias, and turbidites enclosed in hemipelagic deposits suggests a lower-slope environment.

The upper unit consists of 1,000 m of siliceous argillite with thin interbeds of argillaceous conglomerate.

The argillite of the upper unit is believed (Buch, 1984) to have been deposited by fine-grained turbidity flows in relatively deep water, with darker carbonaceous layers deposited by pelagic settling. The calcareous sandstone–sandy limestone may have been deposited by debris flows, and the larger, angular limestone blocks by freefall and rolling from an escarpment. The evidence for the upper unit being Permian is only that it appears conformable with the underlying Lower Permian and is beneath rocks of Early Triassic age (Buch, 1984).

On the 30th parallel the Permian strata are overlain disconformably by a section of argillite with lenses of micrite and intrabasinal conglomerate of Early Triassic (Smithian) age (Buch, 1984). The Triassic is overlain by erosional and angular unconformity by strata of mid-Cretaceous age (Phillips, 1984). The base of the Cretaceous section is marked by a quartzite boulder conglomerate with a siliceous volcanic matrix. A rapid progression from conglomerate to thinly bedded argillite and chert suggests that the Cretaceous basin deepened rapidly up-section.

Mississippian strata in Calamajue Canyon

In Calamajue Canyon (16, Fig. 1; 29°25′N), Hoobs (1985) has described a sequence of bedded chert, flysch, and sandy carbonate rock containing conodonts of medial Mississippian age (Fig. 2). These rocks are overlain by volcanic sequences of both Jurassic and Cretaceous age, and a carbonate/clastic sequence of as yet undetermined age. Here, the apparently oldest part of the section is 1,000 m of chert and shale, and interbedded shale and lithic sandstone, coarsening upward. The uppermost 30 m consists of chert litharenite, pebbly conglomerate, and limestone. The limestone contains crinoid columnals and conodonts. The conodonts include *Gnathodus bilineatus* (Roundy) of early Chesterian age.

These rocks are overlain, apparently with near conformity, by 300 m of basalt, andesite, and epiclastic detritus, which in turn is overlain by 400 m of siliceous volcanic and volcaniclastic strata. Preliminary U-Pb zircon ages are about 155 Ma for the andesite, and 125 Ma for the lower part of the siliceous section (M. Wardlaw, written communication, 1986). A whole-rock Rb/Sr isochron for the fossiliferous upper part of the siliceous section, provided by Sue Gunn of the U.S. Geological Survey (written communication, 1985), indicates 103 ± 4 Ma (Alisitos Formation).

The apparent base of the Cretaceous section is marked by a quartz boulder conglomerate with a siliceous volcanic matrix, as it is on the 30th parallel.

Devonian to Mississippian strata along the Canal Ballenas

Along the coast south from Puerto Calamujue (15, Fig. 1; 29°45′) are nearly continuous exposures of slope and basinal deposits of Paleozoic age. The northern part of this coastal area has been reported in masters theses by Campbell (1985) and Crocker (1987, Fig. 2).

Crocker (1987) describes the northern part of this area as having an average stratigraphic thickness of 2 km, consisting (originally) of thin- and thickly bedded, coarse-grained, bioclastic packstone; black chert; shale; and quartz arenite. To the south (separated by a small tonalite pluton) is a 1-km-thick section of predominantly thinly layered lime-mudstone, coarse-grained bioclastic packstone, and thick boulder and cobble conglomerates with minor bedded chert and argillite. The lime-mudstone suggests carbonate flysch, and is interlayered with intraclastic granule to boulder conglomerate. Some layers contain abundant bioclastic debris.

An anoxic slope-to-basin lithology is suggested outboard of a distally steepening carbonate ramp, with the argillite and chert deposited by hemipelagic settling in an ocean basin, and the carbonate rock representing temporal encroachment of a passive or extensional carbonate platform margin, in part redeposited as carbonate flysch turbidites.

Conodont fragments (Spathognathodiform and Gnathodiform elements) from both the northern and southern parts of the area indicate a Devonian to Mississippian age.

A few kilometers down the coast, Campbell (1985) has mapped four sequences of folded and faulted strata of upper greenschist to lower amphibolite grade. The lowest unit, 1,060 m thick, consists of thick amalgamated chert and slump-folded lime-mudstone, overlain by a thick sequence of argillite. The next unit (1,170 m) is composed of interbedded sequences of impure carbonate rock, chert, argillite, and a minor quartz arenite. Thick-bedded lime-mudstone occurs within the middle portion, and 85 m of pillowed alkaline basalt occurs within the lower portion. The third sequence (1,100 m) is composed primarily of fine-grained clastics interbedded with an abundance of bedded chert and impure carbonate rock. Pillowed alkali basalt within the lower part of the unit reaches a maximum thickness of 400 m. The fourth unit (at least 800 m) is isolated from the other units by a tonalite intrusion. It consists of impure carbonate rock with an intrabasinal pebble to cobble debris flow of carbonate rock and chert clasts. The middle portion of the unit contains impure quartz sandstone. The upper portion is a monotonous sequence of interbedded chert, impure carbonate rock, and minor argillite. Protoliths of the fine clastic and biochemical deposits include bedded cherts, siliceous mudstone, allodapic lime-mudstone, and calcisiltites.

The depositional setting is interpreted as slope to basin. The age of the deposits is unknown.

Midriff islands

Phillips (1964) reported that the metasedimentary rocks exposed on the southwestern coast of Isla Angel de la Guardia (17) closely resembled those that he had visited on the opposite coast (area mapped by Campbell, 1985).

Isla Turner (24, Fig. 1), located at the southeastern corner of Isla Tiburon, consists of bedded chert, flysch, and flysch-like carbonate beds. Limited exposures of a similar sequence occur on the northwestern corner of Isla Tiburon (22), along with a sequence of sandstone. At Punta Onah (23), on the coast of Sonora opposite Isla Tiburon, are exposures of crenulated ribbon chert among other metasedimentary strata. None of these areas have been studied, and no age determinations have been made.

Central Sonora

Miogeoclinal rocks crop out over a considerable area of northern Sonora, west and southwest of Caborca (19). The most recent summary of the work in that area is by Stewart and others (1984). Mississippian and Permian rocks referred to by Rodriquez-Castañeda (1981) as platform facies occur in the vicinity of Hermosillo. Although no fossils have been found southwest of Rancho Pozo Serna (20), similar quartzite/carbonate rock pendants in the Cretaceous granite are exposed farther west. A massive exposure of quartzite just west of Pozo Coyote (21, Fig. 1; latitude 29°37′, longitude 112°22′) and massive exposures of carbonate rock near the coast (22, 22A) suggest that the Miogeoclinal province extends to the gulf.

In the vicinity of Caborca, miogeoclinal strata include rocks of late Proterozoic to Mississippian age. More recently a correlative section has been studied in the Sierra Agua Verde (25), located east of Hermosillo (Stewart and others, 1984). Ketner (1986) believes that the middle Ordovician miogeoclinal quartzite of the Cobachi area (26) is a continuation of the Eureka quartzite of the Owens Lobe area, California. Unpublished work near Cerro Cobachi (26) and Barita de Sonora (27) reveals a sequence of Ordovician to Permian strata of deep-water facies thrust northeastward over an Ordovician to Permian sequence of miogeoclinal facies (Noll, 1981; Ketner, 1983; Poole and Madrid, 1986; Stewart, written communication, 1984–1986). The limits of Paleozoic facies and structures in southern Sonora are not yet well defined, and the metamorphic overprinting of the Laramide-age batholiths are locally very strong.

Northern Sinaloa

Deeper-water-facies rocks believed to be of Paleozoic age occur in at least two areas of northern Sinaloa. Adjacent to Presa Miguel Hidalgo, north of El Fuerte (28), the basement rock consists of thinly bedded chert, carbonate rock, and argillite. Mullan (1978) believed these strata to be contemporaneous with volcanic rocks of rhyolitic and andesitic composition. Attempts to recover adequate zircons from these volcanic strata have so far been unsuccessful, so that this age relation has not been confirmed. Preliminary U-Pb zircon work on a body of two-mica granodiorite believed to be contemporaneous with the volcanic strata indicates a Jurassic age, and a whole-rock Rb/Sr isochron on argillaceous rocks also suggests a Mesozoic age.

Carrillo-Martinez (1971) reported Paleozoic rocks in the area of San José de Gracia (29), near the Chihuahua border, consisting of: a lower sequence of chert, shale, micrite limestone, and quartzite; and an upper sequence of shale and quartzite, which in places is cross-laminated. A calcarenite from the lower sequence yielded the fusulinid *Millerella* (Carrillo-Martinez, 1971), which has a range of Early Mississippian to late Pennsylvanian. In 1984, Gastil and Roldan, revisiting this area (29), found kilometers of exclusively bedded chert exposed along the road approaching San José de Gracia. These cherts are typically in beds of about 5 cm, separated by paper-thin argillite partings. Locally there is quartz arenite in the San José area, and even cobble conglomerate. There are flysch-like fine clastic strata and rare carbonate rock. A small lens-form exposure of calcarenite yielded the conodonts *Streptognathodus*(?) and *Gondolella*, indicating a mid-Pennsylvanian to Early Permian age.

Structure

There has not been a systematic study of the structure of the Paleozoic rocks in peninsular California, nor for that matter, of those immediately on the Sonora side. Mullan (1978) reported a complicated history for the rocks of the El Fuerte area of northern Sinaloa (28).

McEldowney (1968) and Leier-Engelhardt (1986) report that the Paleozoic rocks of the Sierra las Pinta (9), particularly the northern part, are deformed in recumbent folds with subhorizontal axial foliation, and an axial strike of about N80°W with vergence toward the north. Campbell (1985) and Crocker (1987) reported folds along the Ballenas Channel (15–15A), that verge toward the east and northeast. These folds are in sharp contrast to both the style and orientation of folding throughout the western and axial portions of the peninsula, where foliation strike is to the northwest, and inclination is steep to the northeast, generally with a pronounced down-dip stretching lineation (Griffith and Goetz, 1987).

Noll (1981) and Ketner (1986) indicate that the structures of the deeper-water rocks of central Sonora verge to the east. Stewart and others (1987) believe the deep-water Paleozoic rocks were thrust northward over the coeval carbonate and siliclastic rocks in post-mid Permian and pre-Late Triassic time.

Even less is known about the structures of miogeoclinal rocks on either side of the gulf, except those of the Caborca area, which are cut by low-angle "thrusts" presumed to be of Laramide age (Merrian and Eells, 1978; DeJong and Escárcega-Escárcega, 1986). The miogeoclinal rocks were supposedly deposited on more rigid crust than the deeper-water facies, and probably did not deform in the same style. Several workers have suggested that the deeper-water facies were transported eastward and/or northeastward over the miogeoclinal facies in a manner analogous to that of the Antler and/or Sonoman orogeny of central Nevada (Noll, 1981; Poole, 1983; Stewart and others, 1987).

SUMMARY AND CONCLUSIONS

Formations similar in both age and lithology to the miogeoclinal and deep-water Paleozoic facies of the Great Basin are present in northeastern Baja California and in the states of Sonora and Sinaloa (Figs. 1 and 3). The rocks described here all lie

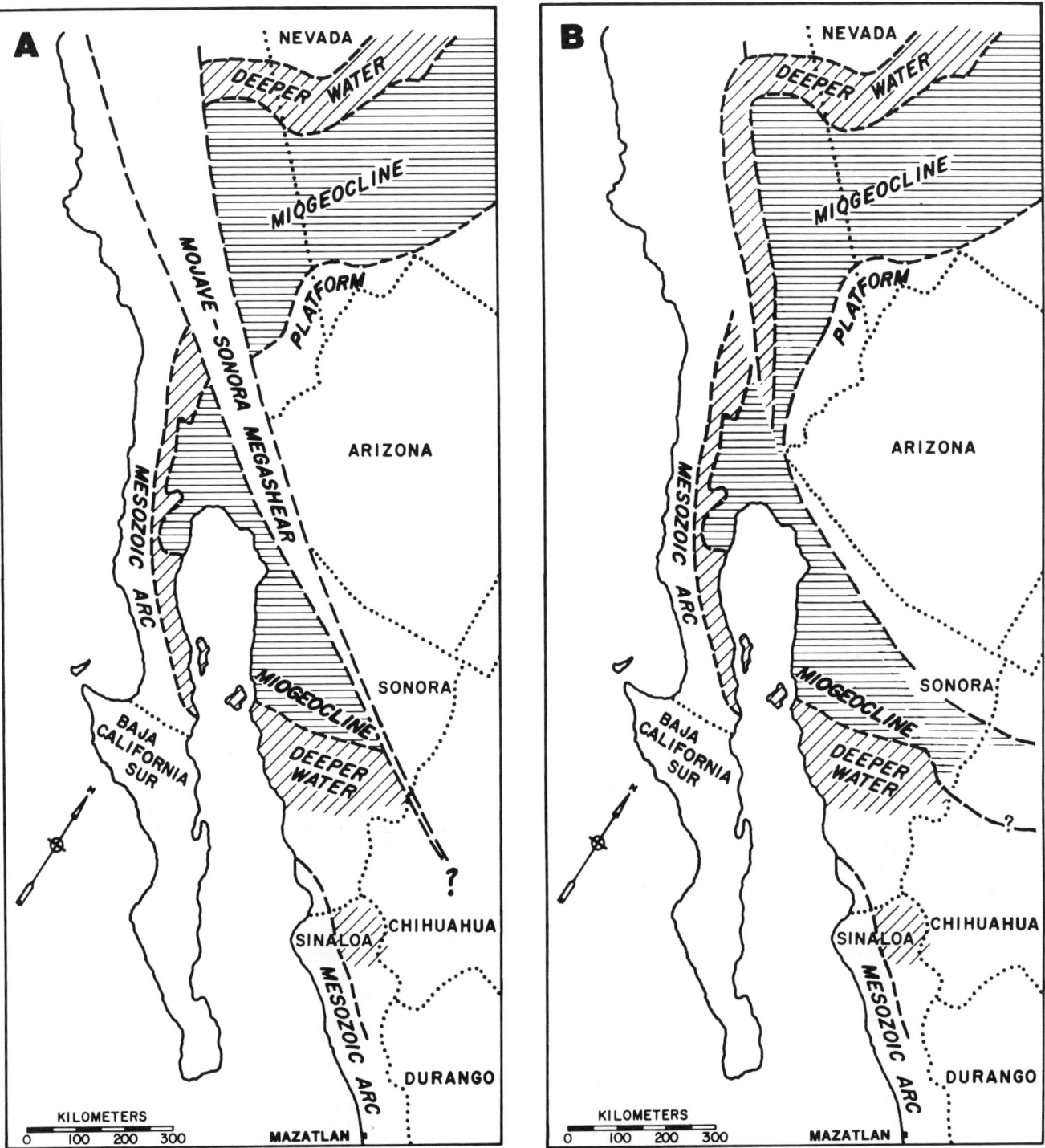

Figure 3. A. Regional tectonic implications of the miogeoclinal/deeper-water Paleozoic facies boundary in peninsular California and adjacent Mexico adopting the Mojave-Sonora Megashear hypothesis. The distributed location for the megashear follows Dickinson, 1981. B. Shows the same facies boundaries without adopting the megashear (see for example Stewart, 1988). Neither diagram has been corrected for displacement along the San Andreas fault.

southwest of the megashear proposed by Silver and Anderson in 1974, and their correlatives to the north are located on the northeastern side. Thus, the position of these rocks is consistent with large left-lateral displacement (Fig. 3A), but could also be explained by a bending strand line around the southwestern corner of the continent (Fig. 3B). The difficulty with the 3A explanation is seen farther north in California. If the megashear concept is correct, we would expect to see the miogeocline/platform boundary of southeastern California offset to the southeast and reappear on the southern side of the shear in Sonora, but the rocks south of the shear are miogeoclinal and deep water (Fig. 1). And, if the miogeoclinal/deep-water boundary that has been mapped east-west across southern Sonora in the past decade (Fig. 1) is offset left-laterally by the megashear it should reappear north of the fault in western Arizona. The absence of these offsets presents problems for the megashear concept (Fig. 3A).

Palinspastic reconstructions based on paleomagnetic data, summarized by Champion and others (1986) and Hagstrum and others (1987), require that peninsular California was located near the middle America trench in Cretaceous time, arriving at the latitudes portrayed in Figure 1 sometime in the early to middle Tertiary. The lateral slip required for the peninsula by this construction is at least 1,000 km. The difficult question is: where is the location of the fault along which the peninsula traveled? The tie of Paleozoic facies across the Gulf of California (Fig. 1) seems to preclude the location of this fault beneath the Gulf of California. Hagstrum and others (1987), referencing Bobier and Robin (1983), place this fault east of the Sinaloa batholith at the latitude of Mazatlan (Fig. 3), and west of El Antimonio in northern Sonora (west of locality 19, Fig. 1). This construction escapes some of the problems inherent in placing the fault beneath the gulf, but appears to face other problems in threading between the terranes of central Sonora. Specifically, it would have to be located far enough east not to cross the miogeocline/deep-water Paleozoic facies boundary east of Hermosillo (25, 26, 27), yet far enough west to place the Caborca terrane (at least 19 and 20) on the eastern side of the fault.

Gastil (1985) made the suggestion that the enigma could be resolved by first moving the peninsula southeast on a left-lateral megashear (mid-Jurassic), and then moving it back right-laterally (after the Cretaceous). This option is, of course, only viable so long as we accept the megashear construction. There is still another possibility. If we prefer the Paleozoic facies boundaries as explained by Figure 3B, but still need the megashear explanation to counter subsequent right-lateral motion, we can place an alternate metashear (of mid-Jurassic age) through the Gulf of California. The post-Cretaceous right-lateral fault would then pass east of the Sinaloa batholith, and beneath the Gulf of California north of the 28th parallel. The Jurassic left-lateral shift would need to approximately balance the early Tertiary right-lateral return. There is no easy solution.

REFERENCES CITED

Anderson, P. V., 1982, Prebatholithic stratigraphy of the San Felipe area, Baja California, Mexico [M.S. thesis]: San Diego, California, San Diego State University, 100 p.

Bobier, C. and C. Robin, 1983, Paleomagnetisme de la Sierra Madres Occidentale dans les etats de Durango et Sinaloa (Mexique); Variations du champ ou rotations de blocs au Paleocene et au Neogene?: Geofis. Int., v. 22, no. 1, p. 57–86.

Brown, A. R., 1980, Limestone deposits of the Desert Divide, San Jacinto Mountains, California, in Fife, D. L., and Brown, A. R., eds., Geology and mineral wealth of the California desert: Santa Ana, California, South Coast Geological Society, p. 284–293.

Bryant, B. A., 1986, Geology of the Sierra Santa Rosa Basin, Baja California, Mexico [M.S. thesis]: San Diego, California, San Diego State University, 75 p.

Buch, I. P., 1984, Upper Permian(?) and Lower Triassic metasedimentary rocks, northeastern Baja California, Mexico: Pacific Section, Society of Economic Paleontologists and Mineralogists Field Trip Guidebook 39, p. 31–36.

Campbell, M. J., 1985, Prebatholithic stratigraphy of the northeastern Sierra la Asamblea, Baja California, Mexico [M.S. thesis]: San Diego, California, San Diego State University, 127 p.

Carillo-Martinez, M., 1971, Geologia de la hoja San Jose de Gracia, Sinaloa: Universidad Nacional Autonoma de Mexico Facultad de Ingenieria, 25 p.

Chadwick, B., III, 1987, The petrology, geochemistry, and geochronology of the Tres Hermanos–Santa Clara region, Baja California, Mexico [M.S. thesis]: San Diego, California, San Diego State University, 195 p.

Champion, D. E., Howell, D. G., and Marshall, M., 1986, Paleomagnetism of Cretaceous and Eocene strata, San Miguel Island, California; Borderland and the northward translation of Baja California: Journal of Geophysical Research, v. 91, p. 11,557–11,570.

Coney, P. J., Jones, D. F, and Monger, J.W.H., 1980, Cordilleran suspect terranes: Nature, v. 288, p. 329–333.

Cooper, G. A., and Arellano, A.R.V., 1946, Stratigraphy near Caborca, northwestern Sonora, Mexico: American Association of Petroleum Geologists Bulletin, v. 30, p. 606–611.

Crocker, J., 1987, Stratigraphy and structure of Paleozoic metasediments south of Bahia Calamajue, Baja California, Mexico [M.S. thesis]: San Diego, California, San Diego State University, 129 p.

D'Allura, J. A., Moores, E. M., and Robinson, L., 1977, Paleozoic rock sof the northern Sierra Nevada; The structure and paleogeographic implications, in Stewart, J. H., Calvin, H. S., and Fritsche, A. B., eds., Paleozoic paleogeography of the western United States: Pacific Section, Society of Economic Paleontologists and Mineralogists, p. 395–408.

DeJong, K. A., and Escarcega-Escarcega, J. A., 1986, Thrust faulting, folding, and backslip in the Cerros Pitiquito near Caborca, NW Sonora, México: Geological Society of America Abstracts with Programs, v. 18, no. 6, p. 100.

DeLattre, M. P., 1984, Lower Permian metasedimentary rocks of Zamora, northeastern Baja California, Mexico, in Frizzel, V., ed., Geology of the Baja California Peninsula: Pacific Section, Society of Economic Paleontologists and Mineralogists, p. 23–29.

Dickinson, W. R., 1981, Plate tectonic evolution of the southern Cordillera, in Dickinson, W. R., and Payne, W. D., eds., Relations of tectonics to ore deposits in the southern Cordillera: Arizona Geological Society Digest, v. 14, p. 113–135.

Dockum, M. S., 1982, Greenschist-facies carbonates, eastern Coyote Mountains, western Imperial County, California [M.S. thesis]: San Diego, California, San Diego State University, 89 p.

Engle, A.E.J., and Schultejohn, P. A., 1984, Late Mesozoic and Cenozoic tectonic history of south-central California: Tectonics, v. 3, no. 6, p. 659–675.

Gastil, G., 1985, Terranes of Peninsular California and adjacent Sonora 1985, in Howell, D. G., ed., Proceedings of the Fourth Circum Pacific Terranes Conference: Circum Pacific Council for Energy and Mineral Resources, Earth Science Series, no. 1, p. 273–283.

Gastil, R. G., and Miller, R. H., 1981, Lower Paleozoic strata on the Pacific plate of North America: Nature, v. 292, no. 5826, p. 828–830.

Gastil, R. G., Lemone, D. V., and Steward, J., 1973, Permian fusulinids from near San Felipe, Baja California: American Association of Petroleum Geologists Bulletin, v. 57, no. 4, p. 746–747.

Griffith, R. C., and Goetz, C. W., 1987, Structural and geochronological evidence for mid-Cretaceous compressional tectonics along a terrane boundary in the Peninsular Ranges: Geological Society of America Abstracts with Programs, v. 19, p. 384.

Hagstrum, J. T., McWilliams, M., Howell, D. G., and Gromme, S., 1987, Mesozoic paleomagnetism and northward translation of the Baja California Peninsula: Journal of Geophysical Research, v. 92, no. B3, p. 2627–2639.

Harris, A. G., Bergstrom, S. M., Ethington, R. L., and Ross, R. J., Jr., 1979, Aspects of Middle and Upper Ordovician conodont biostratigraphy of carbonate facies in Nevada and southeast California and comparison with some Appalachian successions, in Sanberg, C. A., and Clark, D. L., eds., Conodont biostratigraphy of the Great Basin and Rocky Mountains: Brigham Young University Geology Studies, v. 26, pt. 3, p. 7–43.

Hill, R. I., 1984, Petrology and petrogenesis of batholithic rocks, San Jacinto Mountains, southern California [Ph.D. thesis]: Pasadena, California, California Institute of Technology.

Hiner, J. E., 1972, Geology of a portion of the Sierra Santa Rosa, Valle de Chico Quadrangle, Baja California, Mexico [Sr. report]: San Diego, California, San Diego State University, 23 p.

Hoobs, J. H., 1985, Carboniferous island-arc and associated rocks from the Mission Calamuje area, Baja California, Mexico [M.S. thesis]: San Diego, California, San Diego State University, 122 p.

James, A. H., 1973, Structure and stratigraphy of the southern Sierra las Pinta, Baja California, Mexico [M.S. thesis]: San Diego, California, San Diego State University, 56 p.

Ketner, K. B., 1983, Strata-bound, silver-bearing iron, lead, and zinc sulfide deposits in Silurian and Ordovician rocks of allochthonous terranes, Nevada and northern Mexico: U.S. Geological Survey Open-File Report 83-792.

—— , 1986, Eureka quartzite in Mexico?; Tectonic implications: Geology, v. 14, p. 1027–1030.

Leier-Engelhardt, P. J., 1986, Middle Paleozoic strata of the Sierra las Pinta, northeastern Baja California Norte, Mexico [M.S. thesis]: San Diego, California, San Diego State University, 169 p.

Lothringer, C. J., 1984, Geology of a Lower Ordovician allochthon, Rancho San Marcos, Baja California, Mexico, in Frizzell, V. A., Jr., ed., Geology of the Baja California Peninsula: Pacific Section, Society of Economic Paleontologists and Mineralogists, p. 17–22.

McEldowney, R. C., 1970, Geology of the northern Sierra Pinta, Baja California, Mexico [M.S. thesis]: San Diego, California, San Diego State University, 78 p.

Merriam, R. H., and Eells, J. L., 1978, Reconnaissance geologic map of the Carborca Quadrangle, Sonora, Mexico: Universidad de Sonora Departamento geologia Boletin, v. 1, no. 2, p. 87–94.

Miller, R. H., and Dockum, M. S., 1983, Ordovician conodonts from metamorphosed carbonates of the Salton trough, California: Geology, v. 11, p. 410–412.

Mullan, H. S., 1978, Evolution of part of the Nevadan orogeny in northwestern Mexico: Geological Society of America Bulletin, v. 89, p. 1175–1188.

Noll, J. H., 1981, Geology of the Pichacho, Colorado, area, northern Sierra del Cobachi, central Sonora, Mexico: Geological Society of America Abstracts with Programs, v. 3, p. 99.

Pfeiffer, Rangin, F., 1979, Les zones isopiques du Paleozoique inferieur du nordouest Mexicain, temoins du relais entre les Appalaches et la cordillere ouestamericaine: Compte Rendu Academie Sciences Paris, Series D., v. 288, p. 1517–1519.

Phillips, J. R., 1984, "Middle" Cretaceous metasedimentary rocks of La Olvidada, northeastern Baja California, Mexico, in Frizzell, V., ed., Geology of the Baja California Peninsula: Pacific Section, Society of Economic Paleontologists and Mineralogists, p. 37–41.

Phillips, R. P., 1966, Reconnassance geology of some of the northwestern islands in the Gulf of California: Geological Society of America Cordilleran Section Program, p. 59.

Poole, F. G., 1983, Bedded barite deposits of middle and late Paleozoic age in central Sonora, Mexico: Geological Society of America Abstracts with Programs, v. 15, no. 5, p. 299.

Poole, F. G., and Madrid, R. J., 1986, Paleozoic rocks in Sonora, Mexico, and their relation to the southwestern continental margin of North America: Geological Society of America Abstracts with Programs, v. 18, p. 720.

Reed, J. B., 1985, Sedimentology of some Triassic–Middle Jurassic(?) siliciclastic flysch deposits, northwest Baja California Norte, Mexico [M.S. thesis]: San Diego, California, San Diego State University, 265 p.

Rodriguez-Castaneda, J. L., 1981, Notas sobre la geologia del area de Hermosillo, Sonora: Universidad Nacional Autónoma de Mexico Institute Geologia Revista, v. 5, no. 1, p. 30–36.

Ross, R. J., 1977, Ordovician paleogeography of the western United States in Stewart, J. H., Stevens, C. H., and Fritsche, A. E., eds.: Los Angeles, California, Pacific Section, Society of Economic Paleontologists and Mineralogists, p. 19–39.

Sergeeva, S. P., 1962, Stratigraphic distribution of conodonts from the Lower Ordovician of the Leningrad region: Akademiya Nauk SSSR Paleontologicheskiy Zhurnal, v. 146, p. 1393–1395 (in Russian).

—— , 1963, Conodonts from the Lower Ordovician in the Leningrad region: Adademiya Nauk SSSR Paleontologicheskiy Zhurnal, v. 2, p. 93–108 (in Russian).

Silver, L. T., and Anderson, T. H., 1974, Possible left-lateral early to middle Mesozoic disruption of the southwestern North America craton margin: Geological Society of America Abstracts with Programs, v. 6, no. 7, p. 955.

Stewart, J. H., 1981, Paleozoic and uppermost Precambrian Cordilleran geosyncline in the western United States and northwestern Mexico; A review: Geological Society of America Abstracts with Programs, v. 13, no. 2, p. 108.

Stewart, J. H., 1988, Latest Proterozoic and Paleozoic southern margin of North America and the accretion of Mexico: Geology, v. 16, p. 186–189.

Stewart, J. H., 1990, Position of Paleozoic continental margin in northwestern Mexico, present knowledge and speculations: Geological Society of America Abstracts with Programs, v. 22, no. 3, p. 86–87.

Stewart, J. H., and Poole, F. G., 1975, Extension of the Cordilleran miogeosyncline to the San Andreas fault, southern California: Geologial Society of America Bulletin, v. 86, p. 205–212.

Stewart, J. H., McMenamin, M.A.S., and Morales-Ramirez, J. M., 1984, Upper Proterozoic and Cambrian rocks in the Caborca region, Sonora, Mexico; Physical stratigraphy, biostratigraphy, paleocurrent studies, and regional relations: U.S. Geological Survey Professional Paper 1309, 36 p.

Stewart, J. H., and 5 others, 1990, Tectonics and stratigraphy of the Paleozoic and Triassic southern margin of North America, Sonora, Mexico, in Gehrels, G. E., and Spencer, J. E., eds., Geologic excursions through the Sonoran Desert Region, Arizona and Sonora: 86th Annual Meeting, Cordilleran Section, Geological Society of America, Tucson, Arizona, Field Trip guidebook, p. 183–202.

Todd, V. R., Erskine, B. G., and Morton, D. M., 1988, Metamorphic and tectonic evolution of the northern Peninsular Ranges batholith, southern California, in Ernst, W. G., ed., Metamorphic and crustal evolution of the western United States, Rubey Volume VII: p. 894–937.

Webster, G. D., Gastil, G., and Delattre, M., 1985, World's largest crinoid columnals from Baja California Norte: Geological Society of America Abstracts with Programs, v. 17, no. 6, p. 417.

MANUSCRIPT ACCEPTED BY THE SOCIETY APRIL 18, 1990

Printed in U.S.A.

Upper Triassic Barranca Group; Nonmarine and shallow-marine rift-basin deposits of northwestern Mexico

John H. Stewart
U.S. Geological Survey, 345 Middlefield Road, Menlo Park, California 94025
Jaime Roldán-Quintana
Instituto de Geología, Universidad Autónoma de México, Apartado Postal 1039, Hermosillo, Sonora 83000, México

ABSTRACT

The 3,000-m-thick Upper Triassic Barranca Group in the Sierra de San Javier in east-central Sonora, Mexico, is composed, in ascending order, of the Arrayanes, Santa Clara, and Coyotes Formations. The Arrayanes and Santa Clara Formations are composed of fluvial and marine-delta deposits of quartzose and arkosic sandstone, conglomerate, shale, and siltstone; the Santa Clara Formation includes minor amounts of coal and tuff. A sharp contact (perhaps an unconformity) separates the Santa Clara Formation from the overlying Coyotes Formation. The Coyotes consists of alluvial-fan deposits of pebble-to-boulder conglomerate. Paleocurrents were southward during deposition of the Arrayanes and Santa Clara Formations and southwestward during deposition of the Coyotes Formation, assuming that no major post-deposition tectonic rotation has occurred. The Santa Clara Formation has been dated paleontologically as Late Triassic; the age of the entire group is unknown, but is commonly assumed to also be Late Triassic. The Barranca Group in the Sierra de San Javier rests unconformably on a sequence of eugeosynclinal chert, argillite, quartzite, and carbonate rock of Paleozoic age, and is unconformably overlain by the Tarahumara Volcanics, which have been dated no more precisely than latest Triassic to earliest Cenozoic.

The thick, coarse, and laterally variable deposits of the Barranca Group indicate deposition in a basin, or basins, flanked by areas of high relief. Much of the Barranca in Sonora appears to have been deposited in a single basin, which is delineated by the occurrence of major outcrops of the Barranca Group in an east-west–trending belt about 110 km long and 40 km wide. The elongate shape of this basin and the interpretation of flanking areas of high relief suggests a basin of rift origin. If so, the Barranca Group is part of a broad zone of rift-related Upper Triassic sequences in northern Mexico that apparently formed by transtensional and/or extensional faulting.

INTRODUCTION

The Upper Triassic Barranca Group of east-central Sonora, Mexico, is critical in understanding the Mesozoic paleogeography and tectonic evolution of North America. This group records a geologic history of deposition of coarse clastic rocks that contrasts markedly with continental shelf, off-shelf, and oceanic deposits of Paleozoic age in Sonora (Stewart and others, 1984). The Barranca also records tectonic events before the development of a widespread Jurassic magmatic arc and the transform faulting related to the opening of the Gulf of Mexico (Anderson and Schmidt, 1983). A particularly puzzling problem is posed by the scarcity of volcanic material in the Barranca, considering its position near the presumed source region for abundant volcanic detritus of the Chinle Formation in the southwestern United States (Stewart and others, 1986). Abadie (1981) suggested that the

Stewart, J. H., and Roldán-Quintana, J., 1991, Upper Triassic Barranca Group; Nonmarine and shallow-marine rift-basin deposits of northwestern Mexico, *in* Perez-Segura, E., and Jacques-Ayala, C., eds., Studies of Sonoran geology: Geological Society of America Special Paper 254.

Barranca Group lay in a fore-arc basin south of a major arc system that contributed debris northward into the Upper Triassic Chinle basin in the southwestern United States. However, such an interpretation seems unlikely because of the sparsity of volcanic debris in the Barranca Group. Alternately the Barranca may represent a rift-related deposit and part of a widespread group of such deposits in northern Mexico, the Gulf Coast region of the United States, and in the eastern United States.

During our study, we investigated the stratigraphy, provenance, depositional environments, paleogeography, and tectonic setting of the Barranca Group. Our field work was largely in the Sierra de San Javier (Fig. 1) and was done mostly during March 1985. It consisted of reconnaissance field mapping to establish stratigraphic units, measurements of paleocurrent directions, and other detailed studies useful in interpreting depositional environments. Exposures in the Sierra de San Javier area are generally poor, but we concentrated our studies there because the area contains the most complete and least metamorphosed exposures of these rocks in Sonora.

PREVIOUS WORK

Alencaster (1961a) has reviewed the study of Triassic rocks in Sonora. Dumble (1900a, b) applied the name "Barranca Division" to these rocks, which was later modified to "Barranca Formation" by King (1939) and finally to "Barranca Group" by Alencaster (1961a). A threefold subdivision of these rocks was first proposed by Wilson and Rocha (1949), to which Alencaster (1961a) gave the names Arrayanes, Santa Clara, and Coyotes Formations, in ascending order.

The best-studied outcrops of the Barranca Group are in the Sierra de San Javier (Fig. 1). The area includes the town of La Barranca (Fig. 2), for which the group was named, and also the cultural and topographic features for which the Arrayanes, Santa Clara, and Coyotes Formations were named. Wilson and Rocha (1949) and Avila de Santiago (1960) mapped parts of the area. Alencaster (1961a) described the regional stratigraphy of the Barranca Group, using the Sierra de San Javier as the main reference area. The fauna and flora of the Santa Clara Formation have been studied in detail by Alencaster (1961b), Silva-Pineda (1961), Weber (1980, 1985), and Weber and others (1980a, b). Part of the Barranca Group in the Sierra de San Javier is described by Potter and Cojan (1985). The stratigraphy and structure of part or all of the Sierra de San Javier, and the adjacent Sierra El Aliso have been described by Soto-Contreras and Navarro-Martínez (1987), Barrero-Moreno and Domínguez-Perla (1987), Radelli and others (1987), and Bartolini (1988).

STRUCTURAL SETTING

The Barranca Group in the Sierra de San Javier crops out in an irregular, faulted, gently east-southeast–plunging syncline (Fig. 2), defined by southward dips on the northeastern limb and east to northeast dips on the southwestern limb. Considerable structural variability is evident in this syncline. In some places the beds strike at right angles to the general structure, and in an area 4 km northeast of San Javier, overturned beds dip northeast. In this same area, 4 km northeast of San Javier, a few minor folds trend southeast, approximately parallel to the main syncline. The rocks of the Sierra de San Javier are cut by several major southeast-trending, high-angle faults (Fig. 2) having a right-lateral separation of 1 to 2 km. Whether this separation is due to right slip or to normal offset of dipping beds is not known. In many road cuts, normal faults spaced 1 to 10 m apart cut the Barranca Group. Presumably this style of faulting characterizes much of the outcrop area of the Barranca; however, it is commonly obscured by vegetation.

PRE-BARRANCA ROCKS

The Barranca Group rests unconformably on a sequence of chert, shale, siltstone, quartzite, conglomerate, and limestone. These strata are dated, on the basis of graptolites, radiolaria, conodonts, and fusulinids, as Ordovician, Devonian, Mississippian and Pennsylvanian(?), and Permian in age (Bartolini, 1988; J. H. Stewart, unpublished data, 1987). This sequence of Paleozoic rocks apparently includes what Dumble (1900a) suggested were Paleozoic rocks at the "Los Bronces" locality 3 km north-northeast of San Javier. Part or all of this sequence was also included in the Paleozoic by Alencaster (1961a, plate 5), Menicucci and others (1982), Radelli and others (1987), Soto-Contreras and Navarro-Martínez (1987), Barrera-Moreno and Domínguez-Perla (1987), and Bartolini (1988). King (1939), on the other hand, erroneously included these Paleozoic rocks in the Upper Triassic Barranca Group, and this assignment was followed on the 1:250,000-scale map of the Tecoripa sheet (INEGI, 1984).

The most conspicuous unit in the Paleozoic sequence is a 100-m-thick vitreous quartzite that was included in the Upper Triassic Barranca Group by King (1939). This quartzite is now known to be Middle Ordovician in age on the basis of graptolites (discovered by L. A. Navarro-Martínez in 1986 and identified by Clair Carter, written communication, 1987) in a shale within the quartzite. This Ordovician quartzite differs from sandstone and quartzite in the Barranca Group in that it is composed almost entirely of rounded fine- to medium-grained quartz cemented by silica, whereas the quartzite and sandstone of the Barranca Group are locally coarse grained to conglomeratic, feldspathic and/or clay rich, and contain angular grains.

The Paleozoic rocks in Sierra de San Javier are unconformably overlain by rocks of the Barranca Group. The contact was not seen in detail but is recognized by a general lithologic change from chert, siltstone, and fine- to medium-grained vitreous quartzite below to fine- to medium-grained, locally coarse-grained to conglomeratic sandstone or quartzite above. The attitude of the pre-Barranca rocks is generally different from that of the Barranca, and in places, pre-Barranca rocks strike obliquely into the mapped contact. Folding is more conspicuous in pre-Barranca rocks.

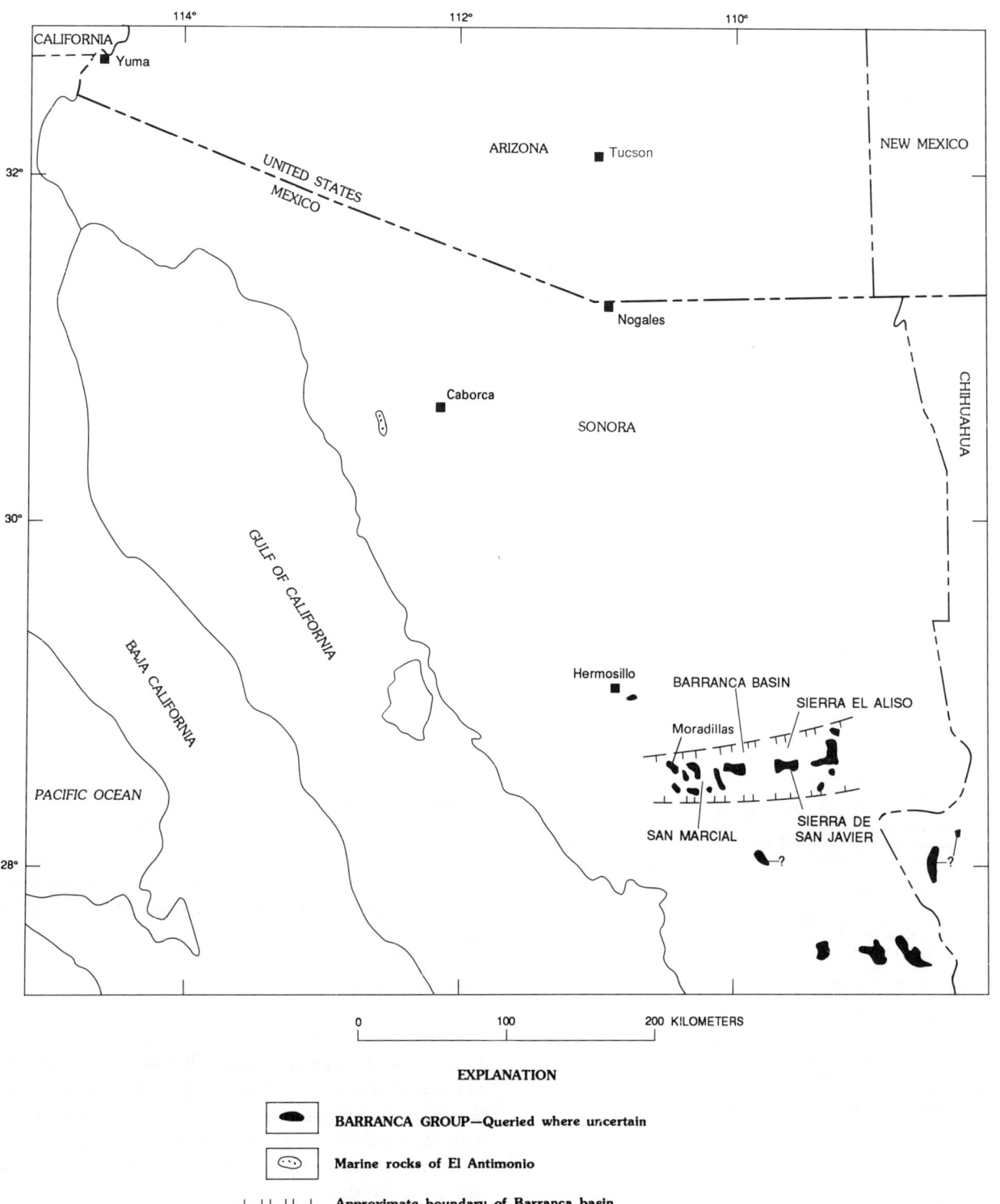

Figure 1. Index map showing outcrops of Upper Triassic rocks in Sonora, Mexico, and location of Sierra de San Javier. Only outcrops that can be clearly identified as the Barranca Group on the basis of paleontologic information or presence of coal or graphite are shown.

EXPLANATION

- Qa — Alluvium (Quaternary)
- Qls — Landslide deposits (Quaternary)
- Tvs — Volcanic and sedimentary rocks (Tertiary)
- TMzi — Intrusive rocks (Tertiary and (or) Mesozoic)
- TMzt — Tarahumara Volcanics (Tertiary and (or) Mesozoic)
- Barranca Group (Upper Triassic)
  - ℞c — Coyotes Formation
  - ℞sc — Santa Clara Formation
  - ℞a — Arrayanes Formation, undivided—Locally divided into
    - ℞au — Upper member
    - ℞am — Middle member
    - ℞al — Lower member
- Pzs — Sedimentary rocks (Paleozoic)—Chert, shale, siltstone, quartzite, conglomerate, and limestone
- Contact
- Fault—Dotted where concealed. Bar and ball on downthrown side
- Syncline
- Strike and dip of beds
  - Inclined
  - Overturned
  - Vertical
- Main road

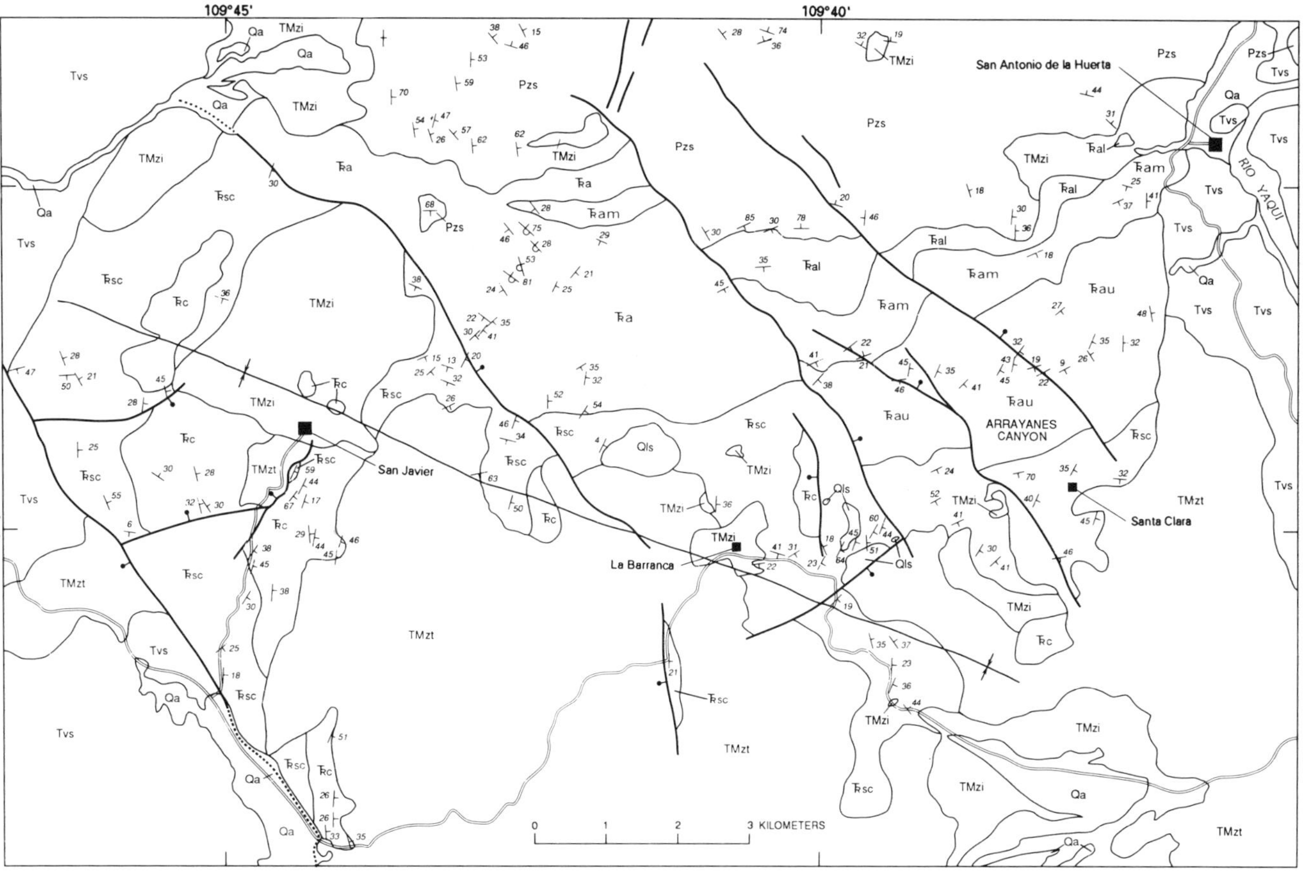

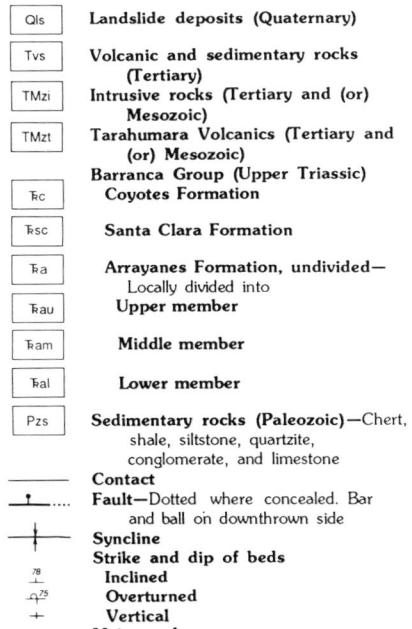

Figure 2. Reconnaissance geologic map of the Sierra de San Javier area. Map based in part on Wilson and Rocha (1949) and Avila de Santiago (1960).

BARRANCA GROUP

The Barranca Group is exposed in the southern part of the Sierra de San Javier over an area about 16 km by 8 km (Fig. 2). It is divided into the Arrayanes, Santa Clara, and Coyotes Formations (Fig. 3).

Arrayanes Formation

The Arrayanes Formation is exposed in an east-west band across the northern part of this area. It is well exposed in the high and rugged central part of the range and is generally poorly exposed in the low hills on the flanks. In the eastern part of the range, it is divided into three members, referred to here informally as the lower, middle, and upper members, whereas in the western part it is undivided.

The lower and upper members of the Arrayanes and the undivided Arrayanes (Figs. 3 and 4) consist of medium gray to light brownish gray sandstone that commonly weathers to very pale orange. The sandstone is composed of fine- to medium-

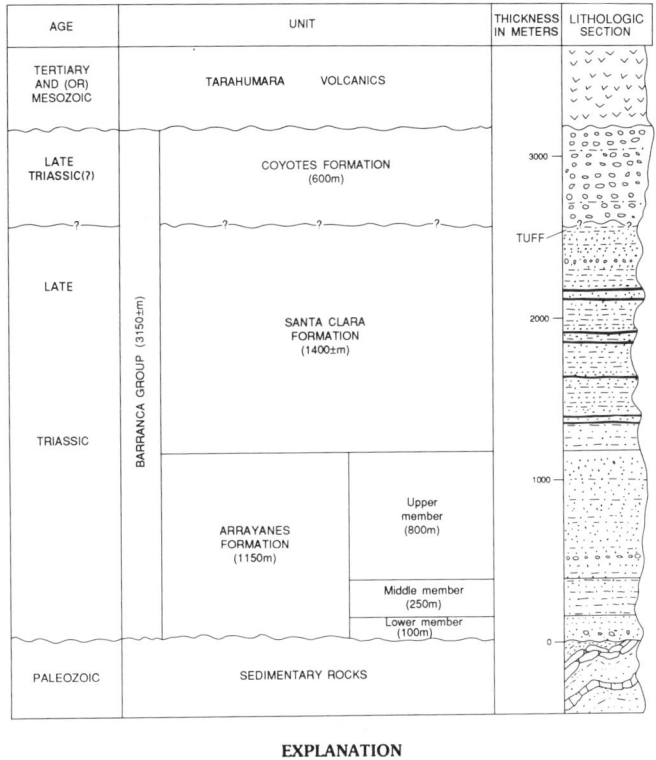

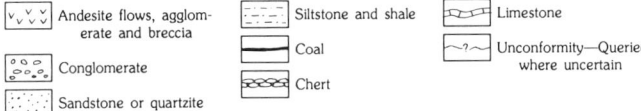

Figure 3. Stratigraphic column of the Barranca Group, Sierra de San Javier. Stratigraphic thicknesses of units and positions of lithologic types are poorly known and are shown diagrammatically.

grained subangular to subrounded quartz and lesser amounts of feldspar. It is generally fairly well to poorly sorted. Coarse to very coarse grained layers are common, and conglomerate is present locally. The conglomerate contains granules, pebbles, and locally cobbles, as much as 14 cm across, of gray and brown quartzite, chert, and sparse white quartz and red chert. The sandstone comprises 1- to 15-m-thick units, many of which are massive and contain only poorly defined internal laminations or thin beds. Other units contain sparse planar and trough cross-strata in units that become finer grained upward (Fig. 4).

Light gray and olive-gray siltstone occurs in very thin beds or in units as much as 20 m thick, interstratified with the sandstone. The siltstone is shaly in places and locally contains plant debris. The amount of siltstone in the Arrayanes is difficult to estimate because of poor exposures. In places, it probably constitutes only one-fourth of the sequence, whereas in other places it may constitute as much as one-half.

The middle member (redbed sequence) of the Arrayanes consists of grayish red siltstone and minor amounts of sandstone (Fig. 4). The predominant red color and abundance of siltstone contrasts markedly with the dominant sandstone and siltstone of the remainder of the Arrayanes that weathers very pale orange. Locally in Arrayanes Canyon, reddish gray siltstone that is similar to the middle member is present within the upper member of the Arrayanes Formation. The middle member is well exposed from about 0.5 to 1 km southwest of San Antonio de la Huerta (Fig. 2). The siltstone commonly contains disseminated very fine to medium-grained sand. It generally contains no internal stratification. About 10 percent of the member consists of yellowish gray sandstone units from 0.2 to 3 m thick interstratified with the dominant siltstone (Fig. 4). Some of these sandstone units are fine grained, others are medium to coarse grained, and still others grade from very coarse grained at the base to fine to medium grained at the top. A few contain granules and small pebbles of chert and quartzite. The sandstone commonly is found in lenticular layers that pinch out along the outcrop; in places, the sandstone fills channels cut into underlying siltstone. Internal stratification in the sandstone units is generally poorly defined, and many individual units appear structureless. Some units contain poorly defined laminae to thin beds; a few contain planar cross-strata, particularly in the upper parts of individual units.

The middle member of the Arrayanes is a well-defined and easily mapped unit in the eastern part of the Sierra de San Javier, but apparently is absent as a distinct lithologic unit in the western part of the area (Fig. 2). Although relations are not entirely clear in the central part of the range, the middle member seems to grade laterally to the west into a sequence of sandstone and siltstone indistinguishable from the lower and upper members of the Arrayanes. Reddish siltstone units in the upper member of the Arrayanes in the eastern part of the Sierra de San Javier apparently grade out to the west into sandstone, because they are not recognized in the western part of the Sierra.

The Arrayanes Formation is not well enough exposed to permit measurement of a detailed stratigraphic section or to make an accurate estimate of thickness. Measurements from generalized cross sections on the map suggest that the formation is about 1,150 m thick and that the lower, middle, and upper members are 100, 250, and 800 m thick, respectively (Fig. 3).

The contact between the Arrayanes and the Santa Clara Formations is fairly well defined in the Santa Clara area, where Wilson and Rocha (1949) mapped it as separating their lower and middle divisions of the Barranca Formation. Here the contact is marked by a fairly conspicuous change from the sandstone-siltstone sequence of the Arrayanes that crops out in the rolling hills to the sequence of siltstone, carbonaceous shale, and sandstone of the Santa Clara Formation that crops out in the valleys. In other parts of the Sierra, the contact is not well defined but is marked by a general change from sandstone-bearing sequences below to siltstone- and shale-rich sequences above. Locally, the contact seems to be gradational across at least 100 m of strata.

Santa Clara Formation

The Santa Clara Formation crops out in an irregular east-west band across the Sierra de San Javier south of the outcrop

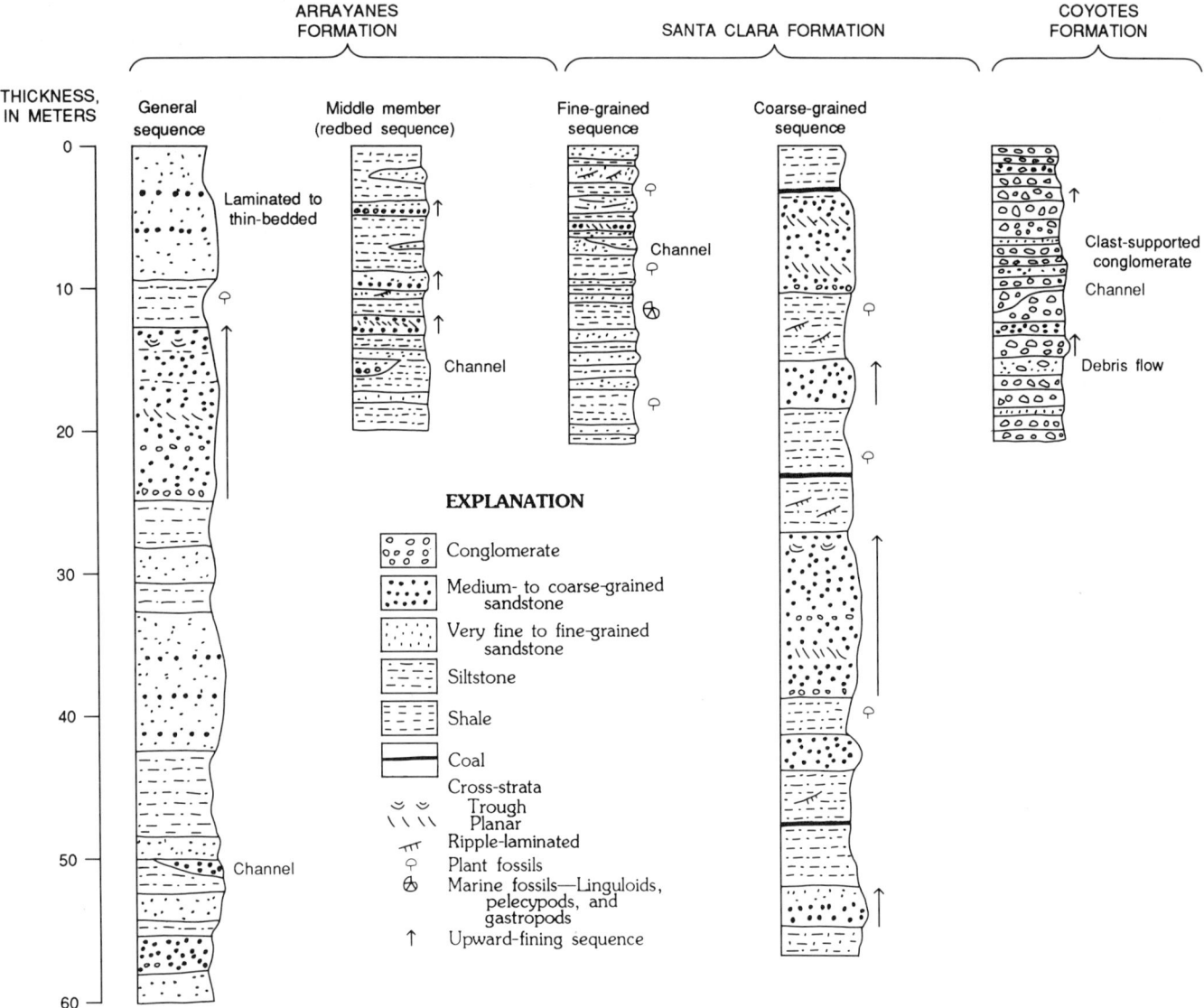

Figure 4. Typical lithologic sequences, shown diagramatically, in the Barranca Group, Sierra de San Javier.

band of the Arrayanes Formation. The Santa Clara Formation crops out in vegetated valleys and low hills and is generally poorly exposed. The best exposures are in roadcuts 2 to 3 km south of San Javier, and 2 km southwest and 5 to 3 km southeast of the town of La Barranca.

The Santa Clara Formation is composed of interstratified siltstone, shale, sandstone, conglomerate, carbonaceous shale, and coal (Fig. 4). The silty and shaly units in the formation are relatively nonresistant. Consequently, fewer cliffs are formed in this formation than in the underlying Arrayanes Formation or in the overlying Coyotes Formation. Two major types of depositional sequences (Fig. 5) are recognized in the Santa Clara Formation: a fine-grained sequence of shale, siltstone, and sandstone, and a coarse-grained sequence of sandstone, siltstone, conglomerate, carbonaceous shale, and coal. These two major depositional sequences commonly are present in megasequences 100 to 300 m thick (Fig. 5), grading from the fine-grained sequence below to the coarse-grained sequence above.

The fine-grained sequence consists of greenish gray and dark gray shale and siltstone, and interstratified yellowish gray and light greenish gray laminated, irregularly laminated, and ripple laminated, very fine to fine-grained sandstone (Fig. 6). Locally, the sandstone contains irregular channels and medium-scale low-angle cross-strata. Tabular layers from 0.5 to 1 m thick of coarse-grained sandstone that form a single set of planar cross-strata are present locally. Plant fossils (Silva-Pineda, 1961; Weber, 1980; Weber and others, 1980a, b) are common in the shale and siltstone. A marine fauna of brachiopods (*Lingula*), pelecypods, and at one locality, an ammonite in shale and siltstone also occur (Alencaster, 1961b). We found two additional faunal localities in

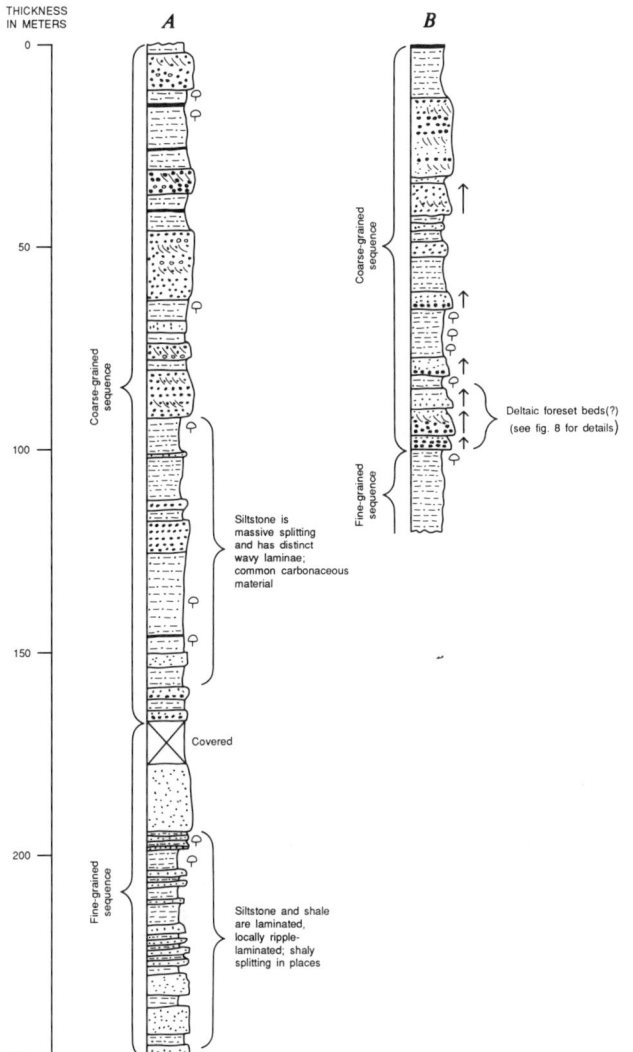

Figure 5. Stratigraphic columns of part of the Santa Clara Formation. Modified from Potter and Cojan (1985). A. Column measured in roadcuts 2.25 km southeast of La Barranca. B. Column measured in roadcuts 1 km east-southeast of La Barranca. See Figure 4 for explanation.

fine-grained sequences: (1) pelecypods identified by N. J. Silberling (written communication, 1985) as *Septocardia*? or *Paleocardia* and a trigoniacean that might be *"Myophorigonia" alasi* Alencaster at 28°34.9'N, 109°39.4'W, and (2) a brachiopod identified by N. J. Silberling (written communication, 1985) as *Lingula,* at 28°34.3'N, 109°45.0'W.

The coarse-grained sequence consists of about equal amounts of sandstone and siltstone, and minor amounts of conglomeratic sandstone and conglomerate, carbonaceous shale, and coal (Fig. 4). The conglomeratic units generally contain clasts ranging in size from granules to small pebbles, although cobbles as much as 4 cm across are present at one locality. Clasts are mostly light gray to black chert; quartzite and white quartz clasts are rare. The sandstone is light to yellowish gray, medium to very coarse grained, fairly well to poorly sorted, and is composed of

Figure 6. Fine-grained sequence of the Santa Clara Formation showing even bedding in roadcut 3 km south of San Javier. Maximum height of outcrop about 3 m.

angular to subrounded grains of quartz and a minor amount of feldspar. Individual units of sandstone range from 1 to 15 m in thickness and commonly form massive units that have little internal stratification (Fig. 7). Some units grade fron conglomeratic very coarse-grained sandstone at the base to medium-grained or even fine-grained sandstone at the top. Other units contain tabular planar and trough cross-strata, although cross-strata are generally sparse in the Santa Clara Formation. At a locality 1 km southeast of La Barranca (Fig. 8), large-scale, very low-angle cross-strata (unit 2, Figs. 8 and 9) are present at the base of a coarse-grained sequence and are thought to be deltaic foreset beds (see section entitled "Depositional environments" below). At another locality (Fig. 10), 1.5 km southwest of La Barranca, a well-defined channel with lateral-accretion bedding of a type believed to form by the lateral migration of a meandering stream (Puigdefabregas and Van Vliet, 1978) is present in a coarse-grained sequence. Siltstone that is interstratified with rocks of coarse-grained sequences is generally greenish gray, massive splitting, and composed of wavy laminae. Plant debris, carbonaceous shale, and coal are common in coarse-grained sequences of the Santa Clara Formation. Wilson and Rocha (1949) document at least nine coal beds near Santa Clara, each locally over 1 m thick.

Several thin beds of tuff occur in the upper part of the Santa Clara Formation along the main road 2 km south of San Javier. Tuff was not noted in the Santa Clara elsewhere in the Sierra de San Javier, but it may have been missed because exposures are generally poor. One tuff bed, which appears to be duplicated by faulting, is about 1.5 to 2 m thick and is composed of angular quartz fragments in an ashy matrix that contains some flattened pumice and black charcoal fragments. A few other tuff layers, 1 to 2 cm thick, occur interstratified with siltstone and sandstone of the Santa Clara Formation down to about 50 m below the main tuff unit. None of the tuff layers in the Santa Clara Formation have been dated radiometrically.

Wilson and Rocha (1949) measured 370 m of the Santa

Clara Formation (their middle division) in an incomplete section in the Santa Clara district. We estimate, based on generalized cross sections, that the thickness could be as much as 1,400 m, but thickness estimates of this unit are not accurate.

We have placed the contact of the Santa Clara and Coyotes at the marked change (perhaps an unconformity) from the sandstone, conglomerate, siltstone, shale, carbonaceous shale, and coal of the Santa Clara Formation to the pebble-and-boulder conglomerate of the Coyotes Formation. As originally defined (Wilson and Rocha, 1949; Avila, 1960; Alencaster, 1961a), the contact of the Santa Clara and Coyotes Formations was placed at the base of the lowest thick quartzite unit above the shaly or silty units in the Santa Clara. We could not consistently map this contact in the Sierra de San Javier, because thick quartzite units apparently are present at several different positions in the Santa Clara and, furthermore, the amount of quartzite in the Santa Clara may increase upward. The lowest occurrence of the pebble-and-boulder conglomerate of the Coyotes Formation is a clearly defined contact throughout the Sierra de San Javier, and we have used this horizon to define the base of the Coyotes Formation.

Using our definition, the contact of the Santa Clara and Coyotes Formations may be an unconformity, although the attitude of bedding in the Santa Clara and Coyotes is generally similar, and the Coyotes everywhere overlie the Santa Clara. The presence of apparently different sequences of rocks from place to place below the Coyotes is evidence that the contact is an unconformity, and suggests that the Coyotes may cut across units of the underlying Santa Clara Formation. This relation is particularly evident between an area 2.5 km west-southwest of San Javier, where thick, cliff-forming quartzite units occur at the top of the Santa Clara; and another area about 2 km south of San Javier, where such thick quartzite units are absent. The amount of lateral variability of rock types in the Santa Clara Formation, on the other hand, has not been established, and some or all of the apparent variability of rock types below the Coyotes could be due to lateral facies changes in the Santa Clara Formation. Another possible indication of an unconformity is in an area 1 to 2 km east of La Barranca where the Santa Clara Formation appears to dip at a much steeper angle than the overall attitude of the Coyotes.

Figure 7. Coarse-grained sequence of the Santa Clara Formation showing massive units in roadcut 1 km east-southeast of La Barranca.

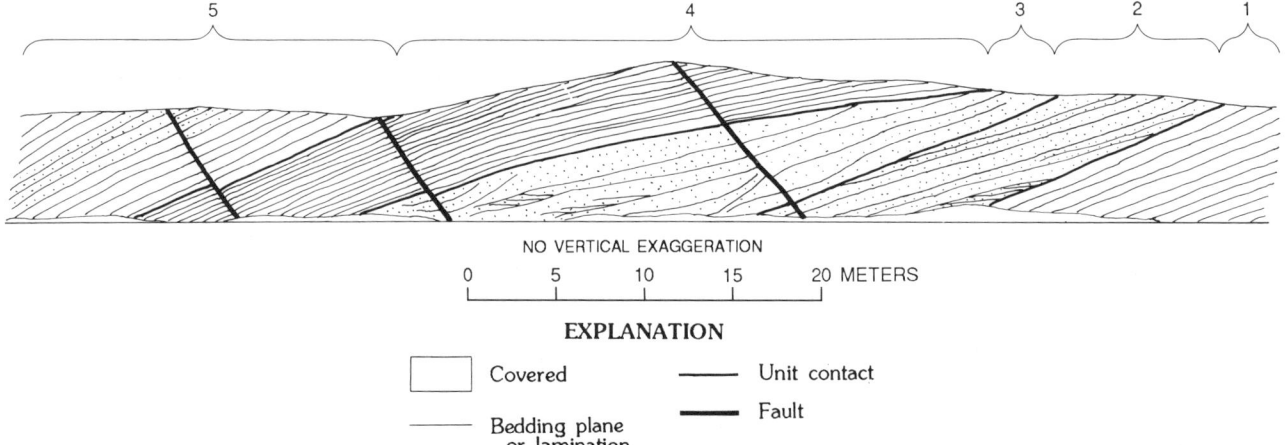

Figure 8. Detailed cross section of possible deltaic foreset beds in part of the Santa Clara Formation, measured in roadcut 1 km east-southeast of La Barranca (see Fig. 5B for location of cross section). Unit 1: siltstone and shale, plant remains common, laminated, some wavy laminae. Unit 2: sandstone, medium- to coarse-grained in lower part grading to fine- to medium-grained in upper part, laminae to very thin beds gently dip into lower contact. Unit 3: sandstone, very coarse grained in lower part to fine grained in upper part, common channel scours from 10 to 30 cm deep in lower part, more evenly bedded in upper part. Some very low angle cross-strata and channels occur in upper part. Unit 4: very fine-grained sandstone in lower part to coarse-grained siltstone in upper part, irregularly laminated in lower half grading upward into more massive strata in upper half, laminae parallel to curving upper contact of unit 2. Unit 5: siltstone, shale, and silty very fine grained quartzite and medial unit of very coarse to very fine-grained sandstone.

Figure 9. Gently dipping sandstone beds of unit 2 overlying siltstone and shale of unit 1. Diagram of these units shown in Figure 8.

The presence of an unconformity in this area, however, remains uncertain because of poor exposures, possible landslide deposits, faulting, and folding.

The contact of the Santa Clara and Coyotes was observed in detail at only one locality—a small exposure in a canyon about 600 m south of San Javier. Here an angular relation of apparently 10 degrees was noted between the Santa Clara and Coyotes. However, this difference in attitude could be due, at least in part, to an initial depositional dip of the pebble-and-boulder conglomerate of the Coyotes on an originally nearly flat surface of the Santa Clara Formation. Our overall conclusion is that a major unconformity is not probable at the Santa Clara–Coyotes contact in the Sierra de San Javier.

Coyotes Formation

The Coyotes Formation crops out extensively in the western part of the Sierra de San Javier and at scattered localities in the eastern part (Fig. 2). It is a resistant formation that forms ridges and peaks.

The Coyotes Formation is dominantly a pebble-and-boulder clast-supported conglomerate composed of angular to subround clasts of quartzite and chert in a fine to very coarse sand matrix (Figs. 4 and 11). Except for localities 4 to 6 km south of San Javier (described below) where limestone and coarse-grained quartzite clasts are found, clast rock types in the Coyotes Formation are similar to rock types in Paleozoic strata that unconformably underlie the Barranca Group. Clasts as much as 20 cm in diameter are common; the largest noted in the Sierra de San Javier is 50 cm. The pebble-and-boulder conglomerate is present in layers from 0.5 to 1.5 m thick, a few of which grade from coarse conglomerate at the base to relatively fine at the top. Channels (Fig. 12) occur rarely.

Minor amounts of matrix-supported conglomerate (Fig. 13), grayish red siltstone, and yellow–gray and grayish red sandstone are interstratified with the clast-supported conglomerate. These generally reddish rocks are, in many places, intergradational with the clast-supported conglomerate. The matrix-supported conglomerate contains clasts of quartzite and chert as large as about 20 cm, set in a grayish red silt to coarse sand matrix.

Finely crystalline limestone clasts as much as 20 cm across occur in a few layers of matrix- and clast-supported conglomerate in outcrops of the Coyotes Formation from 4 to 6 km south of San Javier. A fusulinid from one of these limestone clasts (28°33.1′N, 109°44.1′W) contains *Schwagerina guembeli* Dunbar and Skinner (R. C. Douglass, written communication, 1985).

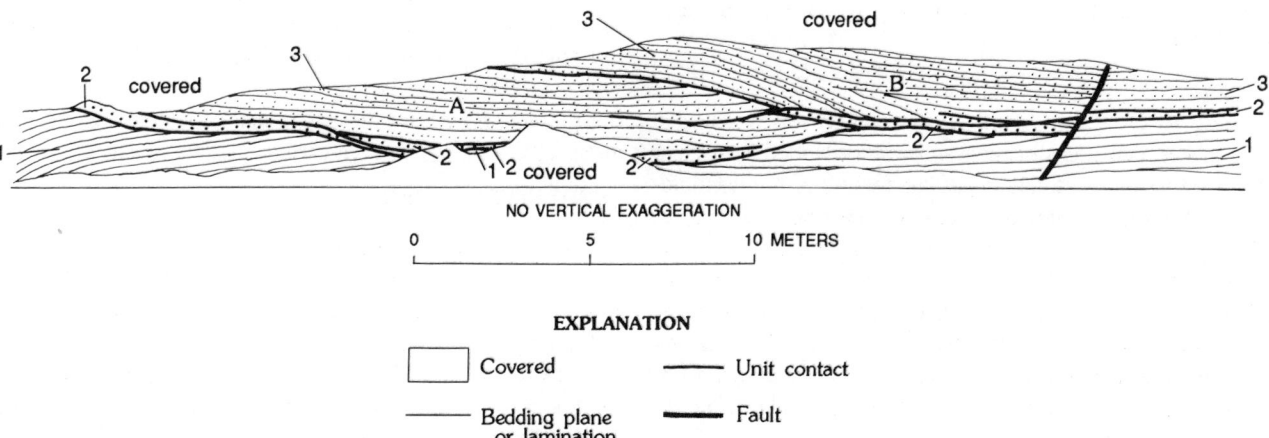

Figure 10. Detailed cross section of part of the Santa Clara Formation in roadcut 1.5 km southwest of La Barranca. Unit 1: silty sandstone, very fine to fine grained, laminated, sparse ripple laminae, plant remains common. Unit 2: sandstone, coarse to very coarse grained, sparse coarse grains to granules of dark gray chert, faint laminae in places. Unit 3: sandstone, medium to very coarse grained, gently dipping cross-laminae. The left-to-right dip of the bedding in unit 3 and the left-to-right change in the position of major channels, from one centered at A to one centered at B, indicate a left-to-right lateral migration of a stream system.

Figure 11. Clast-supported conglomerate in the Coyotes Formation in roadcut 2 km south of San Javier.

Figure 13. Matrix-supported conglomerate in the Coyotes Formation in canyon 1.5 km south of San Javier.

Figure 12. Channel in clast-supported conglomerate in the Coyotes Formation in roadcut 2 km south of San Javier. Arrows point to base of channel surface.

According to Douglass, this form, or a close relative, occurs widely in Permian rocks of the southwestern United States and northern Mexico. The finely crystalline texture of the fusulinid-bearing limestone is similar to fusulinid-bearing limestone in shallow-water sequences of Permian age in Sonora (Menicucci and others, 1982) and dissimilar to the bioclastic turbiditic fusulinid-bearing limestone (Bartolini, 1988) in Paleozoic rocks that unconformably underlie the Barranca Group. The conglomerate layers that contain the limestone clasts also contain a few coarse-grained quartzite clasts that are unlike quartzite in the Paleozoic rocks that underlie the Barranca Group.

The Coyotes Formation is well exposed 1 to 2 km south of San Javier in the canyon of an intermittent stream—a continuation of the stream that passes through San Javier. The Coyotes Formation appears to be about 600 m thick in this area.

The contact of the Coyotes and the overlying Tarahumara Volcanics is a major unconformity. The Tarahumara laps over the Coyotes and rests on the Santa Clara Formation and, locally in the eastern part of the Sierra, on the Arrayanes Formation.

POST-BARRANCA ROCKS

The Tarahumara Volcanics or, locally, Cenozoic volcanic and sedimentary rocks unconformably overlie the Barranca Group (Fig. 2). The Tarahumara Volcanics were named by Wilson and Rocha (1949) in the Santa Clara district. They consist of andesitic flows, agglomerates, breccias, possible intrusive rocks, and local sedimentary rocks. Their age is uncertain. These rocks were considered Cretaceous(?) by Wilson and Rocha (1949) because possible Cretaceous fossils were reported by Dumble (1900b) in supposedly similar rocks in "Arroyo de Obispo, south of the Santa Clara district" and by King (1939), 60 km northeast of the Santa Clara district. The possible Cretaceous fossils at the Arroyo de Obispo locality could not be located. The Tarahumara is intruded by, and is thus older than, 53- to 62-Ma intrusive rocks in the Sierra de San Javier (see below); but on the basis of our present knowledge, the Tarahumara could be latest Triassic, Jurassic, Cretaceous, or even earliest Tertiary in age.

INTRUSIVE ROCKS

Intrusive rocks (quartz diorite to granite) are widely distributed in the Sierra de San Javier. They occur in stocks covering as much as 9 km^2 that form valleys or topographic depressions. Although we did not map different intrusive rock types, much of the rock is fine-grained, dark gray quartz diorite. However, granite that has feldspar phenocrysts as much as 1 cm long forms much of the stock at San Javier. Granodiorite and quartz monzonite (Berlanga, 1971) are also recognized. Many of the intrusive rocks studied in thin section are altered and contain sericite, chlorite, epidote, and opaque minerals.

Ages ranging from 53 to 62 Ma (K-Ar method) were obtained by Damon and others (1983) on four intrusive rocks in or near the Sierra de San Javier.

AGE OF THE BARRANCA GROUP

Abundant and varied flora (Silva-Pineda, 1961; Weber, 1980, 1985; Weber and others, 1980a, b) in the Santa Clara Formation indicate a Carnian age (early Late Triassic). This age is confirmed (Alencaster, 1961b) by a fauna of pelecypods, brachiopods, and a single ammonite in the Santa Clara.

No identifiable fossil material has been obtained from the Arrayanes and Coyotes Formations. The Arrayanes Formation may be Late Triassic in age, because it has a gradational contact with the overlying dated Santa Clara Formation. The Coyotes Formation, however, may rest unconformably on the Santa Clara Formation and, thus, could be significantly younger than the early Late Triassic Santa Clara Formation. It must be older than the 53- to 62-Ma rocks that intrude it.

PETROGRAPHY

Quantitative detrital modes, calculated from point counts of thin sections, were determined on 17 sandstones from the Arrayanes and Santa Clara Formations (Tables 1 and 2; Fig. 14), using the procedures summarized by Dickinson (1985). The modal counts indicate arkosic to quartz-rich sandstones, suggesting a continental block provenance including both stable craton and uplifted basement terranes, again following Dickinson's designations.

The most difficult part of the modal counts was the identification of feldspar, which is partly, and in places apparently entirely, altered to sericite and/or clay minerals. In some sandstone, no feldspar was identified, yet the rock contained deformed grains of sericite and clay minerals that may have been derived from feldspar by postdepositional alteration. Such an interpretation is favored by a clear gradation from rock samples that have only slightly altered feldspar, to samples that have highly altered but still recognizable feldspar, to samples that have no feldspar but abundant grains of sericite and clay minerals. Because of their probable derivation from feldspar, the sericite and clay mineral grains were grouped with feldspar (Tables 1 and 2; Fig. 14). The problem is further complicated by the abundance of clay matrix in some of the rocks. Eleven of the 17 sandstones studied (Table 2) contain over 15 percent matrix, and the maximum is 38 percent. This large amount of matrix may have been derived, at least in part, from feldspar that was altered postdepositionally and then fragmented and deformed to form what now looks like matrix. If so, the original feldspar content of many of the sandstones is higher than indicated.

The sandstones of the Arrayanes and Santa Clara Formations contain grains derived from a variety of granitic, metamorphic, sedimentary, and perhaps volcanic sources. Grains clearly derived from granitic rocks consist of microcline, perthite, and micrographic quartz-feldspar intergrowths. Metamorphic grains include foliate quartzite and quartz-mica aggregates. About three-fourths of the quartz grains have distinct undulatory extinction (>5°), and more than half of the polycrystalline quartz grains contain more than 3 subgrains. An abundance of undulatory quartz and polycrystalline quartz that has more than 3 subgrains is characteristic of a low-rank metamorphic source terrane (Basu and others, 1975). Sedimentary grains consist of chert and lesser amounts of siltstone and sandstone. Sedimentary grains generally constitute a very minor part of the sandstone, although sample 6 is an exception and contains 35 percent chert grains.

The volcanic component, if any, of the sandstone is uncertain. The presence of a few tuff layers in the Santa Clara Formation indicates the likelihood of finding at least some volcanic detritus in the Barranca, yet none was definitely identified in the thin sections studied. Conceivably, some of the sericite-clay mineral grains could be devitrified tuff grains, but no evidence of such an origin was seen. Some of the quartz in the Arrayanes and Santa Clara Formations, particularly the abundant very angular grains, could have been derived from local tuffs, but no clearly defined resorbed quartz or other evidence of a volcanic origin was noted.

TABLE 1. SYMBOLS USED IN MODAL COUNTS (TABLE 2)

A. Interstitial material (I). Percent of total rock. Iron oxide stain generally occurs with the interstitial material and is included with the total percent of interstitial material.
B. Mica (M). Percent of framework grains.
C. Quartzose Grains. Percent of Qt + F + L
  Qt = Qm + Qp
  Qm = Qms + Qma + Qmq + Qmf
  Qp = Qc + Qa + Qf
  Qms = monocrystalline quartz forming single grain
  Qma = monocrystalline quartz forming subgrain >0.0625 mm in size within polycrystalline quartz grain
  Qmq = monocrystalline quartz forming subgrain >0.0625 mm in size within grain of quartz-mica tectonite
  Qmf = monocrystalline quartz forming subgrain >0.0625 mm in size within grain of foliate metaquartzite
  Qc = chert
  Qa = aggregate quartz, grain size <0.0625 mm
  Qf = foliate metaquartzite, grain size <0.0625 mm
D. Feldspar Grains. Percent of Qt + F + L
  F = Fp + Fk + Fm + Fu + Sc
  Sp = plagioclase
  Fk = potassium feldspar
  Fm = microcline
  Fu = undifferentiated feldspar and uncertainly identified feldspar, commonly highly altered to sericite and clay minerals
  Sc = sericite-clay mineral grains
E. Unstable Lithic Fragments. Percent of Qt + F + L
  L = Lst + Lqm + La
  Lst = Siltstone or deformed quartz-sericite aggregates
  Lqm = Quartz-mica tectonite
  La = Shale or argillite
F. Total Lithic Fragments. Percent of Qt + F + L
  Lt = L + Qp
G. Unknown Mineral Grains (U). Percent of total framework

TABLE 2. MODAL COUNTS OF SANDSTONES FROM THE ARRAYANES FORMATION* AND THE SANTA CLARA FORMATION†

Sample No.	1	2	3	4	5	6	7	8	9	10	11	12	13	14	15	16	17
Number of point counts	406	410	409	407	412	416	407	411	403	424	145	430	408	400	399	405	405
(in percent)																	
I	25	12	14	22	6	17	24	15	35	15	28	12	20	21	18	17	38
M	<1	<1	1	<1	0	0	1	0	0	<1	0	0	0	0	0	0	1
Qms	46	53	63	38	68	51	60	61	77	44	99	37	46	86	46	59	62
Qma	2	17	18	10	13	10	8	8	16	12	1	10	6	5	6	15	8
Qmq	0	4	0	0	1	0	3	0	0	0	0	<1	1	0	<1	0	0
Qmf	0	<1	0	0	1	0	0	0	0	0	0	0	0	0	0	0	0
Qc	4	0	<1	2	2	35	<1	6	2	<1	0	1	<1	1	6	21	4
Qa	1	0	2	2	<1	0	0	2	5	3	0	3	8	2	0	3	7
Qf	0	0	<1	<1	0	0	<1	0	0	0	0	0	1	0	0	0	0
Fp	25	1	<1	1	0	0	2	0	0	5	0	8	1	0	4	0	0
Fk	5	1	0	0	0	0	7	0	0	<1	0	3	0	0	8	<1	0
Fm	0	0	0	3	0	0	7	0	0	1	0	2	0	0	5	0	0
Fu	14	14	0	29	0	0	9	0	<1	20	0	19	12	0	23	0	18
Sc	4	7	15	14	12	3	1	20	<1	11	0	8	24	4	0	0	0
Lst	<1	1	0	0	0	0	0	0	0	0	0	<1	2	0	<1	1	<1
Lqm	0	0	<1	0	3	1	4	3	0	4?	0	6	<1	2?	0	1	0
La	0	<1	0	0	0	0	0	0	0	<1	0	<1	0	1	<1	0	<1
U	<1	0	0	0	0	0	1	<1	<1	1	0	0	2	1	1	<1	0
Qm	48	74	81	48	81	61	71	69	93	56	100	47	53	91	52	74	70
Qp	5	0	2	4	3	35	0	8	7	3	0	4	9	3	6	24	11
Qt	53	74	83	52	84	96	71	77	100	59	100	51	62	94	58	98	81
F	48	23	15	47	12	3	26	20	0	37	0	40	37	4	40	0	18
L	0	1	0	0	3	1	4	3	0	4	0	6	2	2	0	2	0
Lt	5	1	2	4	6	36	4	11	7	7	0	10	11	4	6	26	11

Symbols from Table 1
*Samples 1–7
†Samples 8–17

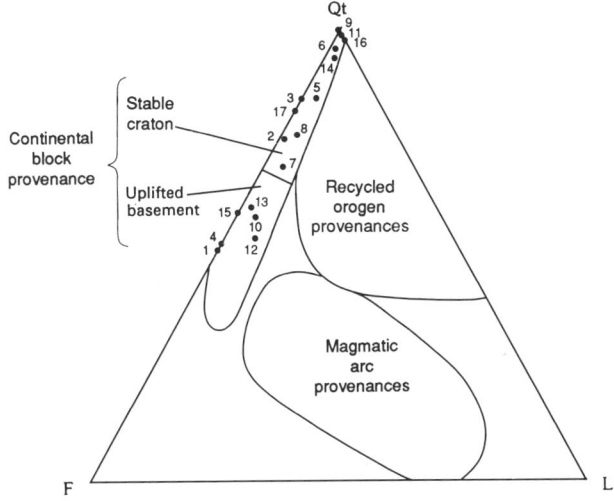

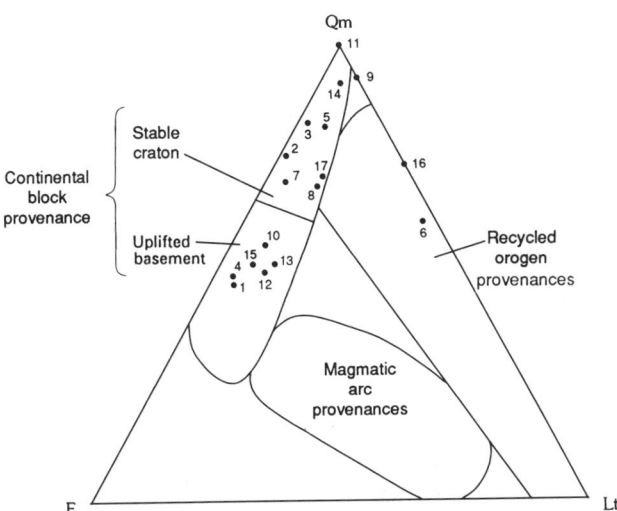

Figure 14. Detrital modes of sandstones of the Arrayanes and Santa Clara Formations. Provenance fields from Dickinson (1985). Symbols from Table 1; sample numbers from Table 2.

PALEOCURRENT DIRECTION

Indications of paleocurrents are puzzlingly scarce in the Arrayanes and Santa Clara Formations. We measured paleocurrent directions on all observed cross-strata that were well enough exposed to determine dip direction. Yet we obtained dip directions on only 16 cross-strata in the Arrayanes Formation. In the Santa Clara Formation, our measurements, combined with those of Potter and Cojan (1985), give a total of 35 readings on cross-strata. Current directions in the Arrayanes and Santa Clara Formation vary but show a general southward direction of flow (Fig. 15). We measured the orientation of 153 flat imbricate pebbles in the Coyotes Formation from several localities from 0.5 to 5 km south of San Javier. The imbrication indicates flow to the southwest or west (Fig. 15). All the measurements in the Barranca Group were corrected for tilt imposed by the structural dip of the rocks, but no attempt was made to correct for rotation caused by plunging folds. An additional problem to keep in mind is that the entire outcrop of the Barranca Group could lie within a rotated block, and measured flow directions would thus be different from true flow directions.

DEPOSITIONAL ENVIRONMENTS

The depositional environments of the Santa Clara Formation are described first because information on this formation is more definitive in defining depositional environments than information on the underlying, and perhaps environmentally similar, Arrayanes Formation. The fine-grained sequences of the Santa Clara Formation (Figs. 4, 5, and 6) consist of shale and siltstone containing common plant debris and a local marine fauna that includes shallow-water near-shore pelecypods and linguloid brachiopods. The presence of the marine fossils, the fine texture of the rocks, and the upward gradation into rocks considered to be of deltaic origin (see below) suggest that the fine-grained sequences are prodelta and delta-front deposits. Associated thin, very fine to fine-grained, laminated sandstone indicates a relatively high-energy influx of clastic material, perhaps during floods when discharge from the delta into the ocean was high.

The fine-grained sequences of the Santa Clara Formation grade upward into coarse-grained sequences (Figs. 4, 5, and 7) containing locally cross-stratified sandstone, wavy laminated siltstone, carbonaceous shale, and coal. Large-scale cross-strata in one of these sequences (Figs. 8 and 9) may represent deltaic foreset beds, whereas small- to medium-scale, cross-stratified sandstone and associated coals are suggestive of fluvial and marsh deposits, respectively. In this interpretation, the upward-coarsening megasequences (Fig. 5) are considered to represent prograding deltaic sequences that start with prodelta fine-grained strata and end with fluvial sandstone and palaudal coal on a largely subaerial delta plain.

The origin of the sandstone and siltstone of the Arrayanes Formation is less clear. Because of the lithic similarity of the sandstone in the Arrayanes to those in the Santa Clara, they are probably mostly fluvial. The amount of deltaic deposits in the Arrayanes, however, is not clear. The red siltstone and sandstone of the middle member of the Arrayanes contain lenticular cross-stratified sandstone, local channel-filling sandstone, and structureless siltstone, suggesting deposition of the sandstone by streams and the siltstone on a flood plain.

The pebble-and-boulder conglomerate that constitutes almost all of the thick Coyotes Formation contrast markedly with the finer-grained sediments of the Arrayanes and Santa Clara Formations. Much of the conglomerate is clast-supported and locally fills channels. Matrix-supported conglomerate and associated sandstone and siltstone also occur in minor amounts. The coarseness, presence of local channels, and great thickness of the

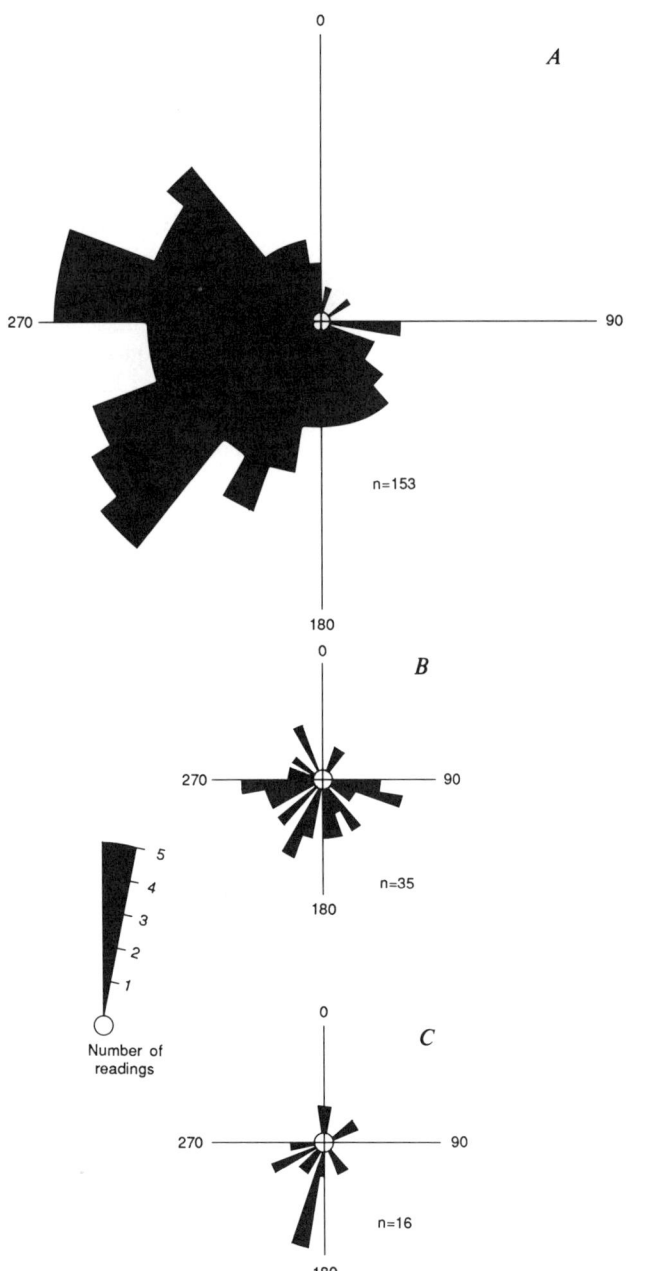

Figure 15. Rose diagrams of paleocurrent directions in the Barranca Group, Sierra de San Javier. A, Coyotes Formation; B, Santa Clara Formation; C, Arrayanes Formation. Paleocurrent directions in the Santa Clara and Arrayanes Formations are based on dip direction of cross-strata and in the Coyotes Formation on orientation of imbricate clasts. For imbricate clasts, direction shown is inferred downcurrent direction 180° from the upcurrent dip direction of the clasts. "n" denotes number of reading. Each reading is on a single set of cross-strata or on an individual imbricate clast.

clast-supported conglomerate all suggest deposition by vigorous streams on alluvial fans adjacent to high mountain ranges. Matrix-supported conglomerates, considered to be debris-flow deposits, are also characteristic of alluvial fans (Miall, 1978).

PALEOGEOGRAPHY AND TECTONIC SETTING

The source terrane of most of the quartz- and feldspar-rich sandstones of the Arrayanes and Santa Clara Formations was probably Precambrian basement similar to plutonic and metamorphic rocks now exposed in Sonora (Cserna, 1970; Damon and others, 1962; Anderson and others, 1979). In detail, grains of micrographic quartz-feldspar that occur in the sandstones of the Barranca Group may have been derived from the micrographic Aibo Granite (1.1 Ga) that is widely exposed in northern Sonora (Anderson and others, 1979; T. H. Anderson, oral communication, 1986). The source of granule- to boulder-size quartzite and chert debris in the Barranca Group, mostly in the Coyotes Formation, is probably primarily Paleozoic quartzite and chert similar to that which underlies the Barranca Group. However, the fusulinid-bearing limestone and coarse-grained quartzite clasts in the Coyotes Formation 4 to 6 km south of San Javier are unlike Paleozoic rocks that underlie the Barranca Group and indicate a more distinct source. The fusulinid-bearing limestone is similar to shallow-water shelf limestone elsewhere in Sonora (Menicucci and others, 1982) and the coarse-grained quartzite is lithologically similar to latest Proterozoic or Lower Cambrian quartzite in Sonora (Stewart and others, 1984).

The lack of volcanic layers or detritus in the Barranca Group, except for a few thin tuff layers in the Santa Clara Formation, is puzzling. The Barranca lies in an area that might be near the source region for widespread volcanic-rich sediments of the Chinle Formation in the southwestern United States (Fig. 16; Stewart, 1969; Stewart and others, 1972, 1986). Abadie (1981) has suggested that the Barranca lies in a fore-arc basin, separated from the Chinle basin by a magmatic arc from which volcanic detritus was fed northward into the Chinle basin. Such a setting seems unlikely, however, because of the sparcity of volcanic layers or detritus in the Barranca. Still, the source of the volcanic detritus in the Chinle Formation and the lack of volcanic detritus in the Barranca are not understood.

The coarse, poorly sorted, and angular detritus of the Barranca Group suggest deposition in a basin, or basins, flanked by areas of high relief. A tectonically active basin is suggested by the abrupt facies and thickness changes, as well as by detritus as much as 50 cm across, in the Coyotes Formation. The arkosic sandstone (by analogy to other occurrences of arkosic sandstone, Dickinson, 1985) suggests a source in fault-bounded basement uplifts. Much of the Barranca in Sonora appears to have been deposited in a single basin that is delineated by the presence of major outcrops of the Barranca Group in an east-west–trending belt about 110 km long and 40 km wide. The elongate shape of this basin and the interpretation of flanking areas of high relief suggest a rift basin. Other outcrops of the Barranca Group (Fig.

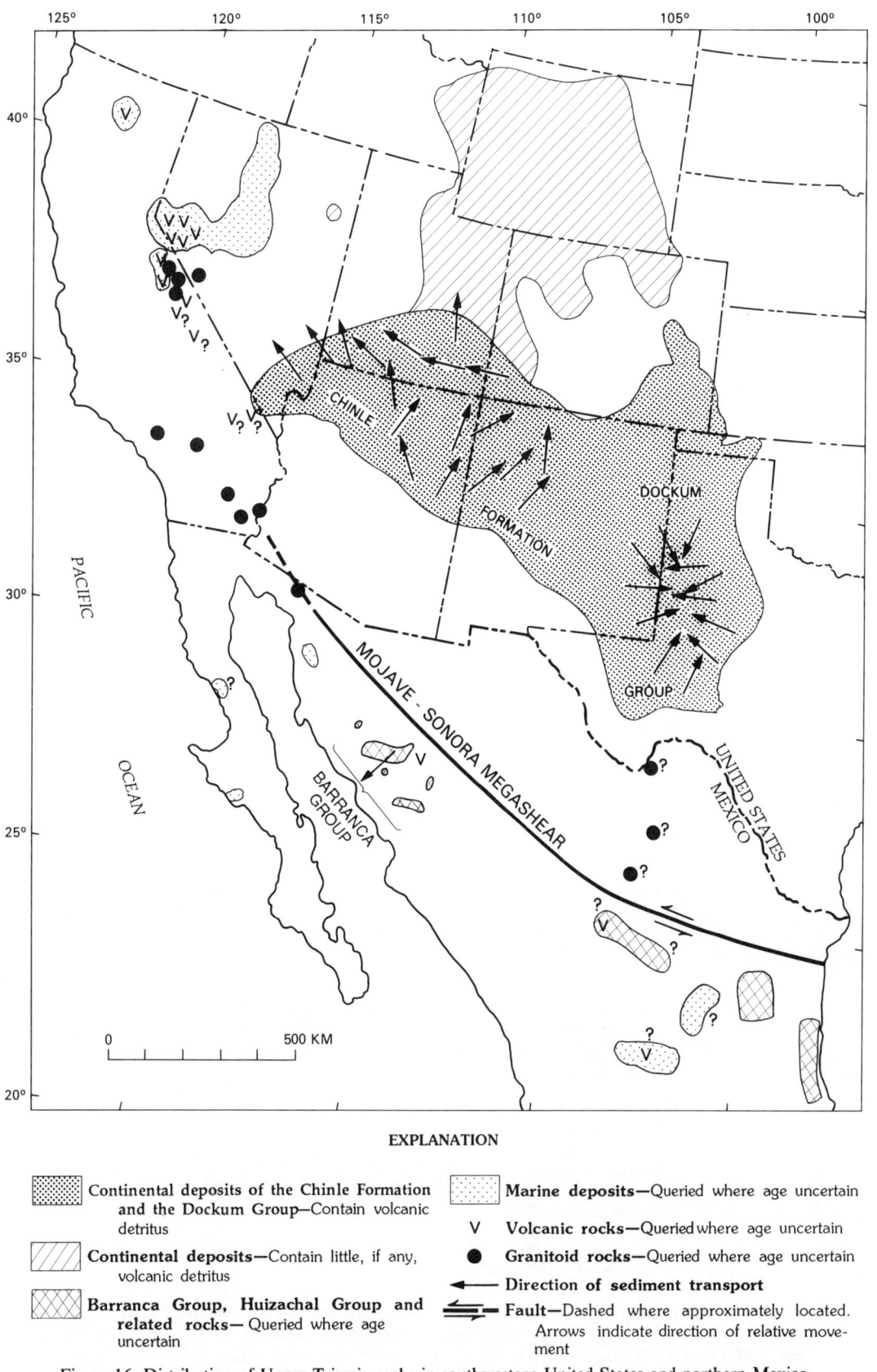

Figure 16. Distribution of Upper Triassic rocks in southwestern United States and northern Mexico (from Stewart and others, 1986).

1) occur outside of this main basin, but the tectonic setting of rocks in these outcrops is unknown.

Deposition in a rift basin best accounts for the alluvial-fan pebble-and-boulder conglomerates of the Coyotes Formation, since they clearly were derived from local sources in an area of high relief. The marine-delta and fluvial deposits of the Arrayanes and Santa Clara Formations, on the other hand, are not so clearly related to a rift basin, and, if they are separated by a major unconformity from the overlying Coyotes Formation, may have had a significantly different tectonic setting. The Arrayanes and Santa Clara Formations, following this line of reasoning, may have been originally more widely distributed and are preserved now only where they were downfaulted and covered by the Coyotes Formation.

We favor, however, the interpretation that all three formations of the Barranca Group are rift-basin deposits because of their interrelated lithology and stratigraphy. All three formations contain chert and quartzite clasts presumably derived primarily from rocks similar to those exposed unconformably below the Barranca Group. The conglomerate is coarsest in the Coyotes

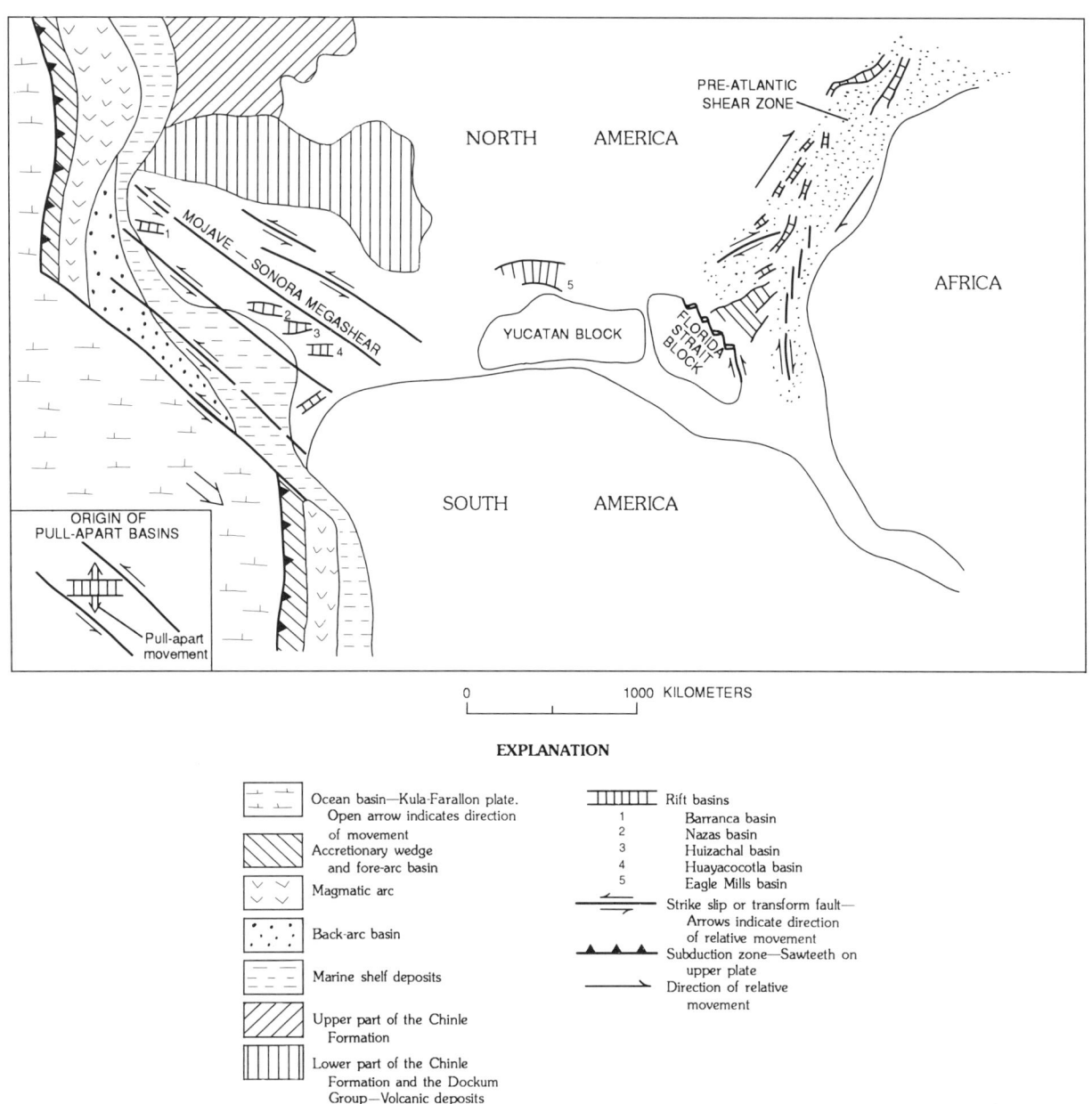

Figure 17. Tectonic model for setting of the Barranca Group. Based in part on Pindell (1985). Insert map shows origin of a pull-apart basin within a left-lateral shear system. In the model, the east-west–elongated Barranca basin (see Fig. 1) and other rift basins containing Upper Triassic rocks in northeastern Mexico (Belcher, 1979) are considered to be pull-apart basins related to shear prior to the separation of the North American, African, and South American blocks.

Formation (clasts as much as 50 cm across), whereas conglomerates in the Arrayanes and Santa Clara contain clasts as much as 14 and 4 cm across, respectively. The basal contact of the Coyotes Formation may be an unconformity, but overall the bedding attitudes in the Coyotes are similar to those in the underlying Santa Clara, and the Coyotes everywhere rests on the Santa Clara Formation. Independent of the Coyotes, the presence of conglomerate and coarse-grained sandstone composed of angular quartz grains in the Arrayanes and Santa Clara Formations indicates a nearby source area of at least moderate relief. If the Arrayanes, Santa Clara, and Coyotes Formations are all rift-basin deposits, the intensity of tectonism and the relief on the bounding faults apparently increased greatly after deposition of the Santa Clara Formation, to account for the marked increase in clast size in the Coyotes Formation.

The basin of the Barranca Group in Sonora is one of a number of Late Triassic and Jurassic rift-related, or possibly rift-related, basins (Fig. 17) extending from Sonora to northeastern Mexico (Huizachal Group; Belcher, 1979), the Gulf Coast region of the southern United States (Eagle Mills Formation; Scott and others, 1961), and the eastern United States (Newark Supergroup; Swanson, 1982; Ratcliffe and Burton, 1985). These basins appear to have formed, at least in part, in an extensional zone or a transtensional shear zone near the boundaries of the North American, African, and South American continents after these continents were assembled to form Pangea in the late Paleozoic and prior to the separation of these continents in the Mesozoic (Pindell, 1985; Stewart, 1988). In northern Mexico, the basins may have formed by transtensional faulting in a wide zone of left-lateral faults (Fig. 17) that are the precursors (as proposed in part by Pindell, 1985) of possible left-lateral faults such as the Mojave–Sonora Megashear in northern Mexico (Silver and Anderson, 1974, 1983; Anderson and Silver, 1979, Anderson and Schmidt, 1983; McKee and others, 1984; Pindell, 1985).

ACKNOWLEDGMENTS

We appreciate the continued help and advice of T. H. Anderson, particularly during the 10-day trip that he took with us in February and March, 1984. During that trip, we looked at many of the outcrops of the Barranca Group and outlined many of the problems that we investigated.

We wish to thank Carlos Gonzales-Leon and Julio Cesar Palomino-Moreno for their expert help in the field. Marco Antonio Bernal kindly let us stay at his company house in San Javier. Manuel Ramirez-Urquijo and Marco A. Montaño-Morales, geologists with the Mina Santa Rosa, lived with us in the house in San Javier and kept up our spirits with their conversation and music. They also helped us gain access to parts of the field area. We appreciate the help of Arnoldo Avila, el presidente of the town of San Javier, and all the people of the town who made our stay enjoyable.

We thank R. J. Madrid for his advice on the petrographic study and for his helpful comments on the manuscript. We also thank N. J. Silberling, T. Calmus, and W. R. Dickinson for reviewing the manuscript.

REFERENCES CITED

Abadie, V. H., III, 1981, Geology of part of the Sierra de Moradillas, Sonora, México [M.Sc. thesis]: Stanford, California, Stanford University, 87 p.

Alencaster, G. de Cserna, 1961a, Estratigrafía del Triásico Superior de la parte central del Estado de Sonora, parte I in Paleontología del Triásico Superior de Sonora: Universidad Nacional Autónoma de México, Instituto de Geología, Paleontología Mexicana número 11, 18 p.

—— , 1961b, Fauna fósil de la Formación Santa Clara (Cárnico) del Estado de Sonora, parte III, in Paleontología del Triásico Superior de Sonora: Universidad Nacional Autónoma de México, Instituto de Geología, Paleontologia Mexicana número 11, 38 p.

Anderson, T. H., and Schmidt, V. A., 1983, The evolution of Middle America and the Gulf of Mexico–Caribbean Sea region during Mesozoic time: Geological Society of America Bulletin, v. 94, p. 941–966.

Anderson, T. H., and Silver, L. T., 1979, The role of the Mojave–Sonora Megashear in the tectonic evolution of northern Sonora, in Anderson, T. H., and Roldán-Quintana, J., eds., Geology of northern Sonora; Geological Society of America Annual Meeting, Guidebook Trip 27; Pittsburgh, Pennsylvania, University of Pittsburgh, and Hermosilla, Sonora, Instituto de Geología, U.N.A.M., p. 59–68.

Anderson, T. H., Eells, J. L., and Silver, L. T., 1979, Precambrian and Paleozoic rocks of the Caborca region, Sonora, Mexico, in Anderson, T. H., and Roldán-Quintana, J., eds., Geology of northern Sonora; Geological Society of America Annual Meeting, Guidebook Trip 27; Pittsburgh, Pennsylvania, University of Pittsburgh, and Hermosilla, Sonora, Instituto de Geología, U.N.A.M., p. 1–22.

Avila de Santiago, G., 1960, Geología de los depósitos de antracita de la Sierra de San Javier y Santa Clara, Municipio de San Javier, Sonora [B.Sc. thesis]: Universidad Nacional Autónoma de México, Facultad de Ingenieria, 35 p.

Barrera-Moreno, E., and Domínguez-Perla, J. E., 1987, Geología de la carta Tónichi (H12D65) con énfasis en el Paleozoico, porción central del Estado de Sonora [B.Sc. thesis]: Hermosillo, Sonora, México, Universidad de Sonora, 93 p.

Bartolini, C., 1988, Regional structure and stratigraphy of the Sierra El Aliso, central Sonora, Mexico [M.S. thesis]: Tucson, University of Arizona, 189 p.

Basu, A., Young, S. W., Suttner, L. J., James, W. C., and Mack. G. H., 1975, Re-evaluation of the use of undulatory extinction and polycrystallinity in detrital quartz for provenance interpretation: Journal of Sedimentary Petrology, v. 45, p. 873–882.

Belcher, R. C., 1979, Depositional environments, paleomagnetism, and tectonic significance of Huizachal red beds (Lowe Mesozoic), northeastern Mexico [Ph.D. thesis]: Austin, University of Texas, 276 p.

Berlanga, G. E., 1971, Geología del Proyecto Aurora, region del bajo Yaqui, Municipio de Onavas, Estado de Sonora [B.Sc. thesis]: México, D.F., Instituto Politecnico Nacional, 69 p.

Cserna, Zoltan de, 1970, The Precambrian rocks of México, in Rankama, K., ed., The Precambrian: New York, Interscience, v. 4, p. 253–270.

Damon, P. E., Livingston, D. E., Mauger, R. L., Giletti, B. J., and Alor, J. P., 1962, Edad del Precambrico "Anterior" y de otros rocas del zocalo de la region de Caborca-Altar de la parte noroccidental del Estado de Sonora, México: Universidad Nacional Autónoma de México, Instituto de Geología Boletín, v. 64, p. 11–44.

Damon, P. E., Shafiqullah, M., Roldan, Q, J., and Cocheme, J. J., 1983, El Batolito Laramide (90-40 m.a.) de Sonora: En Memoria de la XV Convención Nacional de la Asociación de Ings. de Minas, Geol. y Metal. de México, A. C., p. 63–95.

Dickinson, W. R., 1985, Interpreting provenance relations from detrital modes of sandstones, *in* Zuffa, G. G., ed., Provenance of arenites, NATO ASI, Series C; Mathematical and Physical Sciences, v. 148: Dordrecht, Holland, D. Reidel Publishing Co., p. 333–361.

Dumble, E. T., 1900a, Triassic coal and coke of Sonora, Mexico: Geological Society of America Bulletin, v. 11, p. 10–14.

—— , 1900b, Notes on the geology of Sonora, Mexico: American Institute of Mining Engineers Transactions, v. 29, p. 122–152.

INEGI, 1984, Hoja Tecoripa (H12-12), Carta Geología 1:250,000: Instituto Nacional de Estadística, Geografía e Informática (INEGI), Secretaría de Programación y Presupuesto, México, D.F.

King, R. E., 1939, Geologic reconnaissance in northern Sierra Madre Occidental de Mexico: Geological Society of America Bulletin, v. 50, p. 1625–1722.

McKee, J. W., Jones, N. W., and Long, L. E., 1984, History of recurrent activity along a major fault in northeastern Mexico: Geology, v. 12, p. 103–107.

Menicucci, S., Mesnier, H. Ph., and Radelli, L., 1982, Permian, Triassic and Liassic sedimentation (Barranca Formation) of central Sonora, México: Notas Geológicas, Boletín de la Asociaión de Egresados de Geología de la Universidad de Sonora y de la delegación noroeste de la Sociedad Geológica Mexicana, no. 3, p. 2–8.

Miall, A. D., 1978, Lithofacies types and vertical profile models in braided river deposits; A summary, *in* Miall, A. D., ed., Fluvial sedimentation: Canadian Society of Petroleum Geologists Memoir 5, p. 597–604.

Pindell, J. L., 1985, Alleghenian reconstruction and subsequent evolution of the Gulf of Mexico, Bahamas, and proto-Caribbean: Tectonics, v. 4, no. 1, p. 1–39.

Potter, P. E., and Cojan, I., 1985, Description and interpretation of the type section of the Barranca Group east of Rancho La Barranca, Municipio de San Javier, Sonora, *in* Weber, R., ed., Simposio sobre floras del Triásico Tardío, su fitogeografía y paleoecología, III Congreso Latinoamericano de Paleontología, Universidad Nacional Autónoma de México, Instituto de Geología, Memoria, p. 101–105.

Puigdefabregas, C., and Van Vliet, A., 1978, Meandering stream deposits from the Tertiary of the southern Pyrenees, *in* Miall, A. D., ed., Fluvial sedimentology: Canadian Society of Petroleum Geologists Memoir 5, p. 469–485.

Radelli, L., and 9 others, 1987, Allochthonous Paleozoic bodies of central Sonora: Universidad de Sonora, Departamento de Geología Boletín, second epoch, v. 4, no. 1 and 2, p. 1–15.

Ratcliffe, N. M., and Burton, W. C., 1985, Fault reactivation models for origin of the Newark basin and studies related to eastern U.S. seismicity, *in* Robinson, G. R., Jr., and Froelich, A. J., eds., Proceedings of the Second U.S. Geological Survey Workshop on the Early Mesozoic Basins of the Eastern United States: U.S. Geological Survey Circular 946, p. 36–45.

Scott, K. R., Hayes, W. E., and Fietz, R. P., 1961, Geology of the Eagle Mills Formation: Gulf Coast Association of Geological Societies Transactions, v. 11, p. 1–14.

Silva-Pineda, A., 1961, Flora fósil de la Formación Santa Clara (Cárnico) del Estado de Sonora, part 2, *in* Paleontología del Triásico Superior de Sonora: Universidad Nacional Autónoma de México, Instituto de Geología, Paleontología Mexicana número 11, 37 p.

Silver, L. T., and Anderson, T. H., 1974, Possible left-lateral early to middle Mesozoic disruption of the southwestern North America craton margin: Geological Society of America Abstracts with Programs, v. 6, p. 955–956.

—— , 1983, Further evidence and analysis of the role of the Mojave-Sonora Megashear(s) in Mesozoic Cordilleran tectonics: Geological Society of America Abstracts with Programs, v. 15, p. 273.

Soto-Contreras, L. A., and Navarro-Martínez, L. A., 1987, Geología de la carta Tecoripa (H12D64) porción este-central del Estado de Sonora; Nuevas contribuciones Paleozoico [B.Sc. thesis]: Hermosillo, Sonora, México, Universidad de Sonora, 93 p.

Stewart, J. H., 1969, Major Upper Triassic lithogenetic sequences in Colorado Plateau region: American Association of Petroleum Geologists Bulletin, v. 53, no. 9, p. 1866–1879.

—— , 1988, Latest Proterozoic and Paleozoic southern margin of North America and the accretion of Mexico: Geology, v. 16, p. 186–189.

Stewart, J. H., Poole, F. G., and Wilson, R. F., 1972, Stratigraphy and origin of the Chinle Formation and related Upper Triassic strata in the Colorado Plateau region *with a section on* Sedimentary petrology by R. A. Cadigan, *and a section on* Conglomerate studies by W. Thordarson, H. F. Albee, and J. H. Stewart: U.S. Geological Survey Professional Paper 690, 336 p.

Stewart, J. H., McMenamin, M.A.S., and Morales-Ramírez, J. M., 1984, Upper Proterozoic and Cambrian rocks in the Caborca region, Sonora, México; Physical stratigraphy, biostratigraphy, paleocurrent studies, and regional relations: U.S. Geological Survey Professional Paper 1309, 36 p.

Stewart, J. H., Anderson, T. H., Haxel, G. B., Silver, L. T., and Wright, J. E., 1986, Late Triassic paleogeography of the southern Cordillera; The problem of a source for voluminous volcanic detritus in the Chinle Formation of the Colorado Plateau region: Geology, v. 14, p. 567–570.

Swanson, M. T., 1982, Preliminary model for an early transform history in central Atlantic rifting: Geology, v. 10, p. 317–320.

Weber, R., 1980, Megafósiles de Conifera del Triásico Tardío y del Cretácico Tardío de México y consideraciones generales sobre las coníferas Mesozoicas de México; Universidad Nacional Autónoma de México, Instituto de Geología, Revista, v. 4, no. 2, p. 111–124.

—— , 1985, Las plantas fósiles de la Formación Santa Clara (Triásico Tardío, Sonora, México): Estado actual de las investigaciones, *in* Weber, R., ed., Simposio sobre floras del Triásico Tardío, su fitogeografía y paleoecología, III Congreso Latinoamericano de Paleontología, Universidad Nacional Autónoma de México, Instituto de Geología, Memoria, p. 107–124.

Weber, R., Zambrano-García, A., Amozurrutia-Silva, E., 1980a, Nuevas contribuciones al conocimiento de la tafoflora de la Formación Santa Clara (Triásico Tardío) de Sonora: Universidad Nacional Autónoma de México, Instituto de Geología, Revista, v. 4, no. 2, p. 125–137.

Weber, R., Trejo-Cruz, R., Torres-Romo, A., García-Padilla, A., 1980b, Hipótesis de trabajo acerca de la paleoecología de comunidades de la tafoflora Santa Clara del Triásico Tardío de Sonora: Universidad Nacional Autónoma de México, Instituto de Geología Revista, v. 4, no. 2, p. 138–154.

Wilson, I. F., and Rocha, V. S., 1949, Coal deposits of the Santa Clara district near Tonichi, Sonora, Mexico: U.S. Geological Survey Bulletin 962-A, p. 1–80.

MANUSCRIPT ACCEPTED BY THE SOCIETY APRIL 18, 1990

Geological Society of America
Special Paper 254
1991

Depositional environment, petrology, and provenance of the Santa Clara Formation, Upper Triassic Barranca Group, eastern Sonora, Mexico

Isabelle Cojan
Ecole des Mines de Paris, C.G.C.M. Sedimentologie, 35 Rue St., Honoré 77300 Fontainebleau, France

Paul Edwin Potter
H. N. Fisk Laboratory of Sedimentology, Department of Geology, University of Cincinnati, Cincinnati, Ohio 45221-0013

ABSTRACT

The depositional environment and provenance of the Santa Clara Formation of the Upper Triassic Barranca Group in eastern Sonora were studied based on a detailed description of 875 m of section and study of 123 thin sections from both outcrops and cores.

The Santa Clara Formation is a cyclic unit with typical fluvial cycles of 4 to 16 m. Highly altered, quartz-rich sandstone is interbedded with black and gray shale and minor beds of graphite and anthracite, which were formerly mined. Detrital chert and remobilized chert cement are significant features of the sandstones of the Barranca. Diagenesis is extensive, of high rank, and commonly blurs the distinction between primary detrital matrix and chemical cement to produce a predominance of quartz-rich wacke.

A quiescent basin with tropical climate, subdued marginal relief, and no active volcanism is inferred for the Santa Clara, and limited crossbedding measurements indicate a source to the north. Basin type and its relation to other Upper Triassic deposits remain major problems.

INTRODUCTION

The Barranca Group occurs in widely separated outcrops in southeastern Sonora (Fig. 1), where it contains beds of anthracite and graphite. The Barranca Group was first named by Dumble (1900, p. 138–139) in his classic reconnaissance study of Sonora, and was later described by King (1934, 1939). Since then it has received very little study, although studies of the tectonic setting of Sonora commonly mention it (Abadie, 1981; Anderson and others, 1979; Cordoba and others, 1980; Echavarri, 1978; Rangin, 1978). Its thickness is not easy to determine, but Stewart and Roldán-Quintana (this volume) estimate it at more than 3,000 m. Tertiary volcanics discordantly overlie the Barranca Group, which was intruded by Laramide granites. Plant fossils collected from its middle member indicate a Late Triassic age (Weber, 1980, 1985a, b, c; Weber and Garcia, 1985).

The Barranca Group in eastern Sonora consists chiefly of quartzite with beds of granules and pebbles of quartz and chert, quartzitic siltstone, shale, and minor anthracitic coal, which is locally mined for graphite (Alencaster de Cserna, 1961; Pesquera and Carbonell-C., 1960; Wilson and Rocha, 1946; Wilson, 1949). The group has been divided by previous workers into three members: an upper quartzitic unit, a middle anthracite-bearing unit that also contains graphite and carries the name Santa Clara Formation, and a lower quartzitic unit. The graphite was formed from anthracite by heat from intrusions emplaced during the Laramide orogeny. Most exposures of the Barranca have dips of 20° to 40° or more, and many are faulted and intruded by dikes. Fracturing is locally extensive and inhibits measurement of sedimentary structures.

Many questions can be asked about the Barranca Group: In what kind of basin or basins was it deposited? What was its

Cojan, I., and Potter, P. E., 1991, Depositional environment, petrology, and provenance of the Santa Clara Formation, Upper Triassic Barranca Group, eastern Sonora, Mexico, *in* Pérez-Segura, E., and Jacques-Ayala, C., eds., Studies of Sonoran geology: Geological Society of America Special Paper 254.

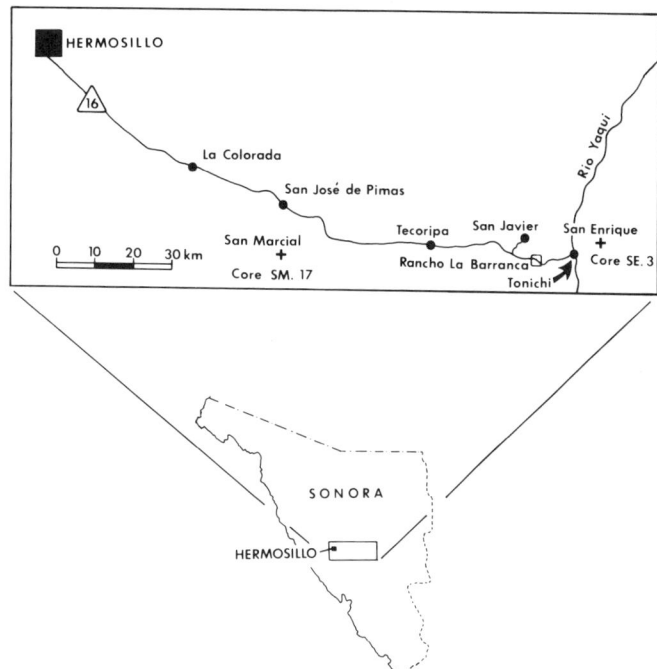

Figure 1. Index map of study area. Type section occurs at Rancho La Barranca.

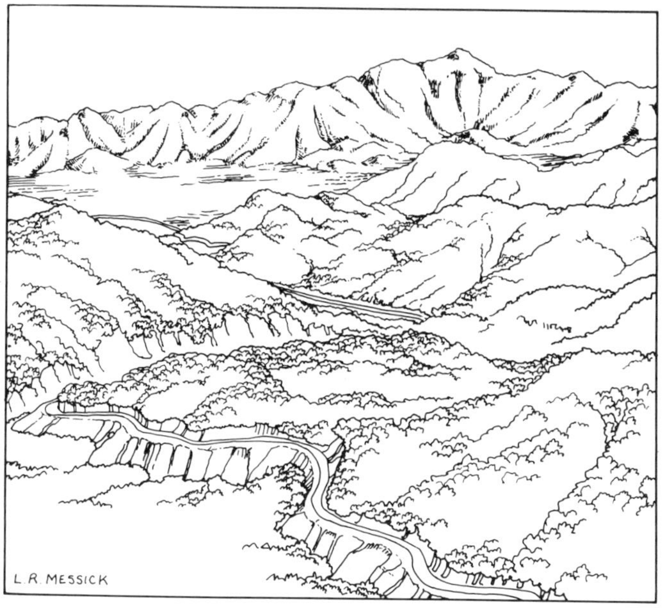

Figure 2. Type section of the Barranca Group along Barranca Canyon as seen from above Rancho La Barranca; Rio Yaqui is beyond the distant mountains to the southeast.

depositional environment? What does the petrology of its sandstones tell us of its source area and its paleoplate setting? How does the petrology of the Barranca compare to that of other Upper Triassic sandstones in nearby parts of North America? We attempt to answer these questions by combining the standard methods of sedimentary petrology with the sedimentological data of Potter and others (1980), who described the Barranca in cores and in outcrop, chiefly at the type section in La Barranca Canyon along Sonora Highway 16. For petrologic study we focused on 380 m of the anthracitic-bearing middle unit, the Santa Clara Formation, in Barranca Canyon along Sonora Highway 16 between 157.5 and 163 km east of Hermosillo. Here, about 1.5 km east of Rancho La Barranca, where Sonora Highway 16 crosses very rugged terrain (Fig. 2), two sections (Figs. 3 and 4) were carefully measured and described in 1980. These outcrops are of the anthracite-bearing middle part of the group. Two of the cores—core SE3 (186.7 m) from San Enrique, and core SM 17 (308.9 m) from San Marcial were also studied petrographically and carefully logged. Because these cores are anthracite-rich, they too may belong to the Santa Clara Formation.

LITHOLOGY AND DEPOSITIONAL ENVIRONMENT

In roadcuts east of Rancho La Barranca (Figs. 3 and 4), the Santa Clara Formation is part of a large homocline, with dips generally eastward between 20° and 40°. The exposure along the highway is exceptionally good—about 70 to 80 percent. Large imposing cliffs of quartzites of the Barranca Group from the north side of the canyon (Fig. 2) and stratigraphically overlie the section exposed along the highway. The highway section contains a number of intrusive dikes and faults; some of the larger faults cut bedding, but most are small bedding-plane faults. A possible large stylolite zone is also present (Fig. 5). All of the cores we studied have faults and dikes in about the same abundance.

The exposures along Sonora Highway 16 show the Santa Clara to consist of quartzite interbedded with hard, dense, dark gray to black shale containing leaf impressions, and some beds of graphite and anthracite. The two measured outcrop sections total 380 m (Figs. 3 and 4), and their proportions of quartzitic sandstone, quartzitic siltstone, shale and anthracite average 54:27:18:1. In core SE-3, these ratios are very similar—55:14:24:7.

Pebbly sandstones are present locally, and consists predominantly of quartz pebbles with black and gray chert (7 to 11 percent) and quartz-rich sandstone pebbles (4 to 14 percent). Virtually all pebbles are rounded to subrounded and occur as lags on bedding planes or in thin beds.

Gray quartzite ranges in thickness from beds of only a few decimeters to units of 19 m, is coarse to fine grained, and contains subrounded to rounded quartz and chert pebbles that are both scattered and in beds. All quartzites are dense and hard. There is also some crossbedding, although it is hard to measure. Perhaps more crossbedding and ripple marks were present originally, but are now obscured by fracturing and shearing. Beds within the quartzite bodies thin upward as grain size fines upward. Plant-bearing, dark, organic rich shales also occur, some as much as 6 m thick. Beds of anthracite and graphite are sheared and fractured, and the thickest is about 0.7 m, although thicknesses of as much as 4 m have been reported elsewhere. No red beds are present in the Barranca in its type section or in the cores that we studied.

Careful descriptions of outcrops and cores permitted us to determine the thickness distribution of the sandstone, siltstone, and shale beds, all of which show log-normal distributions (Fig. 6). Median thicknesses for shales, siltstones, and sandstones are 1.0, 1.5, and 2.3 m, respectively. However, thicknesses of the 80th percentile for shales, siltstones, and sandstones are much larger: 2.8, 3.7, and 6.2 m, respectively. The modal class of sandstone units occurs between 4 and 8 m, a range that includes 32 percent of the thicknesses of all sandstones.

A simple transition matrix (Selley, 1970) was also calculated and, in addition to anthracite, shale, siltstone and sandstone, beds of conglomerate (pebbly sandstones) were recognized (Fig. 7). A strong fining-upward cycle is apparent—sandstone overlies conglomerate beds with transition probability of 1.0 and shale overlies sandstone with transition probability of 0.56. Higher in the cycle, the transitions are less pronounced, and transition probabilities range from 0.3 to about 0.5. The transition probabilities of anthracite to shale, siltstone, and sandstone are roughly of the same value and also of this magnitude. The two cores show a similar fining-upward sequence (Fig. 8). thus, a clear conglomerate-sandstone-shale sequence in the lower part of the cycle changes to a noisier and more random interbedding of thinner units of shale, siltstone and fine sandstone in the upper, low-energy part of the cycle. When total cycle thickness is considered, most are between 4 and 32 m thick—a range that includes 68 percent of all the cycles.

The fining-upward sequences and the sedimentary structures of crossbedding, ripple mark, and rip-up clasts clearly indicate deposition by small to modest meandering streams (Spalletti, 1980, p. 63–85; Davis, 1983, Chapter 8; Collinson, 1986, Chapter 3) on a low-gradient plain, one with abundant muddy swamps and lakes. From a sedimentologic viewpoint, the anthracite-bearing middle part of the Barranca, the Santa Clara Formation, is typical of coal-measure sedimentation such as is found in the Appalachian Basin and described well by Horne and others (1978) and Ferm and Horne (1979). Probably the original peat beds had rather limited extent, perhaps a few tens of square kilometers. The lack of major marine units in eastern Sonora

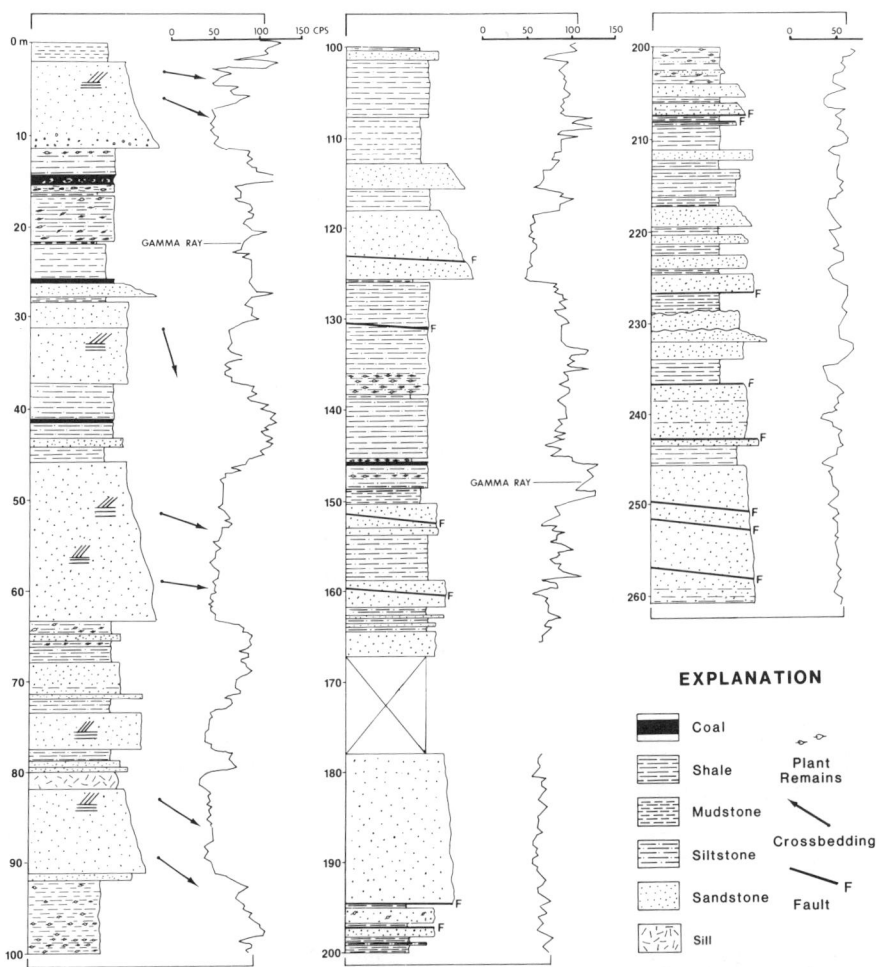

Figure 3. Section exposed along Sonora Highway 16 about 2.5 km east of Rancho La Barranca. Irregular line is gamma ray profile; arrows represent crossbedding orientation.

suggests to us that the deposition was chiefly by streams on a coastal plain. Thus, we concur with Salvador (1987, Fig. 3) that this part of Sonora in Late Triassic time was a low plain—but with a humid climate rather than with appreciable red beds.

The discharge of the rivers to this plain may have been similar to that of the nearby Rio Yaqui. In part, we base this conclusion on the thickness of the median and 80th percentiles of the sandstones (Fig. 6). The sparse available crossbedding indicates that the rives generally flowed to the south: to the southeast in section 1, and to the southwest in section 2. The wide divergence of the two paleocurrent directions may simply be the result of two few measurements but, if not, would indicate a small basin.

SANDSTONE PETROGRAPHY

Petrographic methods. One hundred and twenty-three thin sections were studied, 64 from sandstones and the rest from siltstones and shales. At least 200 point counts per thin section were made for composition, and size was also estimated roughly. Table 1 gives the petrographic variables counted. Some polished sections were used for limited microprobe study of feldspar and matrix. Thin sections from cores SE-3 and SM-17 were studied by Cojan, and those from the outcrop by Potter.

Quartz. Total quartz ranges from as little as 17 to as much as 74 percent in the sandstones, and polycrystalline quartz ranges from 1 to as much as 43 percent. The average values of total monocrystalline and polycrystalline quartz are 48 and 13 percent, respectively. Most of the quartz is angular to subangular. Marginal minor replacement of quartz by fine micas complicates estimation of grain outline on most grains. However, where overgrowths are clear, they nearly always occur on grains with well-rounded outlines (Fig. 9). Thus, at least part of (perhaps most?) the quartz grains had an earlier transport history. Much of the monocrystalline quartz has partial overgrowths that are quite distinctive (Fig. 9), and all of them were later corroded. Much of the monocrystalline quartz is remarkably free of strain, especially considering the high level of pervasive structural deformation in east-central Sonora. The polycrystalline quartz is dominantly anhedral, about 82 percent; polygonal quartz is second in abundance, about 12 percent; stretched quartz, about 4 percent; and other types about 2 percent. It is not always easy to distinguish polygonal quartz from recrystallized chert matrix, nor is it always clear where one grain of polycrystalline quartz ends and another begins.

Detrital chert. Detrital chert is also difficult to distinguish from secondary matrix chert, but distinct detrital grains do occur (Fig. 9). Detrital chert ranges from a trace up to 12 percent and is commonly subrounded to rounded. Its average abundance is 4 percent and it contains fractures of annealed quartz, which were formed in the source region. A few grains of chert also contain secondary mica (see sections on "Mica," "Matrix," and "Cement" for more discussion). There is a gradation between clearly detrital chert grains and those that can only be partially discerned as detrital and structureless, remobilized chert matrix. Considering remobilized chert as all detrital, the initial abundance of detrital chert in the Barranca Group may have been as much as 10 to 16 percent of all framework grains. Cherts of all types in the Barranca certainly deserve much more study.

Feldspar. Feldspar is rare in the Barranca Group and consists chiefly of plagioclase whose polysynthetic twinning is commonly fractured (Fig. 9), and a few grains with Carlsbad twinning. Its occurrence is erratic: most slides totally lack feldspar, but a few have as much as 8 to 10 percent. This variation could be due either to varying degrees of local diagenesis

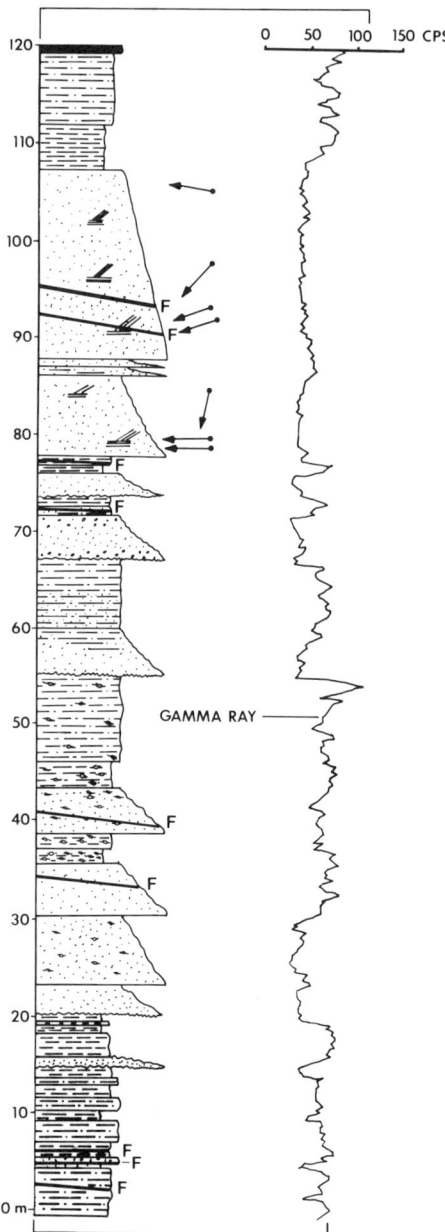

Figure 4. Section exosed along Sonora Highway 16 about 1 km east of Rancho La Barranca. Irregular line is gamma ray profile; arrows represent crossbedding orientation.

Figure 5. Possible styolite zone along Sonora Highway 16 east of Rancho La Barranca.

or to contrasting local source areas. The average value of feldspar is about 1 percent in the quartzites, and it is virtually always much altered. Most plagioclase grains have a cloudy appearance, and in some the polysynthetic twins only form faint, poorly defined ghosts (Fig. 9). Minor calcite replaces the plagioclase, as do some micas. The plagioclase seems to fracture more than the quartz. No microline was observed, nor were there overgrowths on feldspars. Microprobe study by J. B. Maynard of two samples (49 grains were probed in the sample and 50 in another) show all the feldspar to be plagioclase but with much variation. One sample is chiefly albite (Ab 90), whereas the other is much more calcic (Ab 50). Again, one wonders if this is the result of local differences in diagensis or of source area?

Rock fragments. Rock fragments vary widely in abundance, apparently depending on the degree of digenesis. Where digenesis was intense, as is commonly true, rock fragments are hard to identify, and many are probably counted as matrix; where digenesis was weak, however, as many as 42 percent are present, although 10 to 15 percent is far more typical. They average about 8 percent and, in many slides, cannot be identified easily as to type.

There is complete gradation between well-defined argillaceous rock fragments and structureless recrystallized matrix. Where

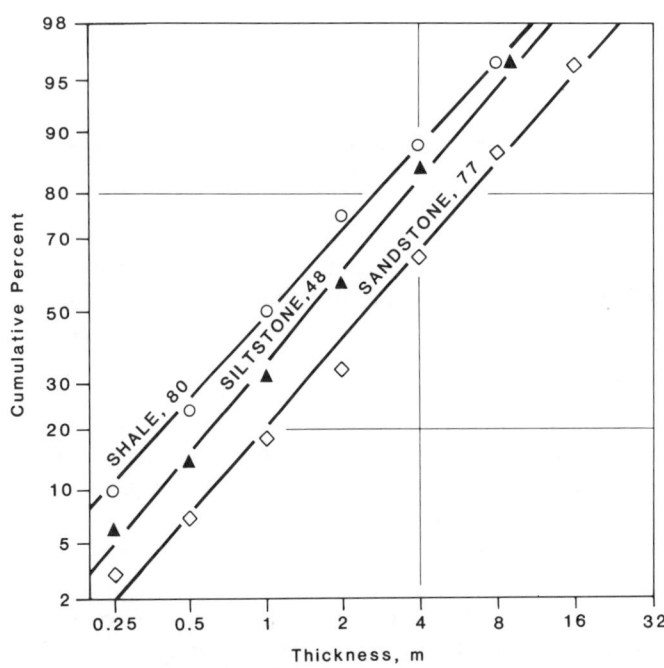

Figure 6. Thickness distribution of shale, siltstone, and sandstone beds; numbers indicate number of units measured.

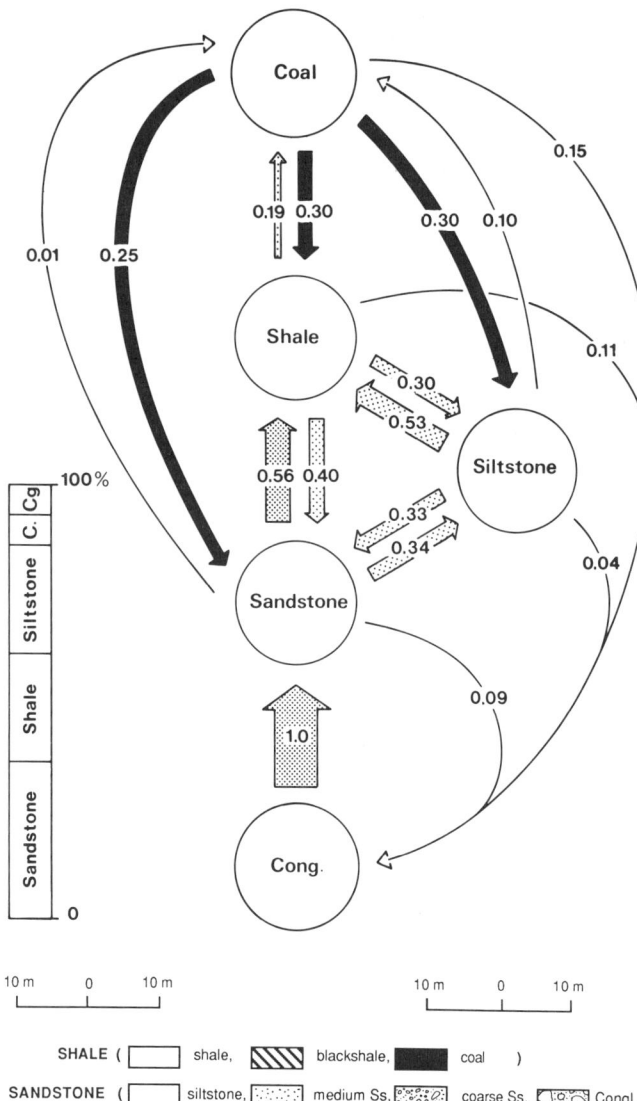

Figure 7. Embedded Markov transition matrix. Lithologic proportions shown in lower left.

alteration is minimal, rock fragments are dominantly argillaceous and now chiefly consist of mosaics of finely recrystallized micas, some with good orientation parallel to original stratification and others with a grid-like interpenetrating fabric that has completely obliterated all original texture. Where rock fragments are not recognizable, well-crystallized irregular mosaics of bright, highly birefringent recrystallized muscovite and chlorite occur. These mosaics also occur between tightly packed quartz grains and represent remobilization and total recrystallization of the original detrital matrix. All other types of rock fragments are rare.

Micas. Micas are present in every slide, but most appear to be secondary rather than detrital and are discussed under matrix and cement below. Flakes of detrital muscovite, biotite, and chlorite are very rare in the sandstones, if present at all.

Matrix and cement. Primary detrital matrix (less than 25 μm) and secondary cement are both troublesome problems for the petrographer who studies the Barranca Group. Commonly, it is difficult to distinguish between original detrital matrix and later-recrystallized cement, except for clear secondary coarse patches of chlorite and muscovite. Thus, the two are discussed together.

The matrix and cement of the sandstones of the Santa Clara Formation average 31 percent and consist almost totally of secondary chert and coarse muscovite followed by chlorite. Most of the chert may have been remobilized from detrital chert, which today forms large irregular masses. Coarse secondary aggregates of micas and chlorites several hundred microns in diameter also occur where diagenesis was high. Muscovites predominate and are perhaps 20 times more abundant than chlorite, some of which have a distinctive Berlin blue color under crossed nicols. Minor, secondary, small (62 μm or less) crystals of muscovite can be found in many slides and even in chert grains.

Some rare carbonate is present, but is possibly the result of outcrop weathering and thus could be caliche. The large size of the micas and the abundant nondetrital chert all point to a secondary origin.

In sum, the diagenesis of the sandstones of the Barranca Group is extensive and locally represents an upper limit of sedimentary analysis, chiefly because of the difficulty of distinguishing detrital framework and matrix from high-rank secondary alteration products. High temperatures related to Cretaceous igneous intrusions and possible concurrent deformation favored pseudoplastic flow of argillaceous rock fragments, argillization of feldspars, and the recrystallization and flow of once-abundant detrital chert. Thus, most of the sandstones are wackes, because they are rich in pseudomatrix (Dickinson, 1970; Morad, 1984).

Classification. Because of extensive diagenesis, classification of the sandstone of the Santa Clara as either arenites or wackes (Fig. 10) has little meaning; however, 61 of the 67 sandstones had more than 15 percent matrix, and only 6 had less (matrix was defined as smaller than 25 μm). At the time of deposition, most of these sands were almost certainly arenites rather than wackes, as judged by the vast majority of modern sands. As a group, about 35 percent of the sandstones studied are quartz-rich, about 45 percent are sublithic sandstones, and about 20 percent are lithic sandstones (Table 2A).

SHALE PETROLOGY

The shales of the Santa Clara Formation vary from black to medium to dark gray and from massive to finely laminated. They also include some that are slightly bioturbated and microbrecciated. In cores, these shales are universally hard and dense and may have tiny filled microfractures. As a group, the shales, which form about 25 percent of the Santa Clara, seem more complex than their associated sandstones and may represent three or four different environments. Plant fossils are common in the shales, especially in the black shales (Weber and Garcia, 1985).

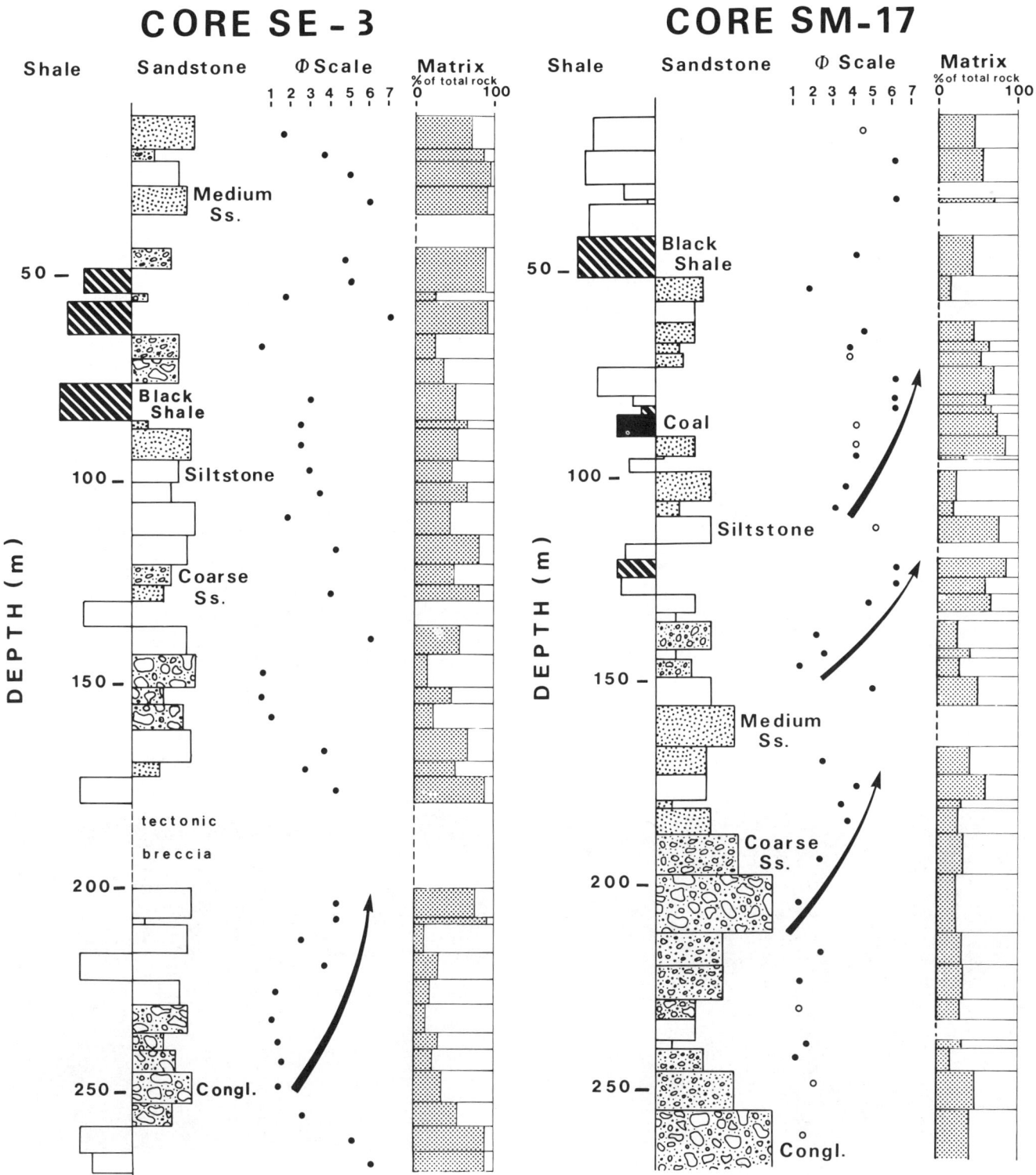

Figure 8. Cores SE-3 and SM-17 both show clear fining-upward cycles. Compare with tree diagram of Figure 7.

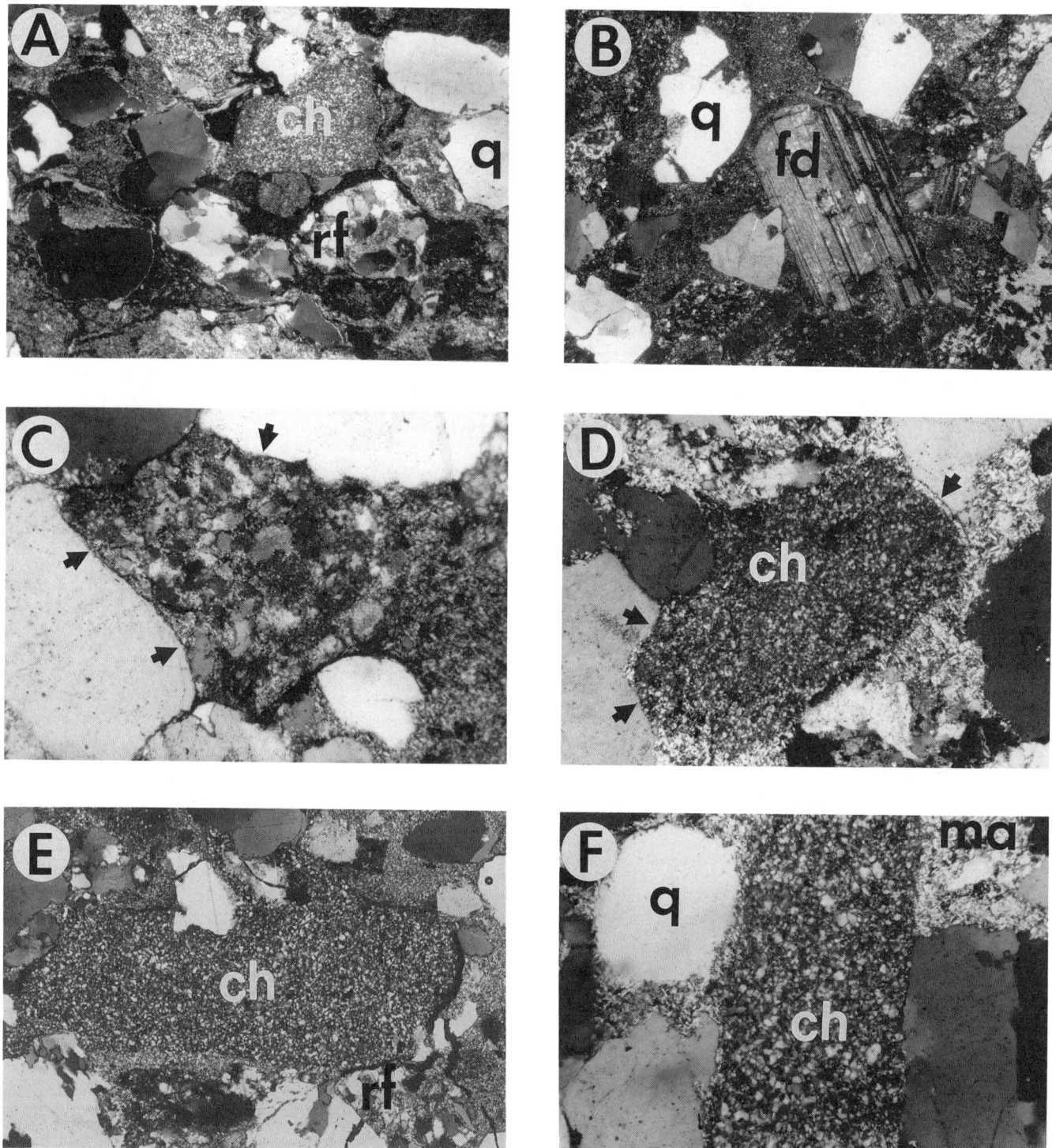

Figure 9A. Photomicrographs of sandstones. (A) Typical appearance in thin section with chert (ch) rock fragment (rf). (B) Reaction pit in quartz and polysynthetic twining in feldspar (fd). (C) Rock fragments. (D) Detrital chert with adjacent recrystallized matrix. (E) Detrital chert and rock fragment. (F) Quartz, detrital chert, and recrystallized matrix.

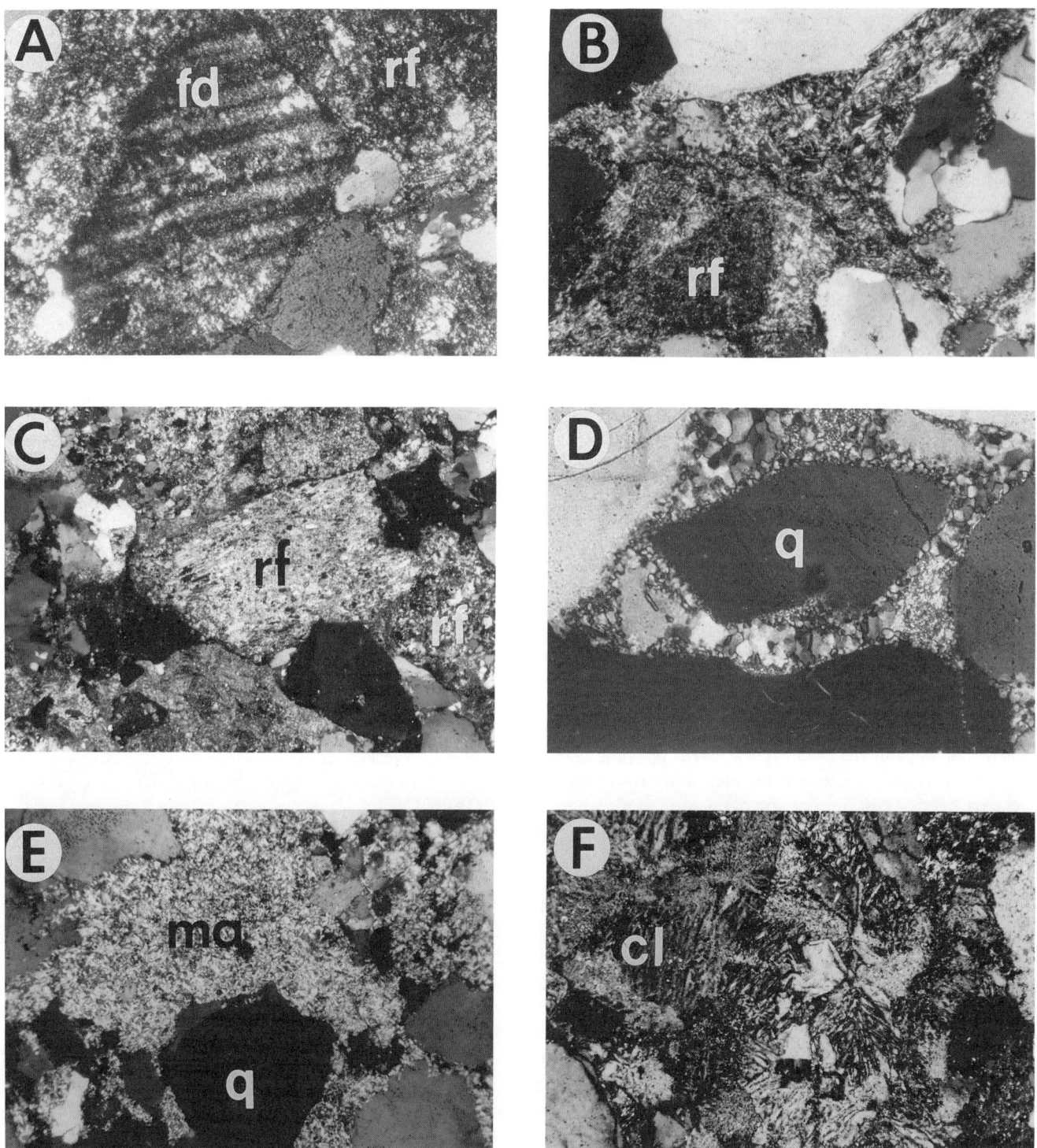

Figure 9B. (A) Highly altered plagioclase. (B) Well-defined rock fragments (rf) and poorly defined ones. (C) Rock fragments. (D) Detrital quartz and recrystallized quartz cement. (E) Pseudo matrix. (F) Chlorite (cl).

TABLE 1. PETROGRAPHIC COMPONENTS COUNTED

Quartz, Tv	
Monocrystalline, Qm	Unit extinction.
Polycrystalline, Qp	Two or more grains; anhedral, polygonal, stretched and fibrous were recognized.
Chert	Distinctly finer grained than polycrystalline quartz.
Feldspar, Ft	
Potash feldspar, Fk	Microcline and orthoclase.
Plagioclase, Fp	Recognized by polysynthetic twinning, Carlsbad twins and lack of yellow stain.
Rock Fragments, Rf	Include plutonic igneous and metamorphic grains, volcanic grains, argillaceous sandstones, siltstone, and carbonate grains.
Mica	Detrital (from form and size) muscovite and chlorite.
Matrix	Smaller than 25 μm, but much, if not most, may be authigenic.
Cement	Calcite and chert and fine-grained micas, none of which are easy to assign to a primary or authigenic origin.
Other	Heavy minerals as well as authigenic opaques.

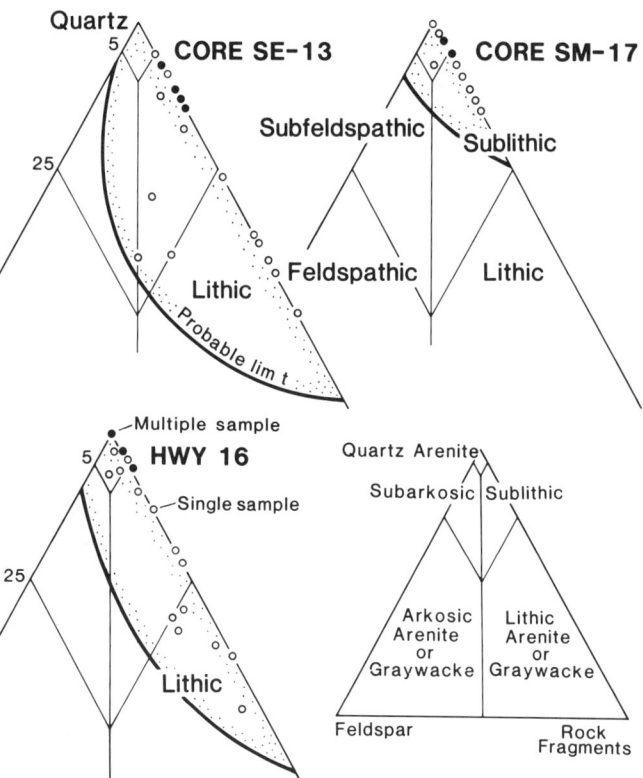

Figure 10. Petrographic classification.

All but a very few of the shales contain some angular quartz silt that is rarely more than 40 μm in diameter (Fig. 11). Its abundance commonly ranges up to 10 to 15 percent. A few argillaceous silt-sized grains are also present, but feldspar and microfossils are both virtually absent. Such silt grains may either form well-defined laminations, or less commonly, are uniformly scattered throughout the shale. A few flakes of white detrital muscovite parallel laminae and are almost always present in unaltered shales. Dark, opaque, organic matter forms scattered blebs and small masses and is concentrated along laminae in the medium to dark gray shales, but forms as much as 85 percent of the black organic-rich shales.

Several variants of the above shales exist. With more silt-sized grains, shales grade into dark argillaceous wackestones. Shales, like their associated sandstones, can be largely altered. When altered, they may be either microbrecciated or massive, extra hard, and seemingly homogenous. Minor to major masses of secondary, light yellow chlorite occur in such shales. This chlorite forms irregular blebs and masses; other secondary low birefringent clays also are present so that finally all primary lamination was destroyed. X-ray analyses of these shales show the most common clay minerals to be iron-rich chlorite and illite. No trace of kaolinite was observed. Another type of shale is a black organic-rich shale with weakly birefringent stringers that form a meshwork parallel to bedding. In a few shales there are euhedral, low-birefringent crystals of uncertain origin (Fig. 11).

The shales of the Santa Clara and Barranca deserve more study, and their petrography should be carefully correlated with field descriptions, possible micro-faunas and plants, and position in the Santa Clara cycle (Fig. 7). Perhaps more than any other lithology, these shales and their fossils could tell us about the environment of deposition and, with application of organic petrology, diagenetic history.

PROVENANCE

Inferring the types of source rocks of the Santa Clara Formation and the Barranca from present sandstone composition is a challenging task because of extensive alteration by high-rank diagenesis associated with the underlying intrusive Laramide granites. This diagenesis has remobilized detrital chert, altered many argillaceous rock fragments and feldspars, and formed much secondary chlorite and muscovite as cement. Recognizing that we have petrographically studied only a small part of the Barranca, we suggest the following.

Inference 1. The original pre-Laramide composition of Barranca sandstones was that of a recycled, chert-rich, lithic or sublithic arenite with somewhat more feldspar and argillaceous rock fragments than at present.

Discussion. Evidence for recylcing comes from the overgrowths on subrounded quartz and rounded to subrounded detrital chert grains, and from chert both as totally remobilized matrix and detrital chert plus hard-to-recognize and highly altered argillaceous rock fragments. Probably, a typical pre-

TABLE 2. PETROGRAPHIC SUMMARY

PART A. PETROLOGIC TYPES (PERCENT)

	Quartz-rich	Sublithic	Lithic	Samples
Core SM-17	61	39	0	23
Core SE-3	0	66	34	21
Outcrop Hwy. 16	42	29	29	23

PART B. TECTONICS BASED ON Q:F:Rf

	Craton Interior	Recycled Orogen	Samples
Core SM-17	61	39	23
Core SE-3	0	100	21
Outcrop Hwy. 16	26	74	23

PART C. TECTONICS BASED ON Qm:F:Lt

	Craton Interior	Recycled	Mixed	Samples
Core SM-17	0	100	0	23
Core SE-3	0	95	5	21
Outcrop Hwy. 16	0	100	0	19

PART D. TYPES OF RECYCLED TERRANES BASED ON Qm:F:Lt

	Quartzose	Transitional	Lithic	Samples
Core SM-17	48	48	4	23
Core SE-3	19	81	0	21
Outcrop Hwy. 16	6	68	26	19

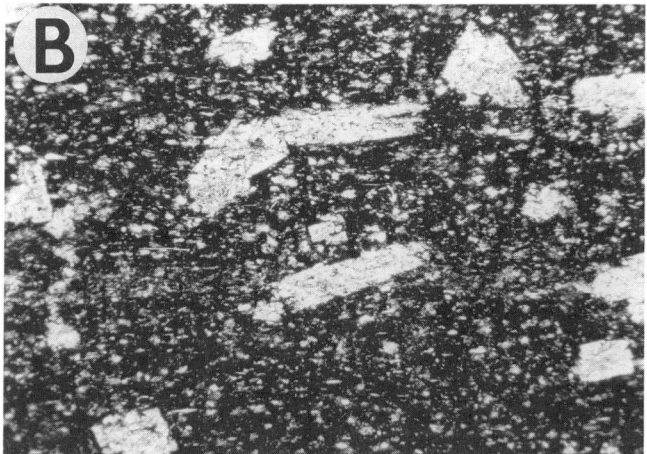

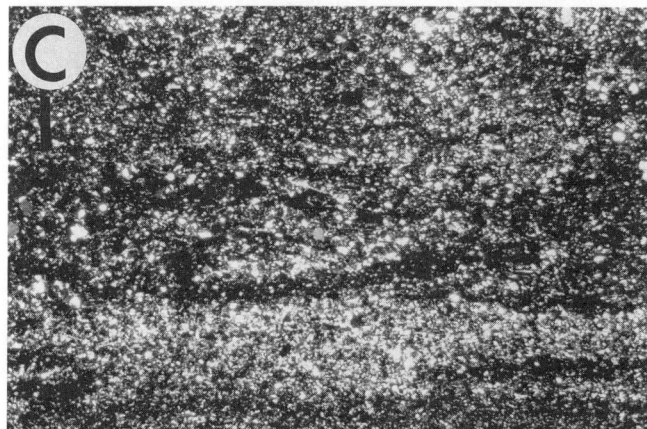

Figure 11. Photomicrographs of shales, all of which are silty and organic rich. Pseudomorphs in photomicrograph need more study.

Laramide Q:F:Rf composition was perhaps 56:10:34. Secondary compaction of rock fragments and argillization of feldspar, chlorite, and mica plus remobilized chert converted an original arenite into the present wacke.

Inference 2. The source consisted chiefly of second-or third-cycle sandstones with some later contributions from plutonic and metamorphic terrains (Figs. 10, 12, and 13; Tables 2B, C and D).

Discussion. Folk (1980, p. 143) recognized a special class of chert-rich lithic arenites that chiefly occur in tectonic belts containing volcanics and cherty geosynclinal sediments, and noted that the Mesozoic sandstones of the Rockies are commonly chert-rich. Using the interpretive scheme of Dickinson and others (1983b, Fig. 1), the Santa Clara shows a provenance dominantly from recycled and mixed terrains. How much of the chert in the Santa Clara is derived form altered volcanics and how much from preexisting earlier Mesozoic or Paleozoic chert sediments? Our petrographic study does not provide a definitive answer, but the presence of as much as 12 percent chert in the Santa Clara shows it to belong to the broad class of chert-rich sandstones associated with subduction-related and oceanic-continental collisions. Consider, for example, the Triassic Vester Formation of central Oregon, which has 20 to 58 percent chert (Dickinson, and others, 1979, Table 3) and the Cadomin Conglomerate of Alberta with 40 to 70 percent chert (Schulthesis and Mounjoy, 1978, p. 315–316). Could some of the chert in the Barranca be

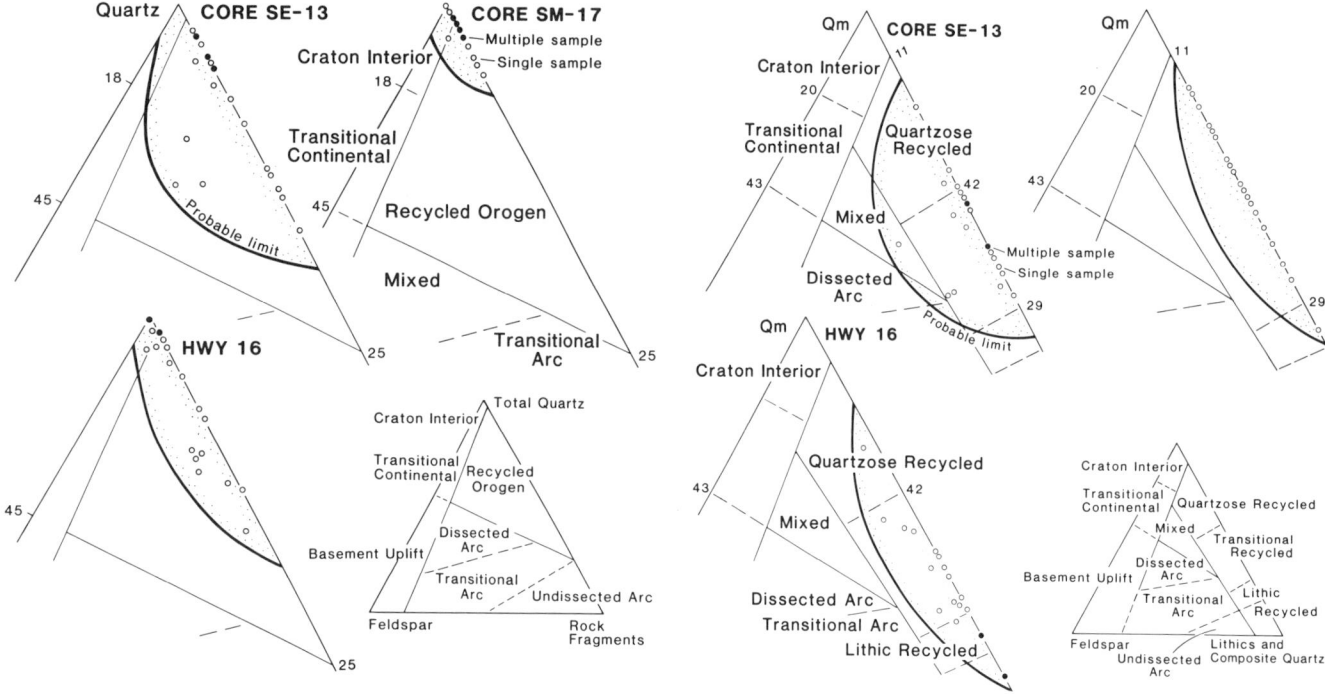

Figure 12. Tectonic classification of sandstones (after Dickinson and others, 1983, Fig. 1). Both triangles A and B suggest that the framework grains of the Santa Clara were chiefly derived from a source with appreciable sedimentary rocks.

relic from a volcanic arc similar to that inferred by Medieta and Partida (1987, Fig. 10) in southern Durango? More detailed petrology of the cherts in the Santa Clara is clearly needed. Progressive reworking and weathering would also help convert chert-rich and lithic conglomerates into quartz-rich conglomerates (Adams, 1979) like those of the Santa Clara. Without doubt, some of these quartz pebbles were derived from quartz veins and perhaps some metamorphic terrains.

Inference 3. Climate was always tropical during the deposition of Santa Clara, and low, vegetation-covered highlands supplied most of its detritus.

Discussion. The abundant anthracite and graphite beds plus the black shales of probable lacustrine origin and the presence of subtropical plant debris (Weber, 1985a, b, c), point to a warm tropical or subtropical climate at the time of deposition. The abundance of chert debris and absence of carbonate clasts suggests that earlier recycling took place under a tropical climate. Low mountains or high lands covered with vegetation supplied sublithic, chert-rich sands and muds from a source to the north, judging by limited crossbedding data.

Regional significance. What do the above inferences tell us about the regional geology of Sonora during Late Triassic time? Regional summaries by Stewart and others (1986, p. 576–578) indicate that the Late Triassic Chinle Formation of Arizona and New Mexico, and the Dochum Group of west Texas (McGowen and others, 1979) are quite different, although both are red beds. The paleocurrents of the Chinle flowed from south to north, and the paleocurrents of the Dochum Group indicate transport to the west (McGowen and others, 1979, Fig. 42). Neither match the attributes of the Barranca of east-central Sonora. Hence, the Barranca had neither a common source nor a common climate with the Chinle and the Dockum, even though they are now separated by only a few hundreds of kilometers. It is also sobering to realize that the contrast—red beds versus coal-graphite beds—is one of aridity rather than latitude. Redbeds could form in the rain shadow of a large mountain range, and coal beds could form on its windward side. In addition, there is the complication of the marine Upper Triassic in northwestern Sonora in Sierra del Alamo (Gonzalez-Leon, 1980).

Does the much discussed Mojave-Sonora Megashear (Anderson and Silver, 1979; Anderson and Schmidt, 1983; Campa and Coney, 1983) provide a possible explanation (Fig. 14)? Because the inferred megashear is a left-lateral fault of Jurassic age and the Barranca lies mostly south of it, the original Barranca basin could have been deposited as far as 800 km to the northwest. Thus, the megashear only seems to add another complication.

Clearly, what is needed are more paleocurrent data from the Barranca in Sonora and adjacent Chihuahua and Durango and careful tectonic studies of each mountain block where it occurs. Certainly the abundance of small-scale deformation in the Barranca indicates very complex regional structure. If so, we need to proceed with care and caution in the study of this widespread and fascinating formation that looms large in the

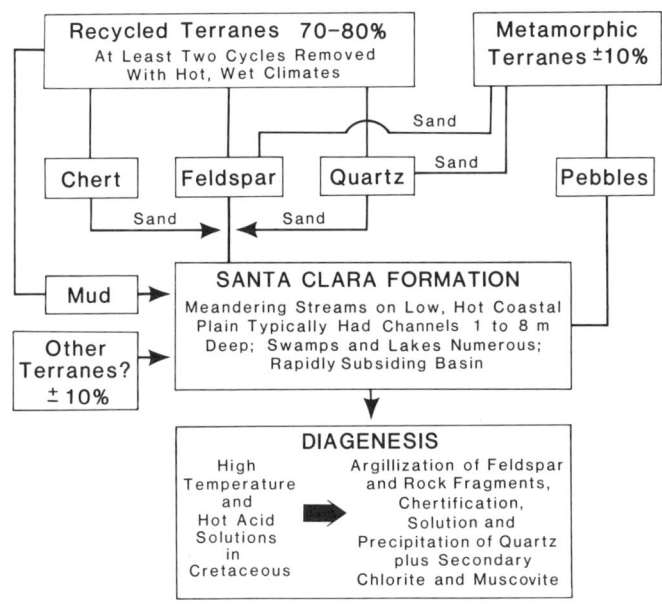

Figure 13. Flow chart for provenance.

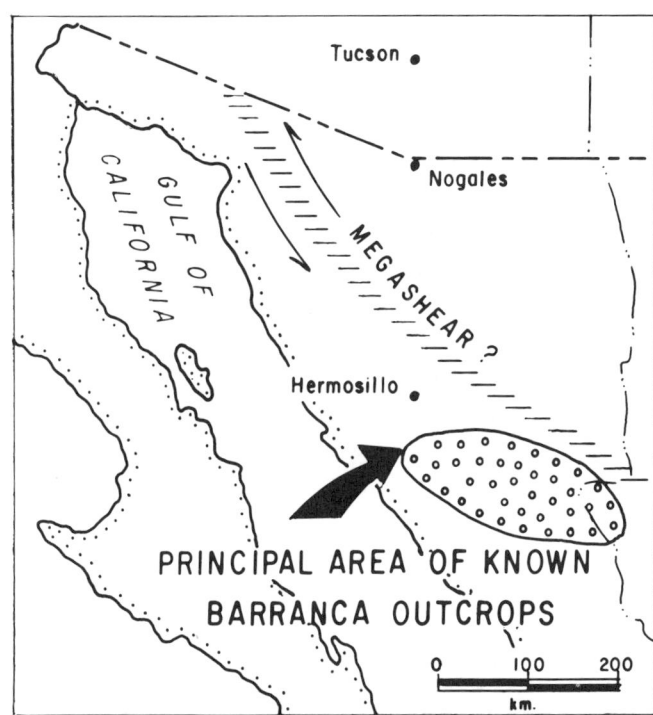

Figure 14. Schematic of Mohave-Sonora megashear and generalized limit of Barranca outcrop as it is today.

geologic evolution of east-central Sonora and far west-central Chihuahua.

As more sections of the Barranca are carefully described and studied petrographically, another question can be addressed: how homogenous is it petrographically? A homogenous petrology probably points to a single large basin, whereas appreciable petrographic inhomogeneity suggests contrasting local sources and small fault-controlled basins. Diversity of paleocurrent directions would also suggest small, tectonically active, structurally independent basins rather than one large single basin. The petrographic variation we see between cores SM-17, SE-3, and the section in Barranca Canyon (Table 2A) points to contrasting local sources in small basins, unless these sites had greatly different diagenetic histories.

ACKNOWLEDGMENTS

We wish to thank both the Comision Federal de Electricidad, Consejo de Recursos Minerales, and the Direction de Mineria, Geologic y Energeticos for their support of our field work and cores in 1980. The Direction de Mineria and Nestor da Silva graciously hosted I. Cojan on a brief visit to the Barranca Group in 1984. Silva also collected core samples for us, and Jesus Aguirre helped us in the field. We also wish to thank E. D. Pittman, who performed the XED analyses; W. D. Huff for his aid with x-ray analyses; J. B. Maynard and L. H. Larsen for their fruitful discussions on the petrography; and J. B. Maynard for his efforts with the microprobe. Amos Salvador, University of Texas, was most helpful, as was K. A. DeJong, University of Cincinnati, and J. H. Steward, U.S. Geological Survey, Menlo Park, California. We thank Lyla Messick for much of the drafting.

REFERENCES CITED

Abadie, V. H., 1981, Geology of part of the Sierra de Moradillas, Sonora, Mexico [M.S. thesis]: Stanford, California, Stanford University.

Adams, J., 1979, Wear of unsound pebbles in river headwaters: Science, v. 203, p. 171–172.

Alencaster de Cserna, G., 1961, Estratigrafia del Triassico Superior de la parte central del Estado de Sonora, in Paleontologia del Triasico Superior de Sonora: Universidad Nacional Autonoma de Mexico, Instituto de Geologia, Paleontologia Mexicana no. 11, 18 p.

Anderson, T. H., and Schmidt, V. A., 1983, The evolution of Middle America and The Gulf of Mexico–Caribbean sea region during Mesozoic time: Geological Society of America Bulletin, v. 94, p. 941–966.

Anderson, T. H., and Silver, L. T., 1979, The role of the Mojave–Sonora megashear in the tectonic evolution of northern Sonora, in Anderson, T. H., and Roldan-Quintana, J., eds., Geology of northern Sonora; Geological Society of America Annual Meeting Guidebook, Field Trip 2: San Diego, California, San Diego State University, p. 59–68.

Campa, M. F., and Coney, P. J., 1983, Tectono-stratigraphic terrains and mineral resource distributions in Mexico: Canadian Journal of Earth Sciences, v. 20, p. 1040–1051.

Collinson, J. D., 1986, Alluvial sediments, in Reading, H. D., ed., Sedimentary environments and facies, 2nd ed.: Oxford, Blackwell Scientific Publications, p. 20–62.

Cordoba, D. A., and others, 1980, Le Mexique Mesogeen et le passage du systeme cordilleran de type Californie, in Les chaines alpines issues de la Tethys 26: Paris, C.I.B., p. 18–29.

Davis, R. A., Jr., 1983, Depositional systems: Englewood Cliffs, New Jersey, Prentice-Hall, Inc., 669 p.

Dickinson, W. R., 1970, Interpreting detrital modes of graywacke and arkose: Journal of Sedimentary Petrology, v. 40, p. 695–707.

Dickinson, W. R., Helmold, K. P., and Stein, J. A., 1979, Mesozoic lithic sandstones in central Oregon: Journal of Sedimentary Petrology, v. 49, p. 501–516.

Dickinson, W. R., and others, 1983, Provenance of North American Phanerozoic sandstones in relation to tectonic setting: Geological Society of America Bulletin, v. 94, p. 222–235.

Dumble, E. T., 1900, Notes on the geology of Sonora: Transactions of the American Institute of Mining Engineers, v. 29, p. 122–152.

Echavirri, Perez, A., 1978, Metallogenic map of Sonora, Mexico–Arizona: Geological Society Digest, v. 11, October, p. 145–154.

Ferm, J. C., and Horne, J. C., 1979, Carboniferous depositional environments in the Appalachian region, *in* Collected papers and fieldguide: Columbia, University of South Carolina, Carolina Coal Group, 760 p.

Folk, R. L., 1980 Petrology of sedimentary rocks: Austin, Texas, Hemphill Publishing Co., 182 p.

Gonzalez-Leon, C., 1980, La Formacion Antimonio (Triasico–Superior–Jurasico Inferior) en la Sierra del Almo, Estado de Sonora: Revista, Universidad Nacional Aautonoma de Mexico, Instituto de Geologia, v. 4, p. 13–18.

Horne, J. C., Ferm, J. C., Caruccio, F. T., and Bagan, B. P., 1978, Depositional models in coal exploration and mine planning in Appalachian region: American Association of Petroleum Geologists Bulletin, v. 62, p. 2379–2411.

King, R. E., 1934, Geological reconnaissance of central Sonora: American Journal of Science, 5th series, v. 28, p. 81–101.

—— , 1939, Geological reconnaissance in northern Sierra Madre Occidental of Mexico: Geological Society of America Bulletin, v. 50, p. 1625–1732.

McGowen, J. H., Granata, G. E., and Seni, S. J., 1979, Depositional framework of the Lower Dockum Group (Triassic): Texas Bureau of Economic Geology Report of Investigations 97, 60 p.

Mendieta, J. A., and Partida-A., R., 1987, Estudio tectonico-sedimentario en el mar mexicano, Estados de Chihuahua y Durango: Boletin de la Sociedad Geologica Mexicana, v. 42, p. 43–88.

Morad, S., 1984, Diagenetic matrix in Proterozoic graywackes from Sweden: Journal of Sedimentary Petrology, v. 54, p. 1157–1168.

Pesquera Velazquez, R., and Carbonell-Cordoba, M., 1960, Geologia y exploracion los depositos de carbon de la region de San Marcial, Estado de Sonora: Consejo de Recursos Naturales y Renovables, Boletin 59, 39 p.

Potter, P. E., and 5 others, 1980, Estudio Formacion Barranca del Centro de Sonora: Hermosillo, Direccion de Mineria Geologia y Energeticos Open-File Report, 32 p.

Rangin, C., 1978, Speculative model of Mesozoic geocynamics, central Baja California to northeastern Sonora (Mexico), *in* Howell, D. G., and McDougal, K. A., eds., Mesozoic paleogeography of the western United States; Paleogeography Symposium, Los Angeles, 2 April, 1978: Society of Economic Paleontologists and Economic Mineralogists Pacific Section, p. 85–106.

Salvador, A., 1987, Late Triassic–Jurassic paleogeography and origin of Gulf of Mexico Basin: American Association of Petroleum Geologists Bulletin, v. 71, p. 419–451.

Schultheis, N. H., and Mountjoy, E. W., 1978, Cadomin Conglomerate of western Alberta; A result of early Cretaceous uplift of the Main Ranges: Canadian Society of Petroleum Geology Bulletin, v. 26, p. 297–342.

Selley, R. C., 1970, Studies of sequence in sediments using a simple mathematical device: Quarterly Journal of the Geological Society of London, v. 125, p. 557–581.

Spalletti, L. A., 1980, Paleoambientes sedimentarios (en secuencias siliclasticas): Assoc. Geologica Argentina, series B., Didactico Complementaria, v. 8, 175 p.

Stewart, John H., and 4 others, 1986, Late Triassic paleogeography of the southern Cordillera: The problem of a source for the voluminous volcanic detritus in the Chinle Formation of the Colorado Plateau region: Geology, v. 14, p. 567–570.

Weber, R., 1980, New observations on the Late Triassic flora of the Santa Clara Formation, Sonora, Mexico: Reading, United Kingdom, IDP International Paleobat. Conference Abstracts, p. 61.

—— , 1985a, Las plantas fosiles de la Formacion Santa Clara (Triasico Tardio, Sonora Mexico); Estado actual de las investigaciones, *in* Weber, R., ed., 3rd Congresso Latinoamericano de Paleontologia, Simposio Sobre Flora del Triasico Tardio, su Fitogeografia y Paleoencologia: Universidad Nacional Autonoma Mexico Instituto de Geologia Memoria, p. 107–124.

—— , 1985b, Helechos nuevos y poco conocidos de la Tafoflora Santa Clara (Triasico Tardio, Sonora) NW-Mexico I Marattiales, *in* Weber, R., ed., 3rd Congresso Latinamericano de Paleontologia, Simposio Sobre Floras de Triasico Tardio, su Fitogeografia y Paleoecologia: Universidad Nacional Autonoma de Mexico, Instituto de Geologia Memoria, p. 125–137.

—— , 1985c, Helechos nuevos y poco conocidos de la Tafoflora Santa Clara (Triasico Tardio, Sonora) NW-Mexico 2, Helectros lepotosporongiados; Cynepteridaceae y Glenicheniaceae, *in* Weber, R., ed., 3rd Congreso Latinamericano de Paleontologia, Simposio Sobre Flora del Triasico Tardio, su Fitografia y Paleoecologia: Universidad Nacional Autonoma de Mexico, Instituto de Geologia Memoria, p. 139–154.

Weber, R., and Garcia, A. Z., 1985, Nuevo de un panorama de la paleoecologia de comunidades de la Tafoflora Santa Clara (Triasico Tardio, Sonora), *in* Weber, R., ed., 3rd Congresso Latinamericano de Paleontologia, Simposio Sobre Flora del Triasico Tardio: su Fitografia y Paleocologia, p. 155–163.

Wilson, I. F., 1949, Coal deposits of the Santa Clara district near Tonichi, Sonora, Mexico: U.S. Geological Survey Bulletin 962-A, p. 1–80.

Wilson, I. F., and Rocha, V. S., 1946, Los yacimientos de carbon de la region de Santa Clara, municipio de San Javier, estado de Sonora: Comite Directivo para la Investigacion de las Recursos Minerales, Bolotin 49, 108 p.

MANUSCRIPT ACCEPTED BY THE SOCIETY APRIL 18, 1990

Paleontology and biostratigraphy of Cretaceous rocks, Lampazos area, Sonora, Mexico

Robert W. Scott
Amoco Production Company, P.O. Box 3385, Tulsa, Oklahoma 74102
Carlos Gonzalez-Leon
Instituto de Geologia, Universidad Nacional Autonoma de Mexico, Apartado Postal 1039, Hermosillo, Sonora 83000, Mexico

ABSTRACT

The Lower Cretaceous marine strata in east-central Sonora are as much as 2,500 m thick and consist of six mappable lithostratigraphic units. This sequence is important not only for delineating the paleogeography of western Mexico, but also for correlating the sections in eastern Mexico with those in western Mexico and Baja California. The Lampazos sequence is near the boundary between the North American and Yaqui blocks, and the stratigraphy is clearly similar to that of the North American block.

Calcareous algae, benthic foraminifers, corals, and rudists described here complement the molluscan assemblage previously reported. Three species of algae are widespread in Caribbean Lower Cretaceous strata. Two additional species of orbitolinids, besides *O. texana,* are documented in the Caribbean province for the first time. These and six other species of foraminifers enable the recognition of Barremian–lower Aptian, lower Albian, and middle-upper Albian. Three corals and two caprinids support the stadial interpretations, which are enhanced by an assemblage of late Aptian ammonites and younger bivalves and gastropods reported previously.

INTRODUCTION

The Lower Cretaceous stratigraphy of the Lampazos area of Sonora, Mexico, is a key to understanding the paleogeography of western Mexico. This section is located near the boundary between the North American and Yaqui blocks, which are separated by the Mojave–Sonora Megashear (Anderson and Schmidt, 1983; Bilodeau and Lindeberg, 1983), or the Texas lineament (Pindell and Dewey, 1982). This section also permits correlation of the Lower Cretaceous strata of eastern Mexico with Baja California. Early studies by Imlay (1939) and King (1939) showed the presence of important and thick sequences of Lower Cretaceous rocks. However, detailed mapping of this area has just begun (González-León, 1988; González-León and others, 1982). New collections of megafossils (Herrera and others, 1984; González-León and Buitrón, 1984) confirm the correlation of this section with the Trinity and Fredericksburg Groups in central Texas (Imlay, 1939). This area documents a succession of environments from nearshore to offshore terrigenous shelf and to a carbonate platform. Consequently, the western shoreline in the paleogeographic reconstructions by Scott (1984a), Bilodeau and Lindberg (1983), and Young (1983) must be shifted westward (Fig. 1). The middle Albian carbonate platform extended southward along the western shoreline. The southern limit of this platform is still indeterminate.

REGIONAL GEOLOGIC SETTING

Lower Cretaceous sedimentary rocks of Sonora occur as scattered outcrops because of the Tertiary volcanic cover and the extensive normal faulting produced by the basin and range deformation (Fig. 2).

Four types of Lower Cretaceous sedimentary sequences can be recognized in the state of Sonora: (1) the Bisbee Group in the northern parts, representing a transgressive-regressive cycle; (2) the shallow marine and fluvial Arroyo Sásabe and El Chanate

---

Scott, R. W., and Gonzalez-Leon, C., 1991, Paleontology and biostratigraphy of Cretaceous rocks, Lampazos area, Sonora, Mexico, *in* Pérez-Segura, E., and Jacques-Ayala, C., eds., Studies of Sonoran geology: Geological Society of America Special Paper 254.

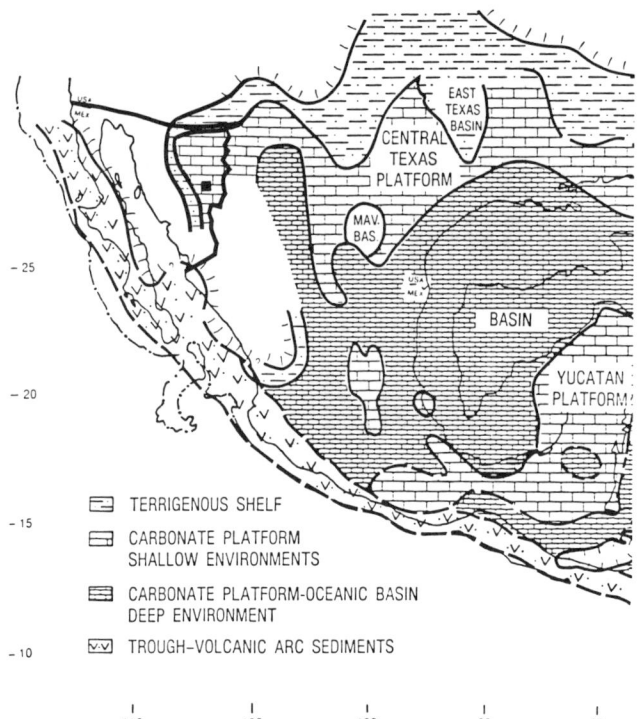

Figure 1. Paleogeography of Aptian–Cenomanian strata in Mexico modified from Scott (1984) to include new data from this study and from Jacques-Ayala and Potter (1987). Boundary between Sonora and Chihuahua is bold N-S line. Hachures show inferred land.

Formations (Jacques-Ayala, 1983; Jacques-Ayala and Potter, 1987) in northwestern Sonora, corresponding to the marginal facies of the Sonora basin; (3) the Lampazos sequence here reported in east-central Sonora, which is most similar to the Lower Cretaceous sequences of Chihuahua and particularly to that recently reported by Ortuño-Arzate (1985); and (4) the thick carbonate Palmar Formation (King, 1939), Sierra Chiltepin sequence (Himanga, 1977), and Potrero Formation (King, 1939), which are known in the Sahuaripa-Arivechi region, 50 km south of this area, the age of which has recently been questioned (Martinez and Palafox, 1985; Minjarez and others, 1985).

In most areas, the base of the Lower Cretaceous sequence is unknown because of the lack of good exposures. However, in northern Sonora, the Bisbee Group unconformably overlies Paleozoic rocks. Near the Lampazos area, the best known pre-Cretaceous rocks are the Upper Triassic Barranca Group (Alencaster, 1961) in central Sonora and a few isolated outcrops of Paleozoic rocks reported by various authors from that same region (King, 1939; Noll, 1981; Poole and others, 1983; Stewart and others, 1984). About 50 km south of the area, Cambrian (Minjarez and others, 1985), Mississippian (Martinez and Palafox, 1985), and Permian (Hewett, 1978; Schmidt, 1978) sedimentary rocks have been reported.

In the Lampazos area, the sequence reported here is the oldest exposed. It has been strongly folded and upthrusted by the Late Cretaceous Laramide orogeny, intruded by acidic plutons, covered by Tertiary volcanic rocks, and affected by normal faulting during Miocene time (Fig. 2).

LITHOSTRATIGRAPHY

The Lampazos sequence is divided into six formations, transitional to each other; from the base to the top, they are: El Aliso, Agua Salada, Lampazos, Espinazo del Diablo, Nogal, and its equivalent, Los Picachos Formation (Fig. 2). Herrera and Bartolini (1983) first reported the stratigraphy from this area and they named the Agua Salada and Espinazo del Diablo Formations; González-León (1988) has mapped the area, as well as reviewed and defined new lithostratigraphic units.

The El Aliso Formation is a 200-m-thick unit, which consists of thin-bedded, light gray, miliolid, calcisphere, algal, bivalve-gastropod fragment wackestone and argillaceous wackestone. Bioturbated clay mudstone is interbedded in the lower part. Thin- to thick-bedded wackestone and orbitolinid packstone predominate in the middle part. The upper 40 m of the formation consists of light yellow, argillic, thin-bedded mudstone-wackestone with thicker intervals of light green shale with oysters.

The Agua Salada Formation transitionally overlies the El Aliso Formation in the central part of the area (Fig. 2), but in the northwestern corner its base is covered. It is 350 m thick, and the lower 100 m consist of black chert in beds 5 to 30 cm thick, with interbedded black shale in beds less than 10 cm thick. Upward, the formation continues with a 30-m-thick sequence of massive brown-green shale with lesser amounts of thin-bedded, fine-grained sandstone. A distinctive, 10- to 15-m-thick unit of thick to massive-bedded oyster wackestone-packstone overlies the shale and in turn is overlain by a 55-m-thick sequence of thin to medium, wavy-bedded radiolarian black chert with *Parahoplites* sp., and minor thin-bedded mudstone-wackestone, nodular cherty mudstone-wackestone, and black shale. The upper part of the Agua Salada Formation is a 150-m-thick sequence of massive black shale with subordinate nodular chert in the middle part of the lower half and nodular, thin-bedded, argillaceous limestone in its upper half, which predominates upward and grades into the Lampazos Formation. The black shale contains fossiliferous beds with late Aptian ammonites (*Dufrenoyia justinae*, *Hypacanthoplites* sp., *Hysteroceras* sp.). The Agua Salada Formation grades to more terrigenous facies in the central part of the area where it reaches 250 m in thickness.

The Lampazos Formation has a thickness of 500 to 600 m. It is formed by 10- to 100-m-thick sequences of thin- to medium-bedded light gray to yellowish, argillaceous and bioturbated mudstone-wackestone with interstratified, 1- to 60-m-thick, yellowish, green, and dark shales with minor intercalations of thin beds of fine-grained sandstone and dark siltstone. Thick beds of oolitic and sandy packstone and grainstone with codiacean algae and miliolids occur locally in the upper parts of the formation and in the limestone sequences, as well as scarce *Toucasia* and orbitolinid biostromes. Poorly preserved bivalves, gastropods, and echi-

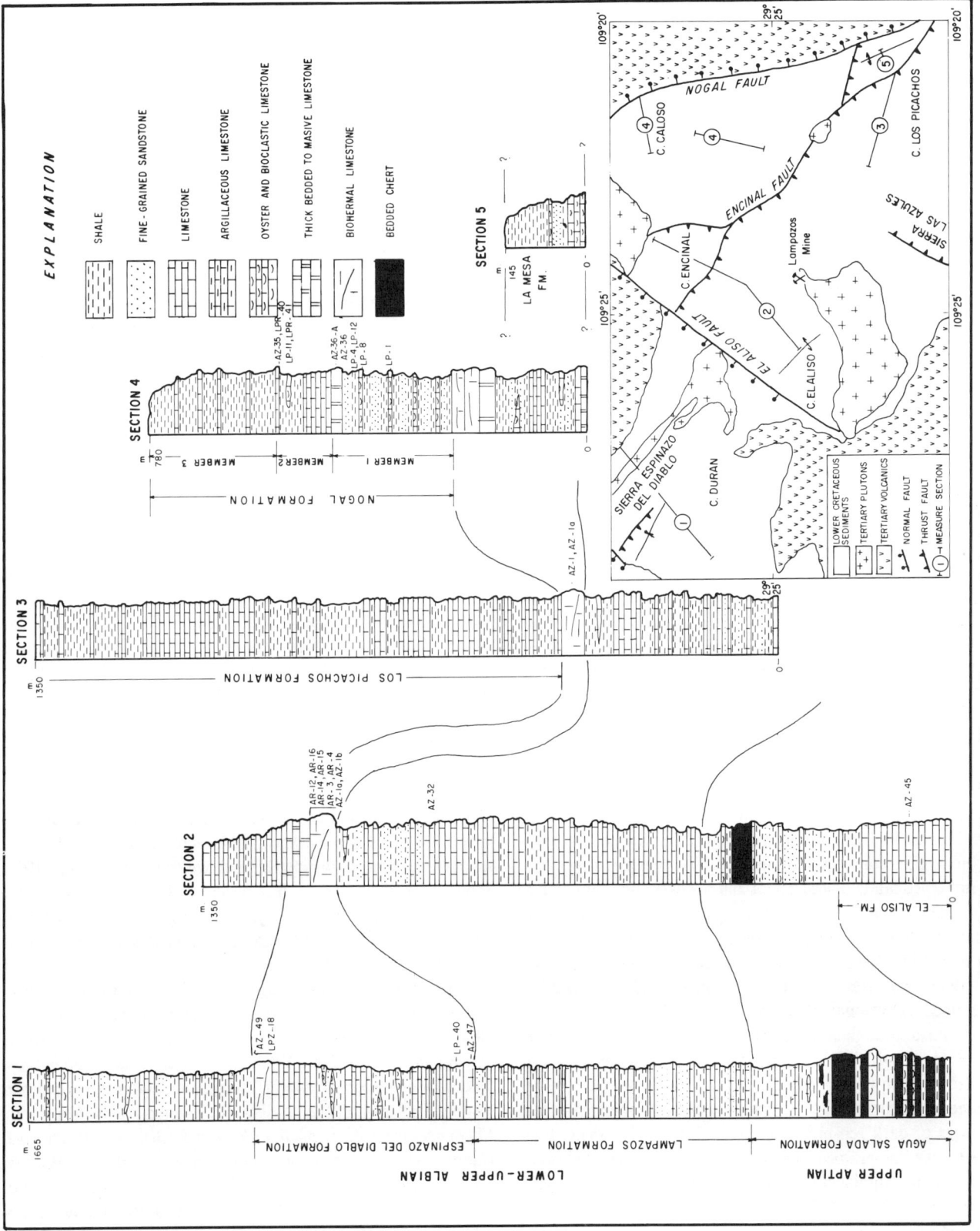

Figure 2. Stratigraphic sections of Lower Cretaceous rocks in the Lampazos area, Sonora. The inset map shows the location of the sections. See Gonzalez-Leon (1988) for a detailed geologic map.

noderms are very scarce. The Lampazos Formation is transitionally overlain by the Espinazo del Diablo Formation.

The Espinazo del Diablo Formation is a distinctive and well-exposed unit throughout the area. It reaches a maximum thickness of 400 m in Sierra Espinazo del Diablo and a minimum thickness of 115 m in Sierra Las Azules. At the first locality (Fig. 2, column 1), the basal part of this formation is a 15-m-thick, massive, light gray limestone with caprinids, requieniids, and colonial corals, which laterally grades to orbitolinid-miliolid wackestone-packstone; above this limestone is a 25-m-thick sequence of calcareous shale with interbedded mudstone-wackestone with *Montlivaltia* sp., *Cladophyllia furcifera* Roemer, *Myriophyllia* sp., and well-preserved bivalves, gastropods, and echinoderms. Above this sequence are 50- to 80-m-thick units of thin- to medium-bedded wackestone-packstone interbedded with similar thicknesses of massive green shale and interbedded siltstone and fine-grained sandstone beds as much as 30 cm thick. The uppermost part of the formation consists of a 30-m-thick massive limestone with a facies similar to the basal limestone. It contains massive corals like *Columnocoenia* sp.; thorough collecting will yield a large number of corals.

In Cerro Encinal, in the north-central part of the area (Fig. 2, column 2), the Espinazo del Diablo Formation is 150 m thick and consists of a bioher25mal structure (or superimposed carbonate buildups) 120 m thick that contains rudists, colonial corals (*Stylosmilia* sp., *Thamnasteria* sp.), bryozoans, gastropods (*Nerinea* sp., *Actaeonella* sp.), algae, and foraminifers. The upper part of the formation consists of thin- to medium-bedded, argillaceous limestone as much as 30 m thick.

The uppermost unit of the Lampazos sequence is the Nogal Formation, which laterally grades into the Los Picachos Formation. The Nogal Formation is divided into three members numbered from base to top. Member 1 consists of interstratified units of massive, dark gray shale; thin-bedded, dark brown, fine-grained sandstone; and minor, red-brown, thin-bedded, arenaceous and argillaceous oyster-limestone. The member contains a well-preserved fauna of gastropods and bivalves and is as much as 190 m thick. Member 2 consists of rudist-algal-orbitolinid biohermal limestone that ranges in thickness from 10 m to 60 to 70 m. It is overlain by thin- to medium-bedded, light gray wackestone; dark gray massive shale with interstratified, thick-bedded, fine-grained sandstone; and lenses of argillaceous oyster-gastropod wackestone-packstone. The top of member 2 is marked by a 3- to 5-m-thick bed of rudist-coral wackestone-packstone with *Texicaprina vivari* (Palmer) and *Caprinuloidea* sp. Member 3 reaches a thickness of 230 m and is composed of interstratified sequences of dark red-brown to yellowish massive shale; thin-bedded, very fine-grained sandstone; and thin-bedded, light blue to yellowish, argillaceous limestone that contains a well-preserved and abundant fauna of ammonites, bivalves, echinoderms, and gastropods.

The Los Picachos Formation reaches 950 m in thickness and is well exposed in the southeastern part of the area. It consists of interstratified 150- to 200-m-thick, monotonous packages of thin-bedded, peloid-miliolid, argillaceous mudstone; thin-bedded, light blue, calcisphere-gastropod mudstone-wackestone; and massive, gray to yellowish shale. In column 2, at the base, is *Caprinuloidea perfecta* Palmer.

A 145-m-thick sequence consisting of massive to thick-bedded oyster-limestone, brown-yellow, fine-grained sandstone, and gray to brown-yellow, massive shale crops out in the southeastern part of the area (Fig. 2, column 5). It is in structural contact with the Los Picachos and Nogal Formations, and is paleontologically and lithologically unrelated to any other formation of the area. It has been named La Mesa Formation (Gonzáles-León, 1988).

Distinct facies can be observed within the Agua Salada, Nogal, and Los Picachos Formations where their outcrops are separated by the El Aliso and Encinal faults. The El Aliso normal fault had lateral displacement during the Laramide deformation, and the Encinal is a thrust fault. The juxtaposition of different sequences is probably the result of the displacements along these faults.

BIOSTRATIGRAPHY

The microfossils and megafossils reported here, together with the mollusks reported by González-León and Buitrón (1984) and Herrera and others (1984), enable recognition of the Aptian and Albian stages (Fig. 3). In addition, this sequence can be correlated with sections elsewhere in the Chihuahua basin (Fig. 4). The ammonite zones of Young (1969, 1972, 1974) and the foraminiferal zones of Moullade and others (1985) can be recognized in the Lampazos area.

The upper Aptian zone of *Dufrenoyia justinae* (Hill) (Young, 1974) is recognized in the Agua Salada Formation by the presence of *D. justinae, Parahoplites* sp., *Hypacanthoplites* sp., *Hysteroceras* sp., and *Cheloniceras* sp. (González-León and Buitrón 1984; Herrera and others, 1984; González-León, 1988). This assemblage zone is distinctive and widespread (Young, 1974; Scott and Kidson, 1977). It is found in the La Peña Shale, the Otates Member of the Tamaulipas Formation, the Pearsall Formation, the Cow Creek Limestone, and the lower member of the Mural Limestone in Arizona.

The ammonite assemblage in member 3 of the Nogal Formation suggests a tentative correlation with Young's (1972) middle to basal upper Albian zones of *Metengonoceras hilli* Böhm, *Oxytropidoceras salasi* Young, *Manuaniceras carbonarium* Young, *Manuaniceras powelli* Young, and *Adkinsites bravoensis* Young. This assemblage consists of *Engonoceras stolleyi* Böhm, *Engonoceras* sp. cf. *E. piedernale* (von Buch), *Engonoceras belviderense* (Cragin) (=uddeni [Cragin] [Scott, 1970a]), *Protengonoceras* sp., *Parengonoceras* sp., *Hoplites* sp., and *Beudanticeras* sp. (González-León and Buitrón, 1984; Herrera and others, 1984; González-León, 1988). This assemblage is found in the Walnut, Comanche Peak, Goodland, and Kiowa Formations in north central Texas and Kansas (Scott, 1970b).

The benthic foraminifers also are age-diagnostic taxa in the

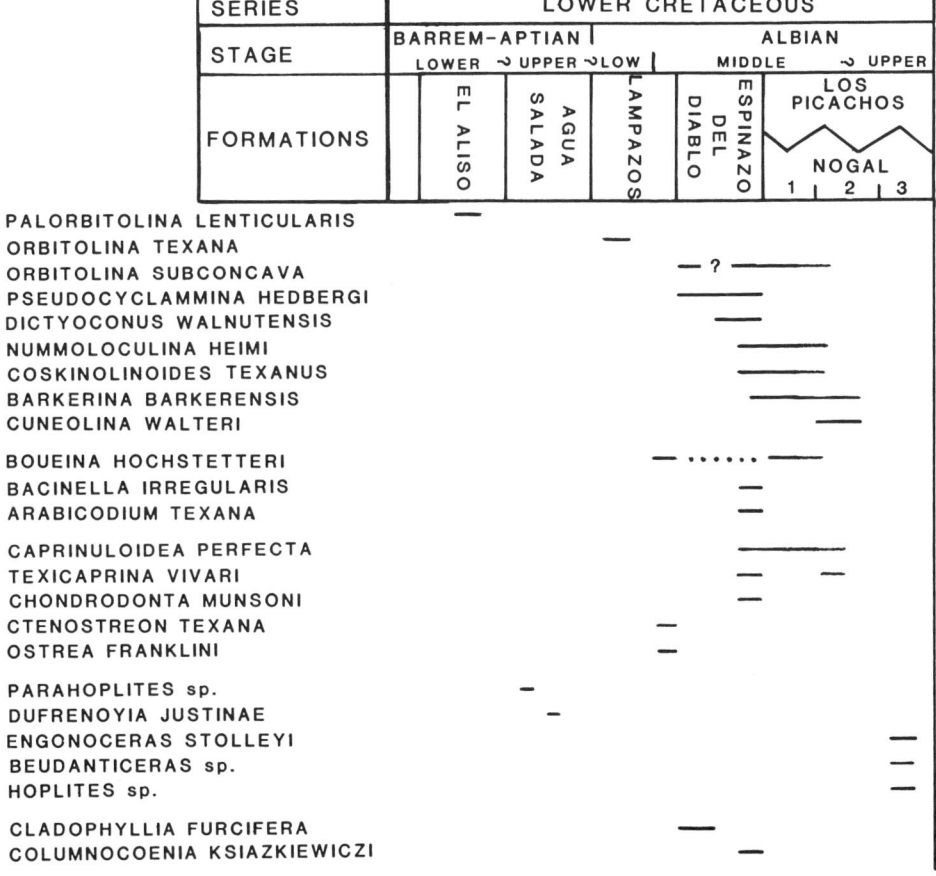

Figure 3. Range chart of important foraminifers (top), algae, bivalves, ammonites, and corals collected from the Lampazos area. Stage and substage boundaries are placed approximately.

Lampazos area. Their importance in the Caribbean region was shown by Maync (1961). *Palorbitolina lenticularis* (Blumenbach) in the El Aliso Formation identifies the upper Barremian–lower Aptian zone of *P. lenticularis* of Moullade and others (1985), which equates approximately with zones III to VI in North Africa (Schroeder and others, 1978). Although this is the first report of *P. lenticularis* in the Caribbean province, it is also present in the Sligo Formation in Texas, indicating that the El Aliso correlates in part with the Sligo and its correlatives.

Orbitolina texana (Roemer) is reported in the Lampazos Formation by González-León and Buitrón (1984); however, our samples have not revealed the diagnostic embryonic apparatus. Their report is quite consistent with the range of this species elsewhere in the Gulf Coast. *O. texana* is characteristic of the lower Albian Glen Rose Formation and its equivalents such as the Mural Limestone (Scott, 1987; Scott and Kidson, 1977). In the Mediterranean province, *O. texana* ranges from uppermost Aptian to basal upper Albian, spanning the zones of *Orbitolina minuta* Douglass, *Simplorbitolina* gr. *manasi* Ciry and Rat, *Simplorbitolina conulus* Schroeder, and *Neorbitolinopsis conulus* (Douvillé) (Moullade and others, 1985). In North Africa, this range zone equates with zones VII to IX of Schroeder and others (1978). However, in the Gulf Coast, *O. texana* occurs with early Albian ammonites; it is not yet recognized in younger zones.

The middle Albian substage is supported by *Dictyoconus walnutensis* (Carsey) (Moullade and others, 1985; Coogan, 1977) and by *Orbitolina subconcava* (Leymerie) (Arnaud, and others, 1981). The range of *O. subconcava* is obscured by taxonomic differences and the inconsistent identification of this species. Moullade and Saint-Marc (1975) group *O. subconcava* with *O. texana*. Fourcade and Raoult (1973) refer to it as *Orbitolina (Mesorbitolina)* sp. A. However, Schroeder (1975), Arnaud and others (1981), and Schroeder and Neumann (1985) separate it and show it ranging from uppermost Aptian to upper Albian. We follow Schroeder's opinion regarding the validity and definition of *O. subconcava* because this taxon is distinct from others and it has stratigraphic utility. These are the first illustrations identified as *O. subconcava* in the Caribbean province, although Azema and others (1985) first reported it from Honduras, and Schroeder (*in* Schroeder and Neumann, 1985) identified some of Douglass's (1960a) specimens from Texas as *O. subconcava*. The species was also figured as *Orbitolina morelensis* n. sp. (Ayala-Castañares, 1960) from Michoacán, according to Schroeder and Neumann (1985).

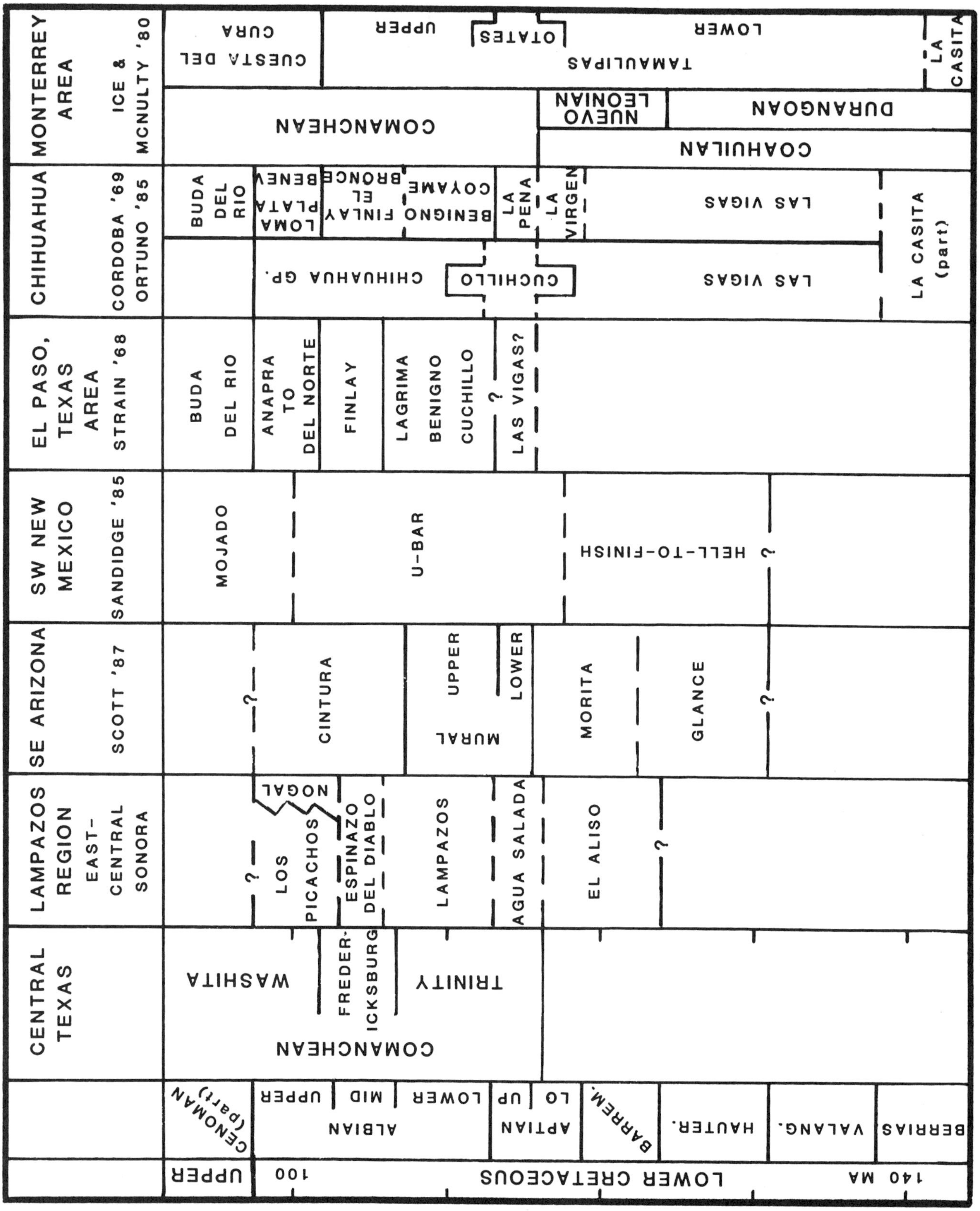

Figure 4. Correlation chart of the Lower Cretaceous strata in the Lampazos area and other well-known areas.

Other middle Albian foraminifers are *Barkerina barkerensis* Frizzel and Schwartz, and *Coskinolinoides texanus* Keijzer (Scott and Kidson, 1977). *Pseudocyclammina hedbergi* Maync and *Nummoloculina heimi* Bonet range from lower to upper Albian. This assemblage is common in units of the Fredericksburg Group in Texas.

The rudists in the Espinazo del Diablo and Nogal Formations also support a middle Albian age. Although no formal zones of rudists in the Gulf Coast Albian have been defined, the ranges are well known. *Caprinuloidea perfecta* Palmer spans the lower-middle Albian (Coogan, 1977), and the genus ranges to the basal Cenomanian (Young, 1982, 1984). *Texicaprina vivari* (Palmer) occurs in the Edwards and Stuart City Limestones (Coogan, 1977; Scott, unpublished). The genus ranges from middle Albian to lower Cenomanian (Young, 1982, 1984).

The corals and other mollusks in the Espinazo del Diablo and Nogal Formations support a middle to upper Albian correlation. *Cladophyllia furcifera* Roemer and *Chondrodonta munsoni* (Hill) are common fossils in the Edwards Limestone. Most of the mollusks in Member 1 of the Nogal Formation (González-León and Buitrón, 1984; Herrera and others, 1984; González-León, 1988) are known from formations of the Fredericksburg Group in central Texas (Stanton, 1947); however, a few gastropods are known only from the Glen Rose Limestone. This occurrence in the Nogal extends the range of *Cassiope branneri* (Hill), *Cassiope burnsi* Stanton, and *Cassiope paluxiensis* Stanton into the middle Albian. The base of *Ceratostreon* [*Exogyra*] *texanum* (Roemer) in the upper part of the Lampazos Formation (González-León and Buitrón, 1984) marks the base of the middle Albian in the Lampazos area. *C. texanum* ranges from the middle Albian *Metengonoceras hilli* Böhm zone to the basal upper Albian *Adkinsites bravoensis* Young zone (Young, 1982).

The mollusks and echinoids in Member 3 of the Nogal Formation (González-León and Buitrón, 1984; Herrera and others, 1984; González-León, 1988) are middle to upper Albian in range. *Protocardia multistrata* (Shumard) (=*Cardium granuliferum* Gabb) ranges from middle Albian to lower Cenomanian (Scott, 1986). *Cyprimeria washitaensis* Adkins occurs in the Washita Group in north-central Texas. The three species of echinoids are known from the Fredericksburg and/or Washita Groups (Cooke, 1946). The middle-upper Albian boundary occurs within this member.

SYSTEMATIC PALEONTOLOGY

Phylum CHLOROPHYCOPHYTA Papenfuss, 1946
Family CODIACEAE Zanardini, 1843

Genus *Arabicodium* Elliott, 1957
Arabicodium texana Johnson, 1968
Figs. 5A–C

Arabicodium texana Johnson, 1968, p. 13–14, pl. 2, figs. 5–8; Johnson, 1969, p. 41, pl. 29, figs. 5–8.
Description. Thallus segments relatively short and thick, length:diameter ratio of 1.2 to 1.7, mean of 1.39; marginal zone of fine tubes oriented about normal to margin, tubes uniform in diameter. Mean outside diameter of the thallus of *A. texana* is 2.07 mm, mean inside diameter is 1.07 mm, and the mean ratio of outside diameter to inside diamter is 2.13, based on measurements of Johnson (1968, Table 3) and measurements of the three new segments. The mean length of segments is 2.82 mm.
Discussion. Arabicodium texana differs from *Arabicodium aegagrapiloides* Elliott (1957), the genotype, by its short stubby thallus segments and by the orientation of the tubes in the marginal area. In *A. aegagrapiloides* these marginal tubes are oblique to the margin and they widen at the margin. Elliott's species is a long, slender codiacean in Lower Cretaceous rocks in Oman.

Arabicodium is distinguished from *Boueina* by its very fine, uniform internal system of tubes and by its simple, elongate thallus segments (Elliott, 1957).
Occurrence. A. texana was found in an ooid grainstone in the upper part of the Espinazo del Diablo Formation in Section 2 (Sample AR-12). The type specimens (USNM 42616) are from ooid grainstone in the Buda Limestone, Pecos County, Texas (Johnson, 1968). The species now ranges from middle Albian to lower Cenomanian.

Genus *Boueina* Toula, 1884
Boueina hochstetteri Toula, 1884
Figs. 5D–G

Boueina hochstetteri Toula, 1884, p. 1319–1324, pl. 7–9; Johnson, 1964, p. 27, pl. 25, figs. 1–6; 1969, p. 42–43, pl. 28, figs. 1–6. (See other references in Konishi and Epis, 1962, p. 69.)
Description. Branching segmented thallus; thickness of wall variable; maximum diameter of segment large, from 2.5 to 3.5 mm; tubes in inner part of cortex diverge outward irregularly, diameter from 27 to 150μm; tubes in outer cortex about normal to surface, diameter from 18 to 74 μm.
Discussion. This, the type species of the genus, has a much larger thallus diameter than *Boueina pygmaea* Pia (1936) and the tubes of the cortex are generally larger. In *B. pygmaea* the tubes range from 12 to 43 μm (Johnson, 1969). Pia (1936) implies that the tubes in *B. hochstetteri* are as much as 150 to 200 μm in diameter, but the Liassic specimens illustrated by Johnson (1964, 1969) have tube diameters from 50 to 148 μm. The new specimens from Lampazos, Mexico, have segment diameters of 2.67 to 3.42 mm, length of 2.58 mm, diameter of inner tubes of 27 to 54 μm, and outer tubes are 18 to 36 μm across. The Buda Limestone specimens (Johnson, 1968, p. 12) have the features of *B. pygmaea* instead of *B. hochstetteri.*

Boueina is similar internally to *Halimeda* Lamouroux; the two have been differentiated by the external shape (Beckmann and Beckmann, 1966).

Boueina is supposed to have a rounded cross section and an unbranched thallus (Pia, 1927). However, the Liassic specimen illustrated by Johnson (1964, p. 25, fig. 6) is branched. The arrangement of the medullary and cortical tubes distinguishes these taxa (Elliott, 1965; Johnson, 1969). Elliott (1965) suggested that *Boueina* and *Arabicodium* be treated as subgenera of *Halimeda.*
Occurrence. The first report of this species was from the upper Neocomian near Pirot, Yugoslavia. In the Lampazos area, it occurs in the Nogal Formation in Section 4 (Samples LP-1, LP-8, AZ-36A), middle-upper Albian, and in the Lampazos Formation in Section 2 (AZ-32), lower Albian.

Genus *Lithocodium* Elliott, 1956
Lithocodium aggregatum Elliott, 1956

Lithocodium aggregatum Elliott, 1956, p. 331, pl. 1, figs. 2, 4, 5; Johnson, 1969, p. 38–39, pl. 27; Scott, 1984b, p. 334, 336, pl. 1, fig. 6; pl. 3, fig. 6.

Description. An irregular nodule as much as 8 mm long encrusting a gastropod consists of dark gray, opaque micrite walls and irregularly intertwined tubes 0.66 to 0.31 mm in diameter and having an irregular polygonal cellular network as much as 0.67 by 0.89 mm across. The tubes branch into smaller tubes at the outer margin.
Occurrence. This species is present in Lower Cretaceous rocks in Europe, the Middle East (Elliott, 1963), and Cuba (Beckmann and Beckmann, 1966). In the Lampazos area, it is in the Espinazo del Diablo Formation, Section 2 (AR-14).

Phylum PROTOZOA von Siebold, 1845
Subphylum SARCODINA Schmarda, 1871
Class RHIZOPODEA von Siebold, 1845
Order FORAMINIFERIDA Eichwald, 1830
Suborder TEXTULARIINA Delage and Hérouard, 1896
Superfamily LITUOLACEA de Blainville, 1825
Family LITUOLIDAE de Blainville, 1825
Subfamily CYCLAMMININAE Marie, 1941

Genus *Pseudocyclammina* Yabe and Hanzawa, 1926
***Pseudocyclammina hedbergi* Maync, 1953**

Pseudocyclammina hedbergi Maync, 1953, p. 101, pl. 16, figs. 1–8.
Discussion. The test of this species has an early planispiral stage and a late uncoiled stage of a few chambers. The periphery tends to be rounded to subacute. The agglutinated wall has a coarse labyrinthic structure and is 111 to 222 μm thick. The last spiral whorl has as many as five chambers in a diameter of 1.7 to 2.2 mm. *P. hedbergi* has been referred to the genus, *Buccicrenata* Loeblich and Tappan (Loeblich and Tappan, 1985), although the aperture, a key feature for classification, is unknown in the type specimens. Until this question is resolved, the original generic assignment is retained. The specimens in the Espinazo del Diablo Formation are fragmented tests of a few chambers cut at oblique or tangential angles. The wall is from 130 to 180 μm thick and composed of coarse-grained particles such as peloids and foraminifers. The whorl diameter is at least 1.15 mm.
Occurrence. This species ranges from Valanginian through the Albian in the Tethyan realm (Simmons and Hart, 1987). In the Lampazos area, it ranges throughout the Espinazo del Diablo Formation (Samples AR-12 in Section 2, and AZ-47, AZ-49 in Section 1).

Superfamily ATAXOPHRAGMIACEA Schwager, 1877
Family CUNEOLINIDAE Saidova, 1981
Subfamily CUNEOLININAE Saidova, 1981

Genus *Cuneolina* d'Orbigny in de la Sagra, 1839
***Cuneolina walteri* Cushman and Applin, 1947**
Fig. 5O

Cuneolina walteri Cushman and Applin, 1947, p. 30, pl. 10, figs. 4, 5a, b.
Discussion. Discrimination among the species of *Cuneolina* is difficult because many were defined by exterior features and the corresponding features are not known in the same detail for each species. Furthermore, thin sections usually reveal only partial sections through the specimens. This Albian species has a triangular to flabelliform, biserial test. The walls consist of uniform micritic particles, and a labyrinthic internal structure. Sutures of the chambers are not depressed and are indistinct. The length is as great as 1 mm, the lateral extent is as much as 1.4 mm, and the thickness is 300 μm. A number of specimens in the Nogal Formation (Section 4, AZ-35, 36, LP-11) have the biserial chamber structure from 196 to 214 μm thick. In transverse sections, the arcuate whorls are 71.4 to 107 μm long and chamberlets are 3.57 to 44.7 μm wide. The length of one test is 2.22 mm. Radial sections show the arcuate, inwardly curved septal faces that do not join the wall of the opposite chamber. This radial structure separates *Cuneolina* from *Pseudotextularia* and *Sabaudia*. Random sections of *Dicyclina* have interior structures like *Cuneolina*, but *Dicyclina* has a planispiral early stage and cyclical chambers in a discoidal test. *Cuneolina walteri* differs from *Cuneolina pavonia* var. *parva* Henson (1948) by its less distinct sutures and by having more chambers per millimeter of radius (10 versus 7 to 8 in *parva*).
Occurrence. C. walteri is found in Lower Cretaceous rocks in the Gulf Coast including member 2 of the Nogal Formation.

Figure 5. A–C: *Arabicodium texana* Johnson; A, B, Espinazo del Diablo Formation (AR-12), longitudinal and transverse sections, 26.2×; C, Nogal Formation (LP-1), transverse and longitudinal views, 16.4×. D–G: *Boueina hochstetteri* Toula, Nogal Formation (LP-1); D, E, G, incomplete segments, 16.4×; F, transverse view showing tubes in inner and outer cortex, 41.3×. H, I: *Dictyoconus walnutensis* (Carsey), Espinazo del Diablo Formation (AR-4), longitudinal view, 16.4× and tangential view, 26.2×. J, K: *Coskinolinoides texanus* Keijzer, Espinazo del Diablo and Nogal Formations (AR-12, AZ-36A), longitudinal and transverse sections, 41.3×. L: *Nummoloculina heimi* Bonet, Nogal Formation (AZ-36A), 41.3×. M, N: *Barkerina barkerensis* Frizzell and Schwartz, Nogal Formation (AZ-36A) and Espinazo del Diablo Formation (AR-3), axial sections, 41.3×. O: *Cuneolina walteri* Cushman, Nogal Formation (LP-11), transverse view normal to growth axis, 41.3×. P, Q: *Orbitolina (Mesorbitolina) subconcava* Leymerie, Nogal Formation (LP-4); P, oblique section showing marginal zone, 26.2×; Q, section through embryonic apparatus, 66.1×. R: *Nautiloculina* sp. within an ooid, Espinazo del Diablo Formation (AR-12), 41.3×.

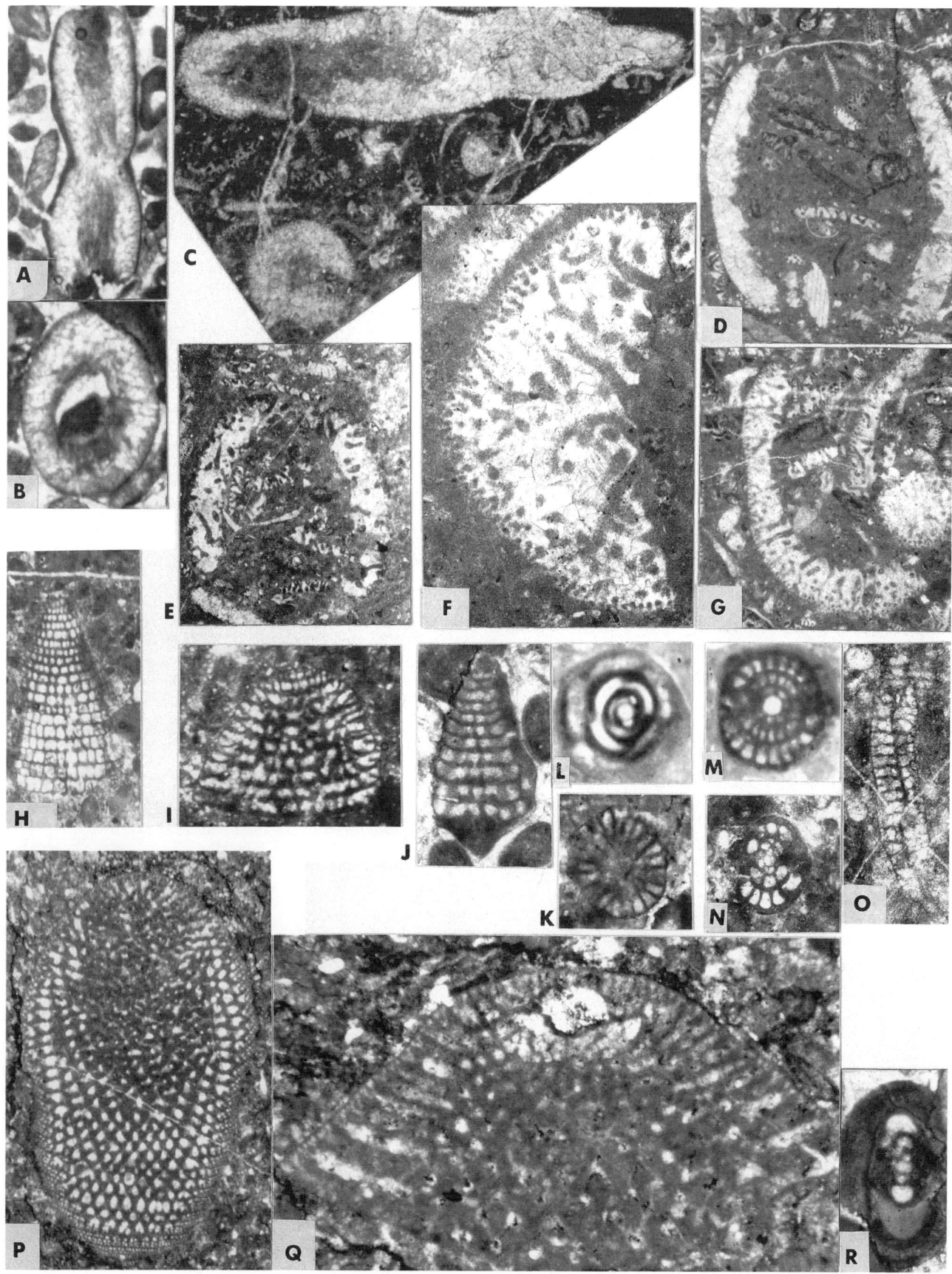

Superfamily ORBITOLINACEA Martin, 1890
Family ORBITOLINIDAE Martin, 1890
Subfamily ORBITOLININAE Martin, 1890

Genus *Palorbitolina* Schroeder, 1963
***Palorbitolina lenticularis* (Blumenbach, 1805)**

Madreporites lenticularis Blumenbach, 1805, pl. 80, figs. 1–6.
Orbitolina lenticularis (Blumenbach), Douglass, 1960a, p. 30–32, pl. 1, figs. 1–26.
Orbitolina (Palorbitolina) lenticularis (Blumenbach), Schroeder, 1963, p. 349–357, pl. 23, figs. 1–9; pl. 24, figs. 1–10; provides intervening references, p. 17.
Palorbitolina lenticularis (Blumenbach, 1805), Arnaud-Vanneau, 1969, p. 16–20, pl. 1, figs. 1–5; Fourcade and Raoult, 1973, p. 239, pl. 1, figs. 4, 5; Schroeder and Cherchi, 1979, p. 575, pl. 1, figs. 1, 2; pl. 2, fig. 3; Gusíc, 1981, p. 192–205, figs. 3, 4; Simmons and Hart, 1987, text-fig. 10.8e, pl. 10.2, figs. 1, 7; Loeblich and Tappan, 1988, p. 166–167, pl. 185, figs. 6–10.

Description. This species is recognized by its megalospheric embryonic apparatus, which has a large, subglobular, central chamber at the apex of the test. This chamber is surrounded by a periembryonic ring, which is subdivided by six to eight radial septa. The outer zones of the central chamber and the periembryonic ring are divided by short vertical partitions. The diameter of the central chamber ranges from 100 to 300 μm; it is 178 μm in the specimen in AZ-45. Schroeder (1963) provides a complete description.

Occurrence. *O. lenticularis* ranges from upper Barremian through lower Aptian and possibly into basal upper Aptian (Schroeder, 1975; Chiocchini and others, 1984; Moullade and others, 1985). It is common in the Tethyan carbonates from Spain to Iran and Oman. In the Lampazos area, it is in the El Aliso Formation (Section 2, AZ-45). This species is also found in the Sligo Formation in Texas.

Genus *Orbitolina* d'Orbigny, 1850
Subgenus *Mesorbitolina* Schroeder, 1962
***Orbitolina (Mesorbitolina) subconcava* Leymerie, 1878**
Figs. 5P, Q

Orbitolina subconcava Leymerie, 1878, pl. E, figs. 7, 8.
Orbitolina (Mesorbitolina) subconcava Leymerie; Schroeder, 1975, p. 124–125; 1979, p. 293–294, pl. 1, fig. 8; Schroeder and others, 1978, p. 246, pl. 2, figs. 2, 4; Schroeder *in* Schroeder and Neumann, 1985, p. 80–82, pl. 37, figs. 1–8; Simmons and Hart, 1987, pl. 10.2, fig. 2.
Orbitolina (Mesorbitolina) texana (Roemer, 1852) in part, Moullade and Saint-Marc, 1975, p. 834–835, pl. 13, figs. 9–11; pl. 14, fig. 1.
Orbitolina (Mesorbitolina) sp. A, Fourcade and Raoult, 1973, p. 239, pl. 1, figs. 8, 9; pl. 2, figs. 1, 2.

Description. Test moderate sized, diameter as much as 8 to 9 mm, conical to convexo-concave; triangular cross sections of chamber passages in radial zone; embryonic apparatus well developed, diameter 0.4 to 0.5 mm (rarely as much as 0.65 mm), deuteroconch and subembryonic chambers subequal and extending laterally around proloculus, marginal septulae inclined towards proloculus; proloculus lenticular to subspherical, diameter about 160 to 220 μm. This species differs from *Orbitolina texana* by its wide, convex deuteroconch, in which the lateral chamberlets are open at the base, and from *Orbitolina aperta* by its shorter deuteroconch with one or two lengths of septulae and by its smaller size of embryonic apparatus and test diameter.

Occurrence. This species is widespread in the Tethyan Realm, particularly in the Mediterranean region in latest Aptian to upper Albian strata (Schroeder, 1975, 1979). *O. subconcava* occurs in the lower Albian Glen Rose Formation in central and west Texas and in Albian rocks in Michoacán according to the synonymies of Schroeder (*in* Schroeder and Neumann, 1985, p. 82). In the Lampazos area *O. subconcava* is found in the base of member 2 of the Nogal Formation (Section 4, LP-4, LP-12). It is questionably identified in the Espinazo del Diablo Formation (section 1, AZ-49).

Subfamily DICTYOCONINAE Moullade, 1965

Genus *Coskinolinoides* Keijzer, 1942
***Coskinolinoides texanus* Keijzer, 1942**
Figs. 5J, K

Coskinolinoides texanus Keijzer, 1942, p. 1016, figs. b–h; Maync, 1955a, p. 89; 1955b, p. 109, pl. 17, figs. 14, 15; Coogan, 1977, fig. 7, pl. 18, figs. 2–5; Applin and Applin, 1965, p. 58, 59, pl. 1, figs. 3, 4, 9, 10; Frizzel, 1954, p. 76; Loeblich and Tappan, 1988, p. 157, pl. 168, figs. 10–14.
Coskinolina adkinsi Barker, 1944, p. 206–207, pl. 35, figs. 1–4.

Description. Agglutinated conical test, height equal to or greater than diameter (0.43 mm maximum height); external profile of cone wall straight to slightly concave; basal, septal margin flat with rounded corners. Early trocoid stage, and later uniserial stage. Later chambers discoidal, subdivided by radial septa, alternately shorter and longer, extending to the center about two-thirds of the radius. Aperture of uniserial stage, multiple round pores in central part of main septa. Locally short buttresses are formed by inward extensions of the basal septum.

Discussion. *Coskinolinoides* does not have the large-diameter central area of vertical pillars that characterizes *Coskinolina* Stache. *Coskinolinella* Delmas and Deloffre has a simple, unpartitioned central area. Maync (1955a, b) noted that *Coskinolina adkinsi* Barker was indistinguishable from *Coskinolinoides texanus* Keijzer.

Occurrence. This species ranges throughout the Fredericksburg Group in central Texas and in equivalent strata in Florida (Applin and Applin, 1965). It may extend into the Washita Group as well (Coogan, 1977). In the Lampazos area, it occurs in the Espinazo del Diablo Formation in Section 2 and possibly in the Nogal Formation in Section 4 (Samples AR-12, AR-3, AZ-36A?).

Genus *Dictyoconus* Blanckenhorn, 1900
***Dictyoconus walnutensis* (Carsey, 1926)**
Figs. 5H, I

Orbitolina walnutensis Carsey, 1926, p. 23, pl. 7, figs. 11a, b; pl. 8, fig. 3.
Dictyoconus walnutensis (Carsey), Maync, 1955a, p. 85–88, pl. 13, 14; includes intermediate references; Douglass, 1960b, p. 257–258, pl. 5, figs. 1–8 (complete synonymy).

Description. Small, conical test, diameter slightly greater than height. Marginal zone with two, rarely three lengths of radial partitions. Central zone with distinct pillars of uniform diameters from base to ceiling of chamber; pillars randomly arranged in basal section. The Lampazos specimens range in height from 0.98 to 1.29 mm and in diameter from 0.84 to 1.24 mm.

Discussion. *Dictyoconus walnutensis pyrenaicus* Moullade and Peybérnes (1975) is distinguished by its smaller dimensions and by an inflection of the chamber floor at the boundary between the marginal and central zones. This subspecies occurs in lower and middle Albian limestones in France and Spain (Peybérnes, 1976).

Occurrence. *D. walnutensis* is widespread in the Gulf region in middle Albian rocks. In the Lampazos area it is found in the Espinazo del Diablo Formation (Samples AZ-1A, Section 3, and AR-4 and AR—3, Section 2).

Superfamily HAPLOPHRAGMIACEA Eimer and Fickert, 1899
Family BARKERINIDAE Smout, 1956, emend.
Hamaoui and Saint-Marc, 1970

Genus *Barkerina* Frizzel and Schwartz, 1950
Barkerina barkerensis Frizzell and Schwartz, 1950
Figs. 5M, N

Barkerina barkerensis Frizzell and Schwartz, 1950, p. 6, pl. 1, figs. 1-6; Hamaoui, 1973, p. 339–343, pl. 1, figs. 1–3; pl. 6, figs. 1, 2; Loeblich and Tappan, 1988, pl. 77, figs. 1–3.
Description. Test small, planispiral, involute, ovoid; chambers subdivided by transverse partitions, very closely spaced in last whorl; sutures indistinct; aperture multiple pores at base of septum. The Lampazos specimens have a spiral diameter from 0.31 to 0.44 mm and an axial width of 0.31 mm.
Occurrence. This species is never abundant in the Fredericksburg Group in central Texas (Frizzell and Schwartz, 1950). In northern Coahuila the base of *B. barkerensis* occurs at the top of the middle Glen Rose Formation, which is correlated with the base of the Telephone Canyon Formation, Fredericksburg Group (Smith and Bloxsom, 1974). R. W. Scott has also collected it from the Del Carmen Formation of Santa Elena Canyon, Big Bend National Park, and the Fort Terrett Formation at several localities in south-central Texas. In the Lampazos area it occurs in the top of the Espinazo del Diablo Formation in Section 2 (Sample AR-3) and in Member 2 of the Nogal Formation in Section 4 (Sample AZ-35, 36, 36A).

Suborder MILIOLINA Delage and Hérouard, 1896
Superfamily MILIOLACEA Ehrenberg, 1839
Family MILIOLIDAE Ehrenberg, 1839
Subfamily MILIOLINELLINAE Vella, 1957

Genus *Nummoloculina* Steinmann, 1881
Nummoloculina heimi Bonet, 1956, emend.
Conkin and Conkin, 1958
Fig. 5L

Nummoloculina heimi Bonet, 1956, p. 402–406, pl. 3-4; Conkin and Conkin, 1958, p. 149–158, pl. 1, figs. 1–25; Saint-Marc, 1974, p. 227, pl. 3, fig. 4.
Discussion. This species is thoroughly described by Conkin and Conkin (1958), and it is widely known in Albian to lower Cenomanian strata of the Gulf basin. A similar Late Cretaceous species is *Nummoloculina regularis* Phillipson. *N. regularis* generally has a smaller diameter, 0.1 to 0.3 mm, compared to the 0.5 to 2 mm diameter of *N. heimi*; *N. regularis* has 6 to 10 whorls and *N. heimi* has 5 to 7 whorls; *N. regularis* has 2 to 4 chambers in the last whorl and *N. heimi* has 10 or more (Saint-Marc, 1974). *Nummoloculina robusta* Torre in the Late Cretaceous has an inflated test and a very thick wall. *Nummoloculina irregularis* Decrouez and Radiocic has a plane of coiling that changes irregularly. The Lampazos specimens range in diameter from 0.53 to 0.98 mm and in thickness from 0.31 to 0.40 mm.
Occurrence. In the Lampazos area *N. heimi* is found in the Espinazo del Diablo Formation in Section 2 (Samples AR-14, AR-3) and Member 2 of the Nogal Formation in Section 4 (Samples AZ-36, AZ-36A).

Phylum COELENTERATA Prey and Leuckart, 1847
Subphylum CNIDARIA Hatschek, 1888
Class ANTHOZOA Ehrenberg, 1834
Order SCLERACTINIA Bourne, 1900
Suborder ASTROCOENIINA Vaughan and Wells, 1943
Family STYLINIDAE d'Orbigny, 1851
Subfamily STYLININAE d'Orbigny, 1851

Genus *Stylosmilia* Milne-Edwards and Haime, 1848
Stylosmilia sp.
Figs. 6A, B

Description. Phaceloid colony, branching angle about 45°; corallite cross section ovate, diameter from 3.4 to 6.3 mm. Epitheca distinct, commonly encrusted by micritic layers. Four cycles of septa, the first two tending to extend to the center of the calyx, several may join the small styliform columella; septal margins dentate; dissepiments form a continuous ring around outer margin of calyx.
Occurrence. This specimen is from the uppermost thick-bedded limestone in the Espinazo del Diablo Formation in Section 2, Sample AR-15.

Suborder FUNGIINA Duncan, 1884
Superfamily THAMNASTERIOIDEA Alloiteau, 1952
Family THAMNASTERIIDAE Vaughan and Wells, 1943, emended Alloiteau, 1952

Genus *Thamnasteria* Lesauvage, 1823
Thamnasteria sp.
Fig. 6C

Description. Colony massive, laminated growth layers; calyces apparently random, slightly depressed. Septa confluent between calyces, approximately 15 to 18 μm thick, arcuate septa, margins dentate or articulate, synapticulae sparsely developed; calyces 3.8 to 4.3 mm diameter and 4.8 to 6.5 mm apart. Columella trabecular with two or three rods. The poor preservation precludes placement in a species.
Occurrence. The specimen is from the upper massive limestone bed of the Espinazo del Diablo Formation at Section 2, Sample AR—16.

Family FAVIIDAE Gregory, 1900
Subfamily FAVIINAE Gregory, 1900

Genus *Cladophyllia* Milne-Edwards and Haime, 1851
Cladophyllia furcifera Roemer, 1888
Figs. 6D, E; Figs. 7A, B

Cladophyllia furcifera Roemer, 1888, p. 8, pl. 1, figs. 4a–b; Wells, 1933, p. 90–91, pl. 8, figs. 5–8.
Description. Corallium branching dichotomously; calyx oval in cross section; distinct epitheca with growth lines; three cycles of septa, 12 extend to center, several join; 12 septa of third cycle short; calycal diameters from 4 to 5 mm. A columella is not developed but pairs of septa join towards the center or the tip of the septum may be enlarged. The epitheca is about as thick as the septa, but it usually is obscured by spar infill. The actual wall can be seen in ultraviolet light (Fig. 7B).
Occurrence. This species is common and widespread in the middle Albian Edwards Formation in Texas (Wells, 1933). In Sonora it was found in the Espinazo del Diablo Formation in Section 1 (Sample LP-4C).

Family PLACOCOENIIDAE Alloiteau, 1952

Genus *Columnocoenia* Alloiteau, 1952
Columnocoenia ksiazkiewiczi Morycowa, 1964
Fig. 6F

Columnocoenia ksiazkiewiczi Morycowa, 1964, p. 67–69, pl. 17, figs. 1–4a, b; pl. 18, figs. 1a, c; text-fig. 16; Morycowa, 1971, p. 96–98, pl. 24, figs. 2, 3; pl. 25, fig. 1; text-figs. 30c, d.

Description. Nodular to hemispherical colony, cerioid, intratentacular budding; three cycles of septa, 12 of which are subequal and extend nearly to columella, these end in pali, the other 12 septa are one-third the length of the first; columella laminar to styliform, connected with one or two septa. The septa continue into intercalycal area, but recrystallization obscures the structure. Septa dentate; sparse tabulae and synapticulae. Calyces moderately sized, 1.72 to 2.34 mm in diameter; spacing between adjacent columellae 2.58 to 3.23 mm. Morycowa (1964, 1971) provides a thorough species description.

Discussion. The size of the calyces, their spacing, and the more oval columella qualifies this Mexican specimen for the subspecies *bucovinensis* Morycowa (1971). The species is known from the Barremian–lower Aptian in the eastern Carpathians of Romania and Poland. It also occurs in Yugoslavia in Barremian-Aptian strata (Turnsek and Mihajlovic, 1981) and in Albian reefal facies in northern Spain (Reitner, 1984). This specimen is also similar to *Stephanocoenia*? *wintoni* Wells (1933, p. 69, pl. 5, fig. 9) from the middle Albian Goodland Formation in north-central Texas. The Goodland specimen appears to be slightly plocoid and the Carpathian specimens are plocoid to locally subceriod. The Mexican specimen appears to be ceriod, but its surface is poorly preserved and it may be altered. This specimen is allocated to Morycowa's species because its calycal structure and size are the same. Well's species has the same size and calycal structure, but is distinctly plocoid. These species should be compared closely.

Occurrence. This coral was collected from the uppermost limestone in the Espinazo del Diablo Formation, Section 1, Sample LPZ-18.

Class BIVALVIA Linne, 1758
Subclass HETERODONTA Neumayr, 1884
Order HIPPURITOIDA Newell, 1965
Superfamily HIPPURITACEA Gray, 1848
Family CAPRINIDAE d'Orbigny, 1850

Genus *Caprinuloidea* Palmer, 1928
Caprinuloidea perfecta Palmer, 1928
Figs. 7C–F

Caprinuloidea perfecta Palmer, 1928, p. 59–60, pl. VIII, fig. 8; pl. IX, figs. 1, 2; text-fig. 6; Coogan, 1977, p. 56, pl. 12, figs. 1, 2, pl. 13, fig. 3; Young, 1984, pl. 2, fig. 9.
Caprinuloidea perfecta gracilis Palmer, 1928, p. 60–61, pl. IX, fig. 3; pl. X, fig. 1; Coogan, 1973, p. 61, pl. 5, fig. 2; Coogan, 1977, p. 56, pl. 13, figs. 5, 6.
Caprinuloidea septata Palmer, 1928, p. 62, pl. IX, fig. 4; pl. X, fig. 3; pl XI, fig. 1; Coogan, 1977, p. 56, pl. 13, fig. 3.
Caprinuloidea costata Palmer, 1928, p. 62–63, pl. XI, figs. 2–5.
Caprinuloidea bisulcata Palmer, 1928, p. 64, pl. XII, fig. 1, 2; Coogan, 1977, p. 56.

Description. Shell large; attached valve elongated and gently curved, subquadrate cross section, anterior-ventral corner angular; free valve spirally coiled. Tooth on attached valve large with few vertical canals, connected to dorsal margin by a thick septum. Teeth on free valve large, anterior tooth triangular, posterior tooth ovate. On attached valve one row of outer pyriform canals and one to two rows of larger polygonal canals; on anterior and ventral wall of free valve one outer row of pyriform canals and one to two rows of polygonal canals, and on dorsal and posterior walls of free valve one row of elongated oval canals.

Discussion. *C. perfecta gracilis* Palmer has a longer and narrower attached valve. Like *C. perfecta* s.s., it has one to two rows of polygonal canals on the attached valve, although Coogan (1977, p. 56) stated that it has two or more rows of polygonal canals and suggested that it be treated as a separate species.

Study by R. W. Scott of the type specimens of *Caprinuloidea perfecta* (CASM 2168), *C. perfecta gracilis* (CASM 2169), *C. septata* (CASM 2171), *C. costata* (CASM 2172, 2173), and *C. bisulcata* (CASM 2174) indicate that each type possesses a single outer row of pyriform canals and one to two inner rows of polygonal canals. The number of the latter varies from thinner to thicker parts of the wall. The distinguishing features used by Palmer to erect these taxa, which are from a 100-ft-thick marine limestone at Soyatlán de Adentro, Jalisco (Palmer, 1928, p. 8, p. 24), should be given no more than subspecies significance. Some may even be phenotypic variations. *C. septata* was characterized by septate body and accessory-socket cavities on the attached valve, but septae are in *C. perfecta* also. *C. costata* was characterized by exterior growth rugae on both valves, but growth lines are also present on *C. perfecta* (Palmer, 1928, pl. 8, fig. 8). *C. bisulcata* was distinguished by a large dorsal groove anterior to the ligament groove and by a large anterior myophore on the attached valve. These features seem to be variable from specimen to specimen and do not have use in separating species.

Although Coogan (1977, p. 56) stated that the attached valve of *C. perfecta* has one inner row of polygonal canals, this cannot be verified on the holotype, nor is it documented in Coogan's illustrations. Palmer (1928, p. 60) described "extensive bifurcation of the vertical plates," which suggests that the number of rows of canals is variable. The single specimen from the lower Albian Glen Rose Formation, which Coogan (1977, pl. 12, fig. 1) illustrated as *C. perfecta,* has all the features of *Coalcomana ramosa* Boehm. By excluding this specimen from *C. perfecta,* its range is limited to the middle Albian.

Occurrence. Coogan (1977) reported new specimens of *C. perfecta* and *C. gracilis* from the middle Albian Stuart City Limestone in the subsurface of south Texas. Coogan's (1973) specimen from the upper Albian Ft. Lancaster Formation near Ft. Stockton, Texas, has canals throughout the cardinal region and may belong to *Kimbleia* instead of *C. gracilis*. In the Lampazos area, *C. perfecta* occurs in the basal part of the Los Picachos Formation and in the Espinazo del Diablo Formation (Section 2, Samples AZ-33, AZ-1a, b, AR-5). In Section 4, it is found in member 2 of the Nogal Formation (Sample LPR—40).

Figure 6. A, B: *Stylosmilia* sp., Espinazo del Diablo Formation (AR-15), 11.5×, note encrusting algae and a tubular organism. C: *Thamnasteria* sp., Espinazo del Diablo Formation (AR-16), 11.5×. D, E: *Cladophyllia furcifera* Roemer, Espinazo del Diablo Formation (LP-4); D, showing fasiculate branches, 3.3×; E, closeup of corallite in Figs. 5A, B, 8.2×. F: *Columnocoenia ksiazkiewiczi* Morycowa, Espinazo del Diablo Formation (LP2-18), 11.5×.

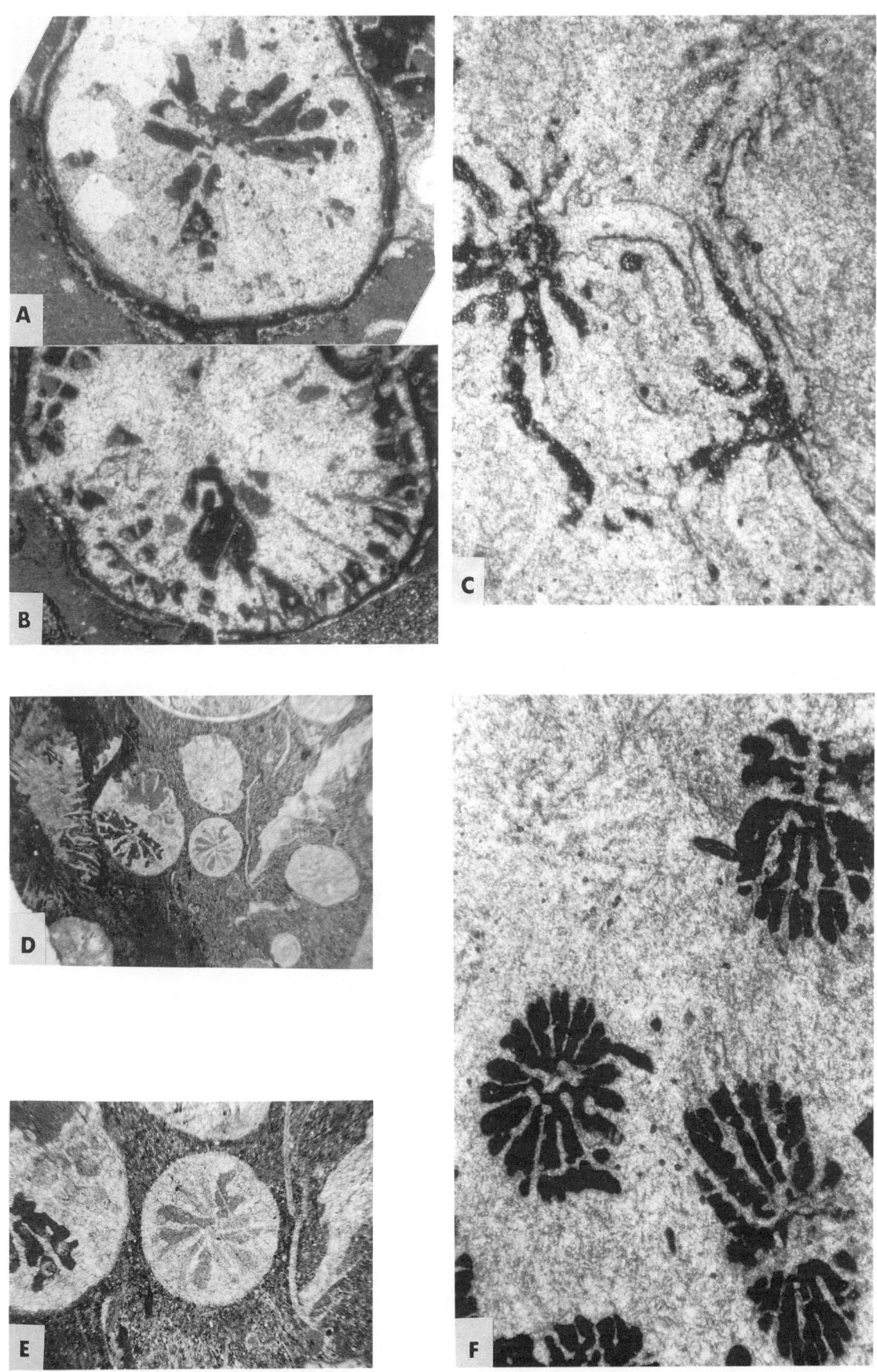

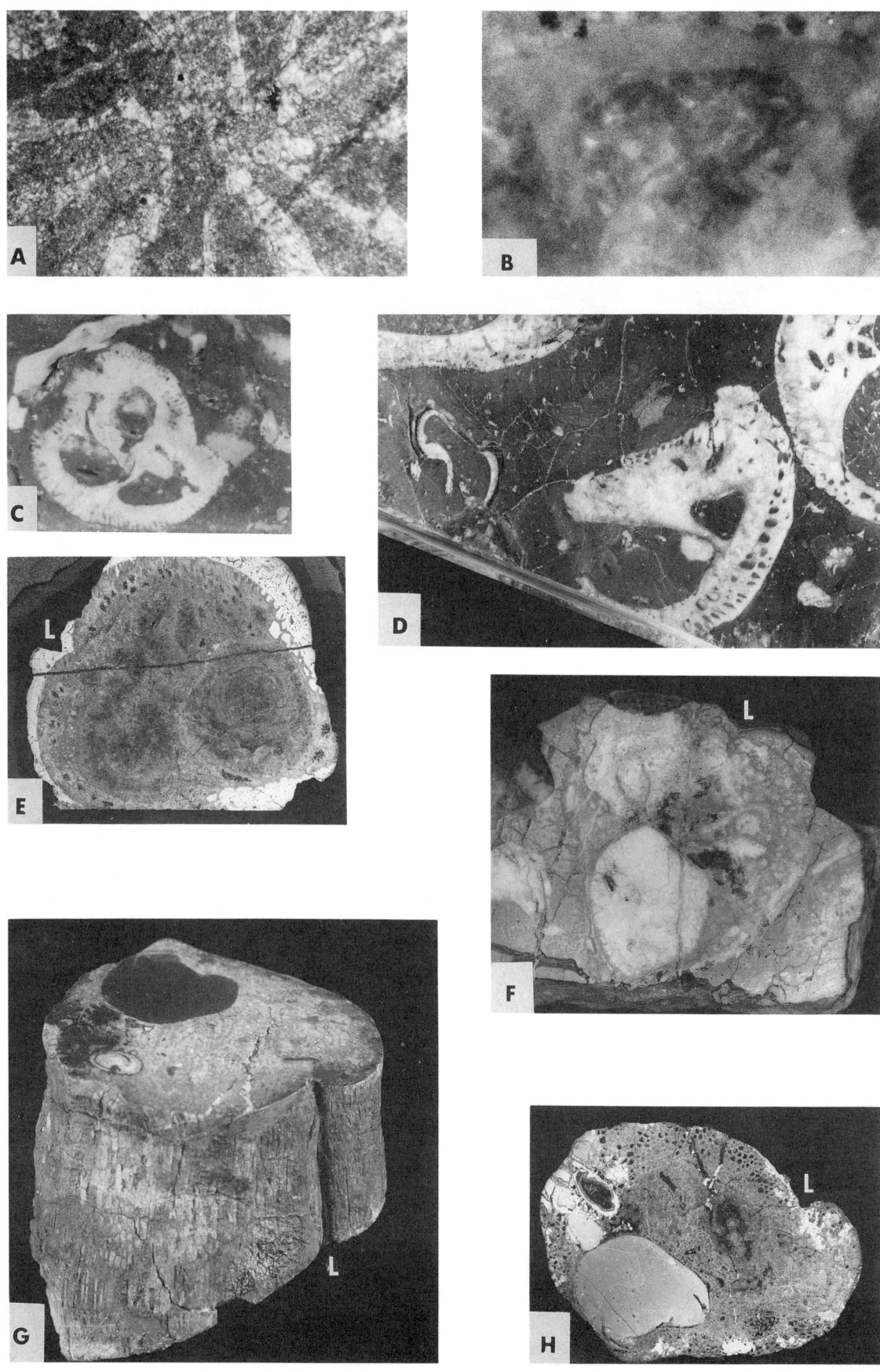

Genus *Texicaprina* Coogan, 1973
***Texicaprina vivari* (Palmer, 1928)**
Figs. 7G, H

Sabinia vivari Palmer, 1928, p. 74, pl. 13, fig. 4; pl. 14, figs. 1–4.
Texicaprina vivari Coogan, 1973, p. 59, pl. 3, figs. 1–6; pl. 4, figs. 1, 2; Coogan, 1977, p. 56, 60, 64, pl. 14, figs. 1, 3, 4; pl. 15, fig. 3; Young, 1984, pl. 3, fig. 3.

Discussion. The new specimen of the attached valve is elongated, gently curved and has a triangular to ovate cross section. Palmer (1928) included the absence of septa in the body cavity and in the outer row of pyriform canals as specific characters. However, the poor preservation of the syntypes (CASM No. 2179, 2180) does not allow verification of the absence of septa. Coogan (1973, pl. 4, figs. 1, 2) included specimens having septate body cavities in this species. The body cavity and accessory cavity are small relative to the valve diameter. This specimen is 57 mm from anterior to posterior and 44 mm from dorsal to ventral.

Occurrences. Coogan (1973, 1977) reported that species from the Edwards Limestone in central Texas ranging from the middle Albian *Oxytropidoceras* zone to the upper Albian Devils River Limestone. It also occurs in the El Abra Limestone in eastern Mexico (Young, 1984). In the Lampazos area, *T. vivari* occurs in member 2 of the Nogal Formation, Section 4 (Sample LPR-41). *Texicaprina*? sp. occurs in the Espinazo del Diablo Formation, Section 3 (Sample AZ-1).

ACKNOWLEDGMENTS

Keith Young, University of Texas at Austin, shared his experiences on the rudist taxonomy and the occurrences of orbitolinids in Mexico. The California Academy of Science permitted study of rudist type specimens collected by R. H. Palmer. We appreciate the helpful reviews by Gloria Alencaster, Jean-Marie Vila, and Ana Luisa Carreño. We are grateful for typing and photography provided by Amoco Production Company.

Figure 7. A, B: *Cladophyllia furcifera* Roemer, Espinazo del Diablo Formation (LP-4); A, septa at center of calyx, 39.2×; B, epitheca in ultraviolet light to distinguish original wall from infilling spar, 82×. C-F: *Caprinuloidea perfecta* Palmer, Espinazo del Diablo Formation (C, of AZ-1; D, of AR—5; E, F, of AZ-33, view into valve), 1×. G, H: *Texicaprina vivari* Palmer, Nogal Formation (LP-41), 1×; L is ligament groove; G, anterior margin is toward viewer; H, looking into valve.

REFERENCES CITED

Alencaster, G., 1961, Estratigrafía del Triásico Superior de la parte central de Sonora: Universidad Nacional Autónoma de México, Instituto de Geología, Paleontología Mexicana 11, pt. 1, 18 p.

Anderson, T. H., and Schmidt, V. A., 1983, The evolution of Middle America and the Gulf of Mexico–Caribbean Sea region during Mesozoic time: Geological Society of America Bulletin, v. 94, p. 941–966.

Applin, P. L., and Applin, E. R., 1965, The Comanche Series and associated rocks in the subsurface in central and south Florida: U.S. Geological Survey Professional Paper 447, 84 p.

Arnaud, A., and others, 1981, Tableau de répartition stratigraphique des grands foraminifères du Crétacé moyen de la region Méditerranéenne: Rapports Commission Internationale Mer Méditerranée v. 27, no. 8, p. 121–123.

Arnaud-Vanneau, A., 1969, Quelques précisions concernant l'appareil embryonnire du genre *Palorbitolina* Schroeder: Revue de Micropaléontologie, v. 12, p. 16–20.

Ayala-Castañares, A., 1960, *Orbitolina morelensis* sp. nov. de la Formación Morelos del Cretacico Inferior (Albiano) en la región de Huetamo, Michoacán, Mexico: Paleontología Mexicana, Mexico, D. F., v. 6, p. 1–16.

Azema, J., and 5 others, 1985, Le Honduras (Amérique centrale nucléaire) et le bloc d'Oaxaca (sud du Mexique): Géodynamique des Caraibes, Symposium Paris, 5–8 Février 1985, Editions Technip, 27, Rue Ginoux, 75015 Paris, p. 427–438.

Barker, R. W., 1944, Some larger foraminifera from the Lower Cretaceous of Texas: Journal of Paleontology, v. 18, p. 204–209.

Beckmann, J.-P., and Beckmann, R., 1966, Calcareous algae from the Cretaceous and Tertiary of Cuba: Schweizerische Paläontologische Abhandlungen, v. 85, 45 p.

Bilodeau, W. L., and Lindberg, F. A., 1983, Early Cretaceous tectonics and sedimentation in southern Arizona, southwestern New Mexico, and northern Sonora, Mexico, in Reynolds, H. W., and Dolly, E. D., eds., Mesozoic paleogeography of west-central United States: Rocky Mountain Section, Society of Economic Paleontologists and Mineralogists, p. 173–188.

Blumenbach, J. F., 1805, Abbildungen naturhistorischer Gegenstände: Göttingen, H. Dieterich, pt. 8, no. 80.

Bonet, F., 1956, Zonificación microfaunistica de las calizas cretácicas del este de México: Asociacion Mexicana de Geólogos Petroleros Boletín, v. 8, p. 389–488.

Carsey, D. O., 1926, Foraminifera of the Cretaceous of central Texas: Austin, Texas University Bulletin 2612, 56 p.

Chiocchini, M., Mancinelli, A., and Romano, A., 1984, Stratigraphic distribution of benthic forminifera in the Aptian, Albian, and Cenomanian carbonate sequences (southern Lazio, Italy): Benthos 1983; Second International Symposium, Benthic Foraminifera, Pau, April, 1983, p. 167–181.

Conkin, J. E., and Conkin, B. M., 1958, Revision of the genus *Nummoloculina* and emendation of *Nummoloculina heimi* Bonet: Micropaleontology, v. 4, p. 149–150.

Coogan, A. H., 1973, New rudists from the Albian and Cenomanian of Mexico and south Texas: Revista del Instituto Mexicano Petróleo, v. 5, p. 51–76.

—— , 1977, Early and Middle Cretaceous Hippuritacea (rudists) of the Gulf Coast: Austin, Texas University Bureau of Economic Geology Report of Investigations 89, p. 32–70.

Cooke, C. W., 1946, Comanche echinoids: Journal of Paleontology, v. 20, p. 193–237.

Córdoba, D. A., 1969, Mesozoic stratigraphy of northeastern Chihuahua, Mexico: New Mexico Geological Society 20th Field Conference Guidebook, p. 91–96.

Cushman, J. A., and Applin, E. R., 1947, Two new species of Lower Cretaceous foraminifera from Florida: Cushman Laboratory for Foraminiferal Research Contribution v. 23, p. 29–30.

Douglass, R. C., 1960a, The Foraminiferal Genus *Orbitolina* in North America: U.S. Geological Survey Professional Paper 333, 52 p.

—— , 1960b, Revision of the Family Orbitolinidae: Micropaleontology, v. 6, p. 249–270.

Elliott, G. F., 1956, Further records of fossil calcareous algae from the Middle East: Micropaleontology, v. 2, p. 327–334.

—— , 1957, New calcareous algae from the Arabian Peninsula: Micropaleontology, v. 3, p. 227–230.

——, 1963, Problematical microfossils from the Cretaceous and Paleocene of the Middle East: Paleontology, v. 6, p. 293–300.

——, 1965, The interrelationships of some Cretaceous Codiaceae: Paleontology, v. 8, p. 199–203.

Fourcade, E., and Raoult, J. F., 1973, Crétacé du Kef Hahouner et position stratigraphique de "*Ovalveolina*" *reicheli* P. de Castro (série septentrionale du Môle néritique du Constantinois, Algérie): Revue de Micropaléontologie, v. 15, p. 227–246.

Frizzell, D. L., 1954, Handbook of Cretaceous Foraminifera of Texas: Austin, Texas University Bureau of Economic Geology Report on Investigations 22, 232 p.

Frizzell, D. L., and Schwartz, E., 1950, A new lituolid foraminiferal genus from the Cretaceous, with an emendation of *Cribrostomoides* Cushman: Missouri School of Mines Bulletin, Technical Series 76, 12 p.

González-León, C., 1988, Estratigrafía y geología estructural de las rocas sedimentarias cretácicas del área de Lampazos, Sonora: Instituto de Geología, UNAM, Revista, v. 7, no. 2, p. 148–162.

Gonzalez-Leon, C., and Buitrón, B. E., 1984, Bioestratigrafía del Cretácico Inferior del área de Lampazos, Sonora, México: Memoria III Congreso Latinoamericano de Paleontología, Universidad Nacional Autónoma de México, Instituto de Geología, p. 371–377.

Gonzalez-Leon, C., Bartolini, C., and Herrera, S., 1982, Geology of the Lampazos area, east-central Sonora: Geological Society of America Abstracts with Programs, v. 14, p. 167.

Gusić, I., 1981, Variation, range, evolution, and biostratigraphy of *Palorbitolina lenticularis* in the Lower Cretaceous of the Dinaric Mountains in Yugoslavia: Palaontologisch Zeitschrift, v. 55, p. 191–208.

Hamaoui, M., 1973, *Barkerina* et formes voisines: Bulletin Centre Recherches Pau-SNPA, v. 7, p. 337–359.

Henson, F.R.S., 1948, Larger imperforate foraminifera of southwestern Asia: London, British Museum (Natural History), 126 p.

Herrera, S., and Bartolini, C., 1983, Geología del area de Lampazos, Sonora [Tesis Profesional]: Hermosillo, Universidad de Sonora, 120 p.

Herrera, S., Bartolini, C., Perez, O., and Buitrón, B., 1984, Paleontología del área de Lampazos, Sonora: Universidad Sonora, Departamento Geología Boletín, v. 1, p. 50–59.

Hewett, R. L., 1978, Geology of the Cerro La Zacatera area, Sonora, Mexico [M.S. thesis]: Flagstaff, Northern Arizona University, 99 p.

Himanga, J., 1977, Geology of the Sierra Chiltepin, Sonora, Mexico [M.S. thesis]: Flagstaff, Northern Arizona University, 99 p.

Ice, R. G., and McNulty, C. L., 1980, Foraminifers and calcispheres from the Cuesta del Cura and lower Agua Nueva(?) Formations (Cretaceous) in east-central Mexico: Transactions of the Gulf Coast Association of Geological Societies, v. 30, p. 403–425.

Imlay, R. W., 1939, Paleogeographic studies in northeastern Sonora: Geological Society of America Bulletin, v. 50, p. 1723–1744.

Jacques-Ayala, C., 1983, Sierra El Chanate, northwestern Sonora, Mexico; Stratigraphy, sedimentology, and structure [M.S. thesis]: Cincinnati, Ohio, University of Cincinnati, 143 p.

Jacques-Ayala, C., and Potter, P. E., 1987, Stratigraphy and paleogeography of Lower Cretaceous rocks, Sierra of Chanate, northwest Sonora, Mexico: Arizona Geological Society Digest, v. 18, p. 203–214.

Johnson, J. H., 1964, The Jurassic algae: Golden, Colorado School of Mines Quarterly, v. 59, no. 2, 129 p.

——, 1968, Lower Cretaceous algae from Texas: Golden, Colorado School of Mines Professional Contributions, no. 4, 71 p.

——, 1969, A review of the Lower Cretaceous algae: Golden, Colorado School of Mines Professional Contributions, no. 6, 111 p.

Keijzer, F. G., 1942, On a new genus of arenaceous foraminifera from the Cretaceous of Texas: Proceedings, K. Nederlandse Akademie Wetensch, v. 45, p. 1016–1017.

King, R. E., 1939, Geological reconnaissance in northern Sierra Madre Occidental of Mexico: Geological Society of America Bulletin, v. 50, p. 1625–1722.

Konishi, K., and Epis, R. C., 1962, Some Early Cretaceous algae from Cochise County, Arizona: Micropaleontology, v. 8, p. 67–76.

Leymerie, M. A., 1878, Description géologique et paléontologique des Pyrénées de la Haute-Garonne: Toulouse, E. Privat., 1010 p., 51 pl.

Loeblich, A. R., Jr., and Tappan, H., 1985, Some new and redefined genera and families of agglutinated Foraminifera I: Journal of Foraminiferal Research, v. 15, p. 91–104.

——, 1988, Foraminiferal genera and their classification: New York, van Nostrand Reinhold Company, v. 1, 970 p., v. 2, 847 pl.

Martínez, J., and Palafox, J., 1985, Estratigrafía del área de Arivechi: Universidad de Sonora Departamento de Geología Boletín, v. 2, no. 1, p. 30–59.

Maync, W., 1953, *Pseudocyclammina hedbergi* n. sp. from the Urgo–Aptian and Albian of Venezuela: Cushman Foundation for Foraminiferal Research Contribution, v. 4, p. 101.

——, 1955a, *Dictyoconus walnutensis* (Carsey) in the Middle Albian Guacharo Limestone of eastern Venezuela: Cushman Foundation for Foraminiferal Research Contributions, v. 6, p. 85–93.

——, 1955b, *Coskinolina sunnilandensis* n. sp. a Lower Cretaceous (Urgo–Albian) species: Cushman Foundation for Foraminiferal Research Contributions, v. 6, p. 105–111.

——, 1961, Foraminiferal key biozones in the Lower Cretaceous of the Western Hemisphere and the Tethys Province: 20th International Geological Congress Mexico, Symposium del Cretácico, v. 1, p. 85–111.

Minjarez, I., Palafox, J., Torres, Y., and Villalobos, R., 1985, Consideraciones respecto a la estratigrafía y estructura del área de Sahuaripa-Arivechi: Universidad de Sonora Departamento de Geología Boletín, v. 2, no. 1, p. 90–105.

Morycowa, E., 1964, Hexacoralla des couches de Grodziszcze (Néocomien, Carpathes): Acta Palaeontologica Polonica, v. 9, no. 1, 114 p.

——, 1971, Hexacorallia et Octocorallia du Crétace Inférieur de Rarau (Carpathes Orientales Roumaines): Acta Palaeontologica Polonica, v. 16, no. 1–2, 149 p.

Moullade, M., and Peybérnes, B., 1975, Biozonation par Orbitolinidés du Clansayésien des Pyrénées franco-espagnoles: C. R. Academic Sciences Paris, Séries D, v. 280, p. 2524–2532.

Moullade, M., and Saint-Marc, P., 1975, Les "Mésorbitolines"; Révision taxinominique importance stratigraphique et paléobiogéographique: Société Géologique France Bulletin, series 7, v. 17, p. 828–842.

Moullade, M., Peybernes, B., Rey, J., and Saint-Marc, P., 1985, Biostratigraphic interest and paleobiogeographic distribution of Early and mid-Cretaceous Mesogean Orbitolinids (Foraminiferida): Journal of Foraminiferal Research, v. 15, p. 149–158.

Noll, H. H., 1981, Geology of the Picacho Colorado area, northern Sierra de Cobachi, central Sonora, Mexico [M.S. thesis]: Flagstaff, Northern Arizona University, 165 p.

Ortuño-Arzate, F., 1985, Evolution sédimentaire mésozoique du bassin rift de Chihuahua le long d'une transversale Aldama–Ojinaga (Mexique) [Thèse Docteur de Troisieme cycle]: Pau, France, Université de Pau et des Pays de L'Adour, 344 p.

Palmer, R. H., 1928, The rudistids of southern Mexico: California Academy of Sciences Occasional Papers, no. 14, 137 p.

Peybernès, B., 1976, Le Jurassique et le Crétacé inférieur des Pyrénées Franco-Espagnoles [Thèse de doctorat és Sciences Naturelles]: Toulouse, France, Université Paul-Sabatier, 459 p.

Pia, J., 1927, Thallophyta, *in* Hirmer, M., ed., Handbuch der Palâobotanik: München and Berlin, p. 31–136.

——, 1936, Calcareous green algae from the Upper Cretaceous of Tripoli (North Africa): Journal of Paleontology, v. 10, p. 3–13.

Pindell, J., and Dewey, J. F., 1982, Permo-Triassic reconstructions of western Pangea and the evolution of the Gulf of Mexico/Caribbean region: Tectonics, v. 1, p. 179–211.

Poole, F. G., Murchey, B. L., and Stewart, J. H., 1983, Bedded barite deposits of middle and late Paleozoic age in central Sonora, Mexico: Geological Society of America Abstract with Programs, v. 15, p. 299.

Reitner, J., 1984, Microfazielle, palökologische und paläogeographische analyse ausgewählter vorkommen flachmariner karbonate im basko-kantabrischen strike slip fault-becken-system (Nordspanien) an der wende von der Unterkreide zur Oberkreide [Ph.D. thesis]: Tubingen, Eberhard-Karls-Universität, 211 p.

Roemer, F., 1852, Die Kreidebildung von Texas und Ihre organische einschüsse: Bonn, Adolph Marcus, 100 p.

——, 1888, Ueber eine durch die haeufigkeit hippuriten-artiger chamiden ausgezeichnete fauna der oberturonen kreide von Texas: Palaeontologische Abhandlungen, Vierter Band, heft 4, p. 281–296.

Saint-Marc, P., 1974, Etude stratigraphique et micropaléontologique de l'Albien, du Cénomanien et du Turonien du Liban: Notes et Mémoires Moyen-Orient, v. 13, p. 7–298.

Sandidge, M. H., 1985, Aptian–Albian ammonoids of the Oyster Limestone Member of the U-Bar Formation, Big Hatchet Mountains, southwestern New Mexico: Newsletter of Stratigraphy, v. 14, p. 158–168.

Schmidt, T. G., 1978, Geology of the northern Sierra El Encinal, Sonora, Mexico [M.S. thesis]: Flagstaff, Northern Arizona University, 80 p.

Schroeder, R., 1963, *Palorbitolina,* ein neues subgenus der gattung *Orbitolina* (Foraminifera): Neues Jarbuch Geologie Paläontologie Abhandlungen, v. 117, p. 346–359.

——, 1975, General evolutionary trends in Orbitolinas: Revista Española de Micropaleontologia, numero especial, enero 1975, p. 117–128.

——, 1979, Les Orbitolines de l'Aptien; Définitions, origine et évolution; Colloque sur l'Urgonien des pays meditérranéens, Grenoble, September, 1979: Géobios, Mémoire spécial, no. 3, p. 289–299.

Schroeder, R., and Cherchi, A., 1979, Upper Barremian–lowermost Aptian orbitolinid foraminifers from the Grand Banks continental rise, northwestern, Atlantic DSDP Leg 43, Site 394, Initial reports of the Deep Sea Drilling Project: Washington, D.C., U.S. Government Printing Office, v. 63, p. 575–583.

Schroeder, R., and Neumann, M., ed., 1985, Les grands foraminifères du Crétacé moyen de la région Méditerranéenne: Geobios, Lyon, mémoire spécial no. 7, 160 p.

Schroeder, R., Cherchi, A., Guellal, S., and Vila, J.-M., 1978, Biozonation par les grands foraminifères du Jurassique supérieur et du Crétacé inférieur et moyen des séries néritique en Algérie du Nord-Est: Actes du VI Colloque Africain de Micropaléontologie, Annales des Mines et de la Géologie, Tunis, no. 28, v. 2, p. 243–253.

Scott, R. W., 1970a, Paleoecology and paleontology of the Lower Cretaceous Kiowa Formation, Kansas: Lawrence, University Kansas Paleontological Contributions, Article 52 (Cretaceous 1), 94 p.

——, 1970b, Stratigraphy and sedimentary environments of Lower Cretaceous rocks, southern Western Interior: American Association Petroleum Geologists Bulletin, v. 54, p. 1225–1244.

——, 1984a, Mesozoic biota and depositional systems of the Gulf of Mexico–Caribbean region: Geological Society of Canada Special Paper 27, p. 49–64.

——, 1984b, Significant fossils of the Knowles Limestone, Lower Cretaceous, Texas, *in* Proceedings, 3rd Annual Research Conference: Gulf Coast Section, Society of Economic Paleontologists and Mineralogists Foundation, p. 333–346.

——, 1986, Paleobiology of Early Cretaceous protocardiids, Caribbean Province: Journal of Paleontology, v. 60, p. 1186–1211.

——, 1987, Stratigraphy and correlation of the Lower Cretaceous Mural Limestone, Arizona and Sonora: Arizona Geological Society Digest, v. 18, p. 327–334.

Scott, R. W., and Kidson, E. J., 1977, Lower Cretaceous depositional systems, West Texas, *in* Bebout, D. G., and Loucks, R. G., eds., Cretaceous carbonates of Texas and Mexico: Austin, University of Texas Bureau of Economic Geology Report Investigation 89, p. 169–181.

Simmons, M. D., and Hart, M. B., 1987, The biostratigraphy and microfacies of the Early to mid-Cretaceous carbonates of Wadi Mi'aidin, Central Oman Mountains, *in* Hart, M. B., ed., Micropalaeontology of carbonate environments: Chichester, England, Ellis Horwood Ltd., p. 176–207.

Smith, C. I., and Bloxsom, W. E., 1974, The Trinity Division and equivalents of northern Coahuila, Mexico, *in* Geoscience and man: Baton Rouge, Louisiana State University, v. 8, p. 67–76.

Stanton, T. W., 1947, Studies of some Comanche pelecypods and gastropods: U.S. Geological Survey Professional Paper 211, p. 1–256.

Stewart, J. H., McMenamin, M.A.S., and Morales-Ramirez, J. M., 1984, Upper Proterozoic and Cambrian rocks in the Caborca region, Sonora, Mexico; Physical stratigraphy, biostratigraphy, paleocurrent studies, and regional relations: U.S. Geological Survey Professional Paper 1309, 36 p.

Strain, W. S., 1968, Cerro de Muleros (Cerro de Cristo Rey): West Texas Geological Society Publication 68-55, p. 82.

Toula, F., 1884, Geologische untersuchengen im westlichen Teile des Balkans und in den angrenzenden Gebieten: Akademie Wissenschaften Wienna, Mathematik-Naturforschung, Sitzungsbericht, v. 88, pt. 1, p. 1279–1348.

Turnsek, D., and Mihajlovic, M., 1981, Lower Cretaceous Cnidarians from eastern Servia: Slovenska Akademija Znanosti in Umetnosti Academia Scientiarum et Artium Sloveniča, Razprave Dissertationes, v. 23, no. 1, 1–54 p.

Wells, J. W., 1933, Corals of the Cretaceous of the Atlantic and Gulf Coastal Plains and Western Interior of the United States: Bulletins of American Paleontology, v. 18, p. 83–286.

Young, K., 1969, Ammonite zones of northern Chihuahua: 20th New Mexico Geological Society Field Conference Guidebook, p. 97–101.

——, 1972, Cretaceous paleogeography; Implications of endemic ammonite faunas: Austin, University of Texas Bureau of Economic Geology Geological Circular 72-2, 13 p.

——, 1974, Lower Albian and Aptian (Cretaceous) ammonites of Texas, *in* Geoscience and man: Baton Rouge, Louisiana State University, v. 8, p. 175–228.

——, 1982, Cretaceous rocks of central Texas; Biostratigraphy and lithostratigraphy, *in* Maddocks, R. F., ed., Texas Ostracoda: University of Houston Department of Geosciences, p. 111–126.

——, 1983, Mexico; The Phanerozoic geology of the world; 2, The Mesozoic (B): Netherlands, Elsevier Science Publishers B. V., p. 61–88.

——, 1984, Biogeography and stratigraphy of selected middle Cretaceous rudists of southwestern North America: Memoria III Congreso Latinoamericano de Paleontología, p. 341–360.

MANUSCRIPT ACCEPTED BY THE SOCIETY APRIL 18, 1990

Printed in U.S.A.

Geology and chemical composition of the Jaralito and Aconchi batholiths in east-central Sonora, Mexico

Jaime Roldán-Quintana
Instituto de Geología UNAM, Apartado Postal 1039, Hermosillo, Sonora 83000, México

ABSTRACT

The Aconchi–El Jaralito area is 100 km northeast of Hermosillo, and about 200 km southwest of Douglas, Arizona. The intrusive rocks of the region are divided into three units on the basis of their field relations, morphology, mineralogy, chemical composition, and isotopic ages. These units are two granitic batholiths and a group of porphyritic rhyolite stocks. The oldest batholith is the granitic-granodioritic El Jaralito batholith, which has the largest exposure area. Lithologically it varies from a true granite to quartz monzonite, quartz diorite, and granodiorite. The rocks of this batholith are considered I-type or Cordilleran granites. The K/Ar ages reported for El Jaralito batholith range from 51.8 to 69.6 Ma. The second batholith is the Aconchi batholith, which has a smaller exposure area. It is lithologically simple, consisting almost exclusively of alkali granites, including abundant pegmatites. This is considered an S-type granitoid, with primary muscovite and red garnet. One K/Ar date for this unit is 35.96 ± 0.70 Ma. Chemical analyses show no clear separation between the two batholiths. There is an overlap in the values of some of the oxides. The youngest intrusive rocks in the area are two porphyritic rhyolite stocks, which have quartz, feldspar, and biotite phenocrysts set in a quartz-feldspar matrix. The chemical composition of the stocks is similar to the alkali granites of the Aconchi batholith. Cataclastic deformation was observed within both batholiths in isolated localities; however, the structure of these deformed rocks was not studied in detail. Chemical and isotopic data for these batholiths remain incomplete; it is highly recommended that future chemical data include trace-element concentrations.

INTRODUCTION

The presence of a batholithic belt in western North America is well known (Bateman and Clark, 1974; Bateman and Busacca, 1983; Hamilton and Bradley, 1967; Henry, 1975). The intrusive rocks of Sonora represent an extension of this belt into Mexico, where it continues along the Pacific coast. Batholithic rocks are widely exposed in northwestern Mexico, in the states of Baja California, Sonora, and Sinaloa.

In central Sonora the intrusive rocks have extensive exposures estimated to be about 2,500 km^2. The batholith represents an intrusive complex consisting of different lithologies that vary from granites sensu stricto to quartz monzonites, granodiorites, and diorites. A broad period of intrusion, which ranges between 40 and 90 Ma, has been documented by Damon and others (1983a). These rocks are of great economic importance because of their well-known association of porphyry copper (Cu-Mo) deposits (Sillitoe, 1976), as well as W and Fe skarn deposits (Pérez, 1985). Contact and vein Ag-Pb-Zn deposits associated to these plutons are economically less important (Roldán, 1979). In central Sonora, these intrusions are also important in the genesis of graphite deposits (Vassallo, 1985).

Most of the isotopic studies of these intrusive rocks have been done by Damon and others (1981, 1983a, b) using K/Ar dating. Anderson and Silver (1974) reported U/Pb ages for some plutons in Sonora, and Anderson and others (1980) described the age and distribution of lineated plutons in northern and central parts of the state.

This work was done despite the lack of systematic field studies of the Late Cretaceous–early Tertiary batholiths (40 to 90 Ma), which were termed "Sonoran Batholith" by Damon and others (1983a). In this chapter, I describe the geology and chemi-

Roldán-Quintana, J., 1991, Geology and chemical composition of the Jaralito and Aconchi batholiths in east-central Sonora, Mexico, *in* Pérez-Segura, E., and Jacques-Ayala, C., eds., Studies of Sonoran geology: Geological Society of America Special Paper 254.

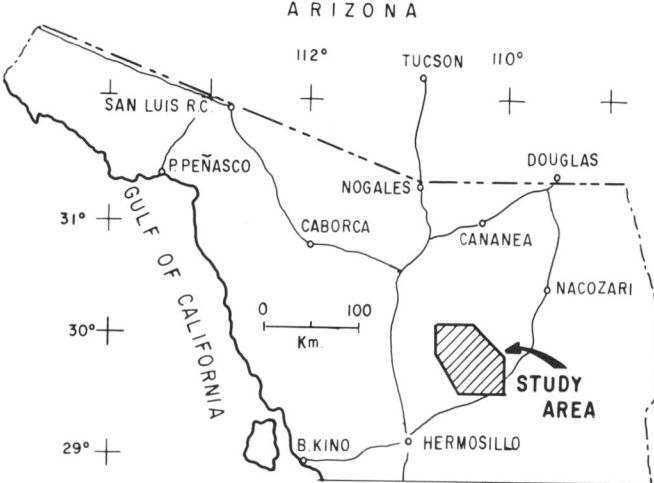

Figure 1. Location map of the Aconchi–El Jaralito area in east-central Sonora.

cal composition of a part of this batholith in the Aconchi–El Jaralito area in east-central Sonora (Fig. 1). This region was selected for study because it appears to be representative of other exposures of this batholith. The work described here shows that the "Sonoran Batholith" consists of at least two magma pulses, Laramide (51.9 to 69.6 Ma) and mid-Tertiary (near the Eocene-Oligocene boundary, 36 Ma).

Within the study area, several investigations of economic geology have been done. Pedrazzini (1961) studied beryl deposits associated with granitic pegmatites in the southern part of the Sierra de Aconchi. Geology and tungsten mineralization of the El Jaralito District were investigated by Mills and Hokuto (1971), and a metallogenetic study covering a portion of the area was done by Chávez (1978). Peabody (1979) studied in more detail the geology of the tungsten-bearing skarn in the region of El Jaralito. More recently, Rodríguez (1984) and Castillo (in preparation) described the geology of two areas in the northwestern part of the region. All previous studies are indicated on Figure 2.

Reconnaissance mapping for the study described here covered about 3,300 km^2 (Fig. 2), and required six months time during a 3-year period (1985–1987). The main objective of this chapter is the geology of intrusive rocks in the area.

GEOLOGIC FRAMEWORK

The Aconchi–El Jaralito area is located 100 km northeast of Hermosillo, the capital of Sonora, and about 200 km southwest of Douglas, Arizona (Fig. 1). It is in the subprovince of Elongated Ranges, part of the Sierra Madre Occidental physiographic province (Raisz, 1964). The region is characterized by northwest-trending ranges, bordered on their flanks by normal faults with the same orientation. It is near the southern margin of the Precambrian North American craton, within the Basin and Range Province. The ranges are separated by valleys filled with clastic rocks.

The Sierras de Aconchi and El Jaralito (Fig. 2) form a horst flanked by the grabens of the Sonora River to the east and the San Miguel River to the west. The highest point of the area is located south of Rancho El Llano, with elevations of 2,180 m above sea level. In the area of El Jaralito there are elevations of 1,500 m above sea level.

In the northern portion of the area, Precambrian gneisses, schists, and arkosic metasediments are intruded by granite and granodiorite, 1.6 to 1.8 Ga (Anderson and Silver, 1979), and by ultramafic dikes and stocks (Castillo, in preparation). Paleozoic(?) clastic and calcareous sediments, most of them altered by contact metamorphism, generally occur as roof pendants in batholiths, such as in Los Amoles and El Jaralito (Fig. 2). Paleozoic age was assigned to them on the basis of lithology (Roldán, 1989), but quartzite and conglomerate in the northwestern portion of the area could well be late Proterozoic.

Rodríguez (1984), tentatively assigned a Jurassic age to volcanic and volcaniclastic rocks in the Sierra del Cobre, in the northwest corner of the area. Isolated outcrops of Cretaceous rocks are present in the San Felipe area, where calcareous rocks contain early Albian fossils. These rocks are interbedded with andesitic tuffs and volcaniclastic sediments. In the Los Amoles–Rancho Los Taraises area, similar sequences also appear to be Cretaceous.

Tertiary volcanic rocks and continental deposits are widespread in east-central Sonora. The volcanic rocks include andesites, rhyolites, rhyolitic tuffs, and basalts exposed in the northern and southeast portion of the area. The oldest volcanic rocks may be coeval with the El Jaralito batholith. Continental red beds, arkosic sandstone, and conglomerate cover large areas in the valleys of the Sonora and San Miguel Rivers.

The most important structural features of the area are northwest-southeast normal faults that border horst and graben structures, which characterize this part of Sonora. Some of these faults cut late Tertiary clastic deposits and control hot springs along the Agua Caliente fault.

BATHOLITHIC ROCKS

In the Aconchi–El Jaralito area, intrusive rocks consist of two granitic batholiths and two porphyritic rhyolite stocks. Younger andesitic stocks were not studied in detail. Intrusive rocks cover more than half of the area under discussion. Of two batholiths, the El Jaralito is the older and has the larger exposure area. This pluton displays great lithologic variety, ranging from granite, quartz monzonite, quartz diorite to granodiorite. Isotopic ages for rocks of El Jaralito batholith span from 51 to 69 Ma (Gastil, written communication, 1986; Mead, 1982; Anderson and others, 1980).

The younger Aconchi batholith is exposed over about 200 km^2 and consists exclusively of alkali granite. It hosts abundant pegmatites, as dikes (200 m long) and stocks (50 m in diameter).

The only published K/Ar date is 35.96 ± 0.70 Ma (Damon, *in* Roldán, 1979). Both batholiths are bordered by intrusive contacts in a few places only. Most contacts are younger faults, mainly late Oligocene normal faults. The actual size of the batholiths must be larger than indicated by present outcrops.

El Jaralito Batholith

The informal name "El Jaralito batholith" is proposed for an intrusion in the Sierra El Jaralito. Some of its best exposures occur around the El Jaralito mine, designated the type locality. This intrusion covers an area of about 750 km^2, and extends south beyond the mapped area.

The El Jaralito batholith forms the cores of northwest-southeast ranges, generally more than 1,000 m above sea level. The slopes of the ranges are gentle, except southwest of Mazocahui, where the Rio Sonora has carved a canyon into the granite.

The El Jaralito batholith is part of the regional "Laramide Sonoran Batholith" of Damon and others (1983a). Besides the large outcrop in Sierra El Jaralito, there are exposures of this pluton northeast of Rayon, east of Opodepe, and northeast of Baviácora (Fig. 2).

The El Jaralito batholith intruded Precambrian rocks near the La Lobera and La Ramada Ranches. Around El Jaralito and west and southwest of Baviácora, it intruded Paleozoic(?) rocks. Along these contacts, important tungsten skarn deposits occur (Peabody, 1979). Other skarns are host to economically important deposits of iron, lead, silver, and zinc sulfides.

South of Baviácora and northeast of Mazocahui, the El Jaralito batholith intruded andesitic volcanic rocks, causing extensive silicification and kaolinization. In most places, the batholith is in fault contact with Tertiary clastic rocks. Many of these faults are of regional extent. For example, the Agua Caliente fault extends for more than 50 km along the eastern flank of Sierra El Jaralito.

In the study area (Fig. 2), the most common lithologies are granite and granodiorite. Alkali granite, quartz diorite, and quartz monzonite are less abundant (Fig. 3). The contacts between the different facies of the batholith were not studied in detail, but at least some are intrusive.

Granite occurs throughout the area, but is specially abundant along the road from Mazocahui to Ures, the El Jaralito area, and in the vicinity of Rancho Los Taraises. Granodiorite is abundant east and northeast of Mazocahui, where it was possible to separate it from the granites (Fig. 2). Granodiorites were also identified in the area of El Jaralito and to the north of Mazocahui.

In the area east and northeast of Mazocahui, gray and black xenoliths are included in the granodiorite. Near the Granja El Mariscal, these xenoliths are tabular and 2 to 25 cm long, preferentially oriented N20°W. In thin section the xenoliths were classified as fine-grained diorite.

Microscopically, the rocks of the El Jaralito batholith have hypidiomorphic (equigranular to porphyritic) holocrystalline textures. Micrographic textures are also common. There are signs of cataclastic deformation in some samples. They include brecciation of quartz, an incipient lineation of the opaque minerals, and folding or brecciation of the feldspars at their margins. The quartz in almost all the samples shows undulatory extinction.

Twenty-two thin sections of the El Jaralito batholith were point counted, 600 points per sample (Fig. 3). The essential minerals are quartz (approximately 30 modal percent, average of five samples); microcline and orthoclase, 30 percent; and albite-oligoclase, 41 percent. The most common accessory minerals are biotite, 4.5 percent; hornblende, 3.5 percent; sphene, 1.5 percent; zircon, 1 percent; and apatite as minor constituent (Fig. 4A). Some thin sections contain muscovite, possibly an alteration product of biotite. The opaque minerals are hematite, magnetite, and pyrite. The most common secondary minerals are chlorite, sericite, calcite, and epidote.

Figure 3 shows that the lithologic variability of the El Jaralito batholith is due mainly to the relative proportions of K-feldspar and plagioclase because the quartz content remains between 20 and 42 percent. The El Jaralito batholith corresponds to the I-type or Cordilleran granites, according to the classification of Chappell and White (1974).

Other samples of the Sonoran batholith reported by Damon and others (1983a), and of the Sinaloa batholith (Henry, 1975) correspond to the granodioritic facies of the El Jaralito batholith (Fig. 3).

Aconchi granitic batholith

The informal name "Aconchi granitic batholith" designates an intrusive body in the Sierra de Aconchi and Sierra de Los Locos. It encompasses an area of about 200 km^2, smaller than that of the El Jaralito batholith.

The Aconchi batholith is an alkali granite (Fig. 3) with abundant associated pegmatites and aplites, as well as dikes or small intrusions of irregular shapes. The batholith is homogeneous, in contrast to the El Jaralito batholith.

The Aconchi batholith is higher topographically than the El Jaralito batholith, 2,180 m above sea level in the Picacho Alto—the highest point in Sierra de Aconchi. To the north, the altitude diminishes to 1,300 m in Rancho El Llano. In outcrops, rocks of the Aconchi batholith are distinctly lighter than those of the El Jaralito batholith. Dikes and small bodies of porphyritic or fine-grained andesite intrude the Aconchi batholith.

The granites of the Aconchi batholith generally are porphyritic to pegmatitic. Graphic textures were observed megascopically and microscopically. Near Rancho El Llano, quartz and microcline crystals as large as 20 cm were found.

Primary muscovite and red garnet are characteristic of the Aconchi batholith. These minerals provide a good field criterion to differentiate the rocks of the Aconchi batholith from the El Jaralito batholith. Pegmatites are an important constituent; south of Rancho El Torreón, they are more voluminous than muscovite granite. In the southern part of the Sierra de Aconchi, beryl was mined in pegmatites (Pedrazzini, 1961). The two-mica granite in

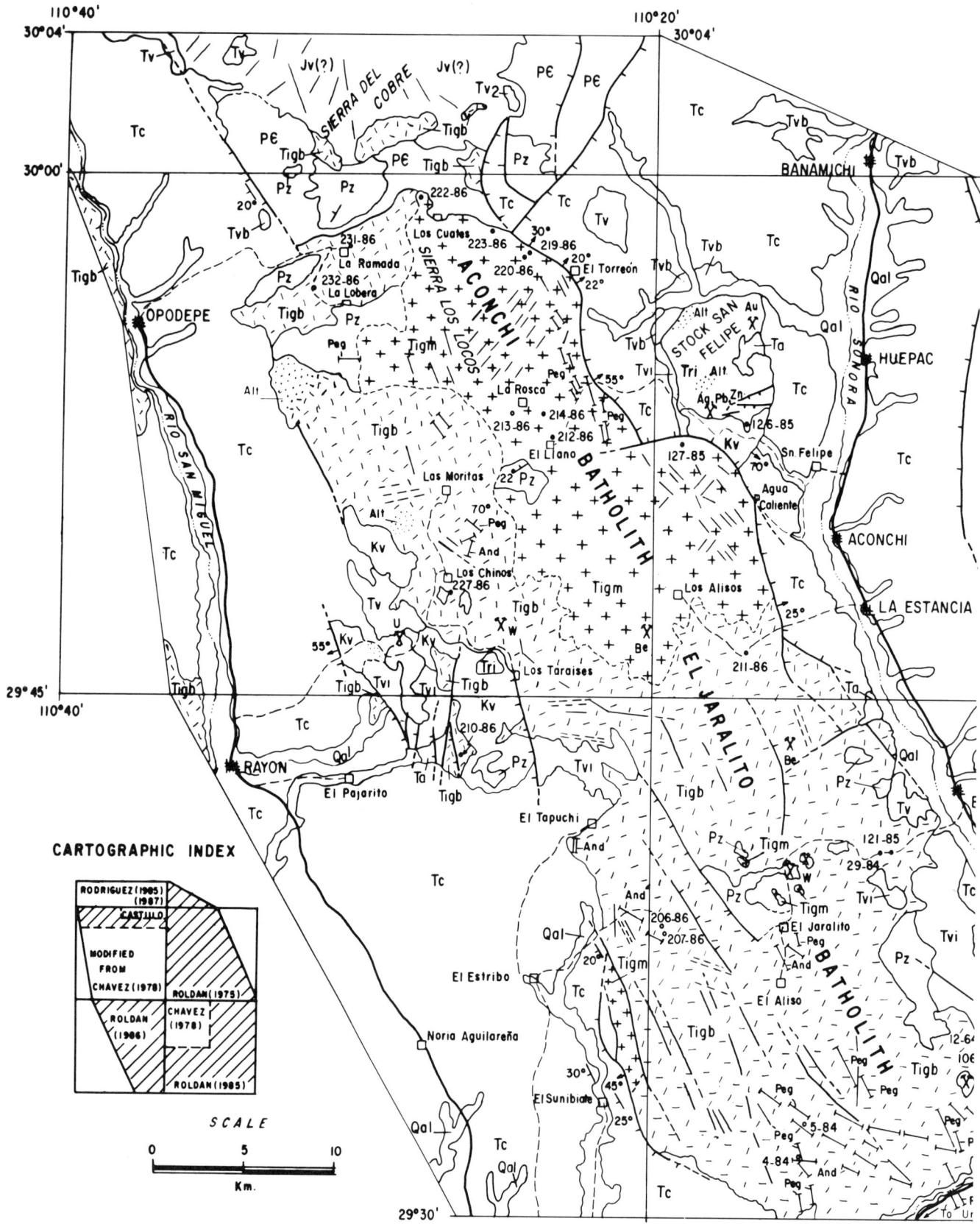

Figure 2. Generalized geologic map of the Aconchi–El Jaralito region and surrounding areas.

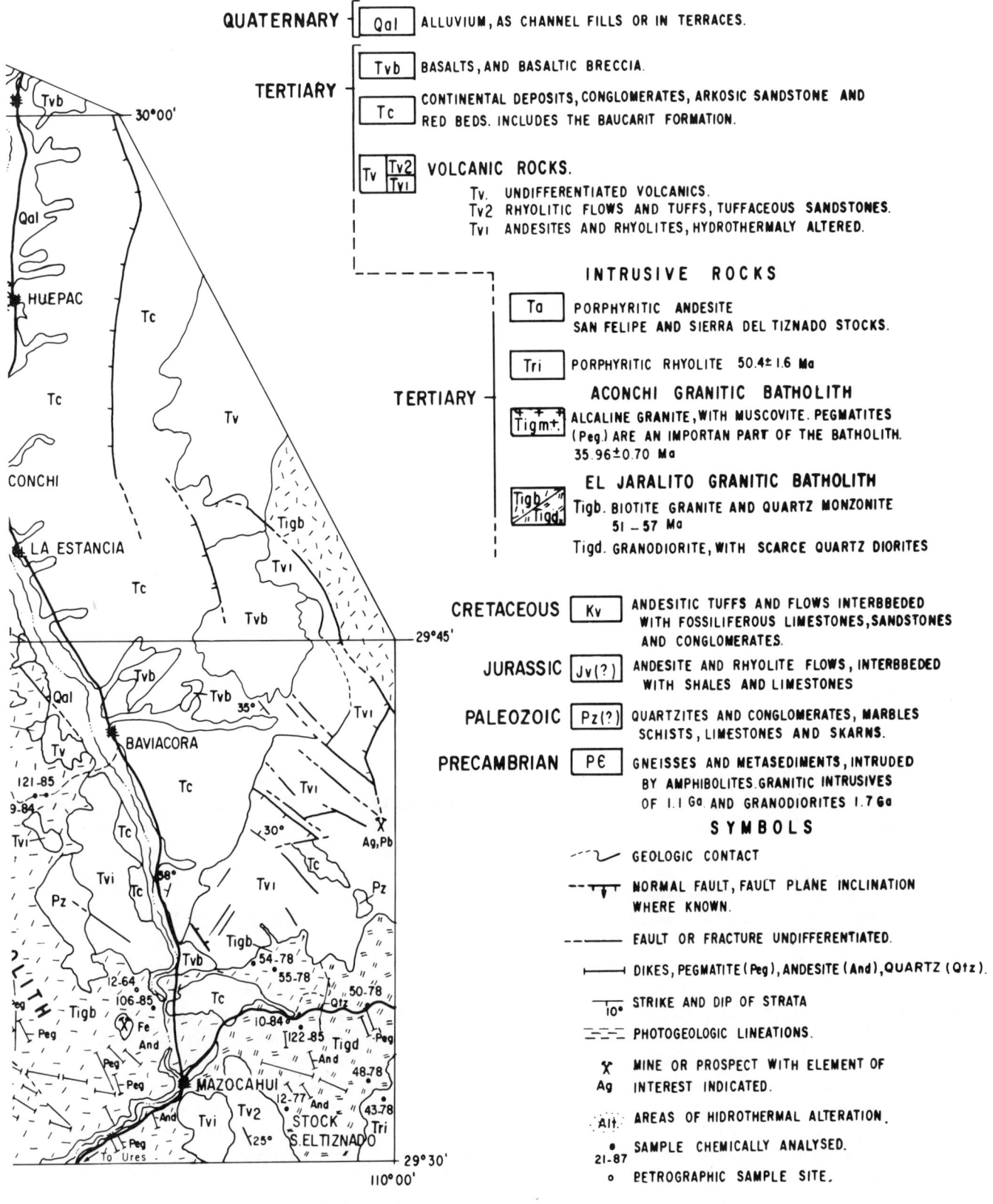

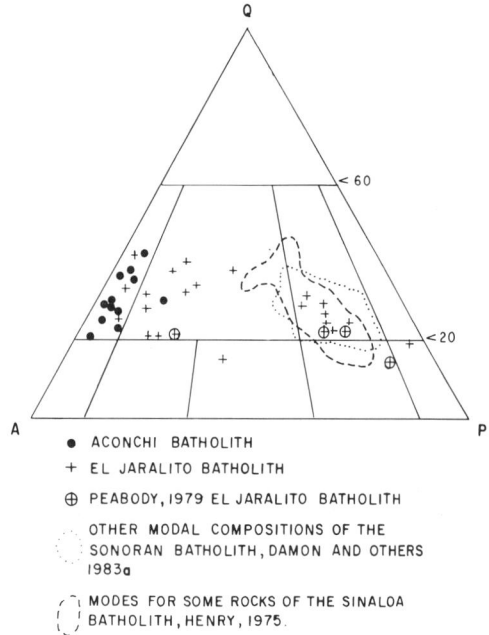

Figure 3. Modal compositions for 37 samples from El Jaralito and Aconchi batholiths. Classification of Streckeisen (1974). Six hundred points per sample were counted. (Q) quartz, (P) plagioclase, (A) alkali feldspar and albite.

the Aconchi batholith is fine grained, and muscovite generally predominates over the biotite. In Arroyo Peñasco (2.5 km northwest of Rancho El Torreón), pegmatitic lenses grade into the alkali granite with abundant red garnet. The relative ages for the two batholiths are confirmed by the younger radiometric age for the Aconchi batholith and numerous muscovite-garnet pegmatite and aplite dikes that intrude the El Jaralito pluton in most of its exposure area. Small bodies of muscovite granite intrude the El Jaralito batholith in the El Jaralito area.

The Aconchi intrusion appears to cut the El Jaralito batholith, although actual contacts have not been observed. The contact inferred within a few hundred meters south of the Sierra de Aconchi, west of Sierra Los Locos, is interpreted from aerial photographs. To the northeast of the Sierra de Aconchi, the batholith intrudes Precambrian gneisses and metasediments. East of Rancho El Llano, it is in fault contact with Tertiary clastic sediments or Cretaceous rocks (Fig. 2).

Of 12 point-counted thin sections of rocks of the Aconchi batholith, only one is granite sensu stricto; the others are alkali granites according to their modal compositions (Fig. 3). The alkali granites have a hypidiomorphic holocrystalline texture, commonly porphyritic. Their mineralogy is simple, with microcline, orthoclase, albite, and quartz as major constituents (Fig. 4B). The principal accessory minerals are muscovite and red garnet. Zircon and opaque minerals occur as minor constituents. The most common secondary minerals are sericite, chlorite, and epidote. In a few places, like that near Rancho Los Cuates in the northern end of Sierra Los Locos, this pluton displays cataclastic deformation where the rock is brecciated and approaches being a true mylonite.

PORPHYRITIC RHYOLITE STOCKS

Two rhyolitic stocks are exposed (Fig. 2), one north of the Sierra de Aconchi and another in the southeast corner of the map area. The stock in the southeast corner of the map area intrudes granodiorites of the El Jaralito batholith, but most of it is outside of the mapped area. Similar rocks crop out northwest of San Felipe.

Microscopically it shows phenocrysts of orthoclase, corroded quartz, and biotite. Contrasting with the batholiths, the porphyries have fine-grained groundmass as an important constituent. No point counting was done on these rocks.

Hydrothermal alteration and spatially related Ag, Pb, and Zn veins are associated with both intrusions. Silicification and quartz-sericitic alteration occur mainly in the San Felipe stock (Roldán, 1979).

Andesitic stocks and dikes were not studied in any detail; field relations suggest they are the youngest intrusions (Fig. 2). Andesitic dikes strike northwest and are well exposed along the road from Mazocahui to Ures (30 km southwest of Mazocahui and outside the geologic map area).

CHEMISTRY OF THE BATHOLITHS AND STOCKS

Eleven chemical analyses of the El Jaralito batholith were obtained (Table 1). An AFM diagram showing the chemistry of the Aconchi and El Jaralito batholiths is presented in Figure 5. The CIPW norms show unusually high values of corundum (samples 54-78, 50-78, and 55-78). This could be due to the presence of modal mica (Barker, 1983) or the effect of hydrothermal alteration (Charoy, 1986).

From Figure 6, the variation of SiO_2 versus the oxides of Na, Ca, Mg, Fe, and Al for the El Jaralito batholith form smooth curves. It is also evident that some overlap between the two batholiths exists above 71 percent SiO_2. However, from field relations and geochronology it is clear that the overlap does not have a genetic meaning.

Table 2 lists four chemical analyses of the Aconchi batholith. From the Harker variation diagram (Fig. 6), it is clear that the rocks of the Aconchi batholith are generally richer in SiO_2 than those of the El Jaralito batholith. In the four samples reported for the Aconchi batholith, content varies from 71 to 74 percent. This is a narrow interval compared with El Jaralito batholith, which span from 54 to 73 percent.

The Aconchi batholith is similar in chemistry to some middle Cretaceous–early Tertiary peraluminous granites described from the southwestern United States (Table 3). It is evident that more chemical analyses are needed for a better understanding of the chemistry of the Aconchi batholith.

For the porphyritic rhyolite stocks, the two chemical anal-

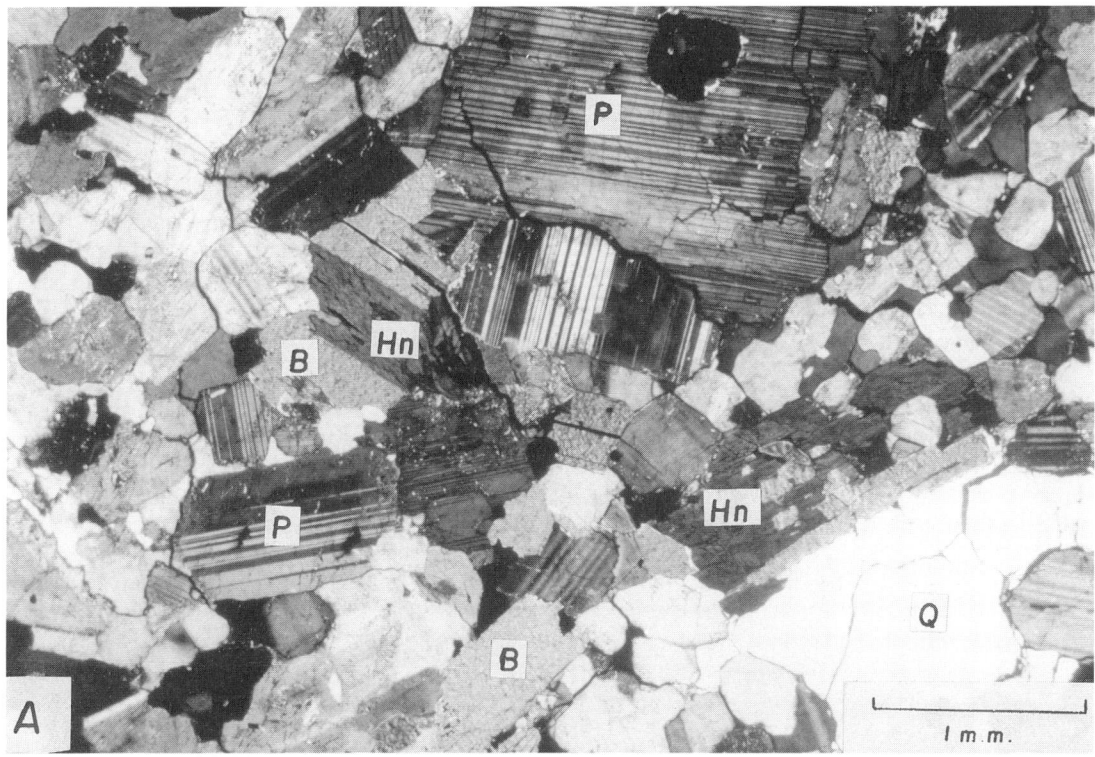

Figure 4. Photomicrographs of rocks from the batholiths. A. Granodiorite from the El Jaralito batholith: (P) plagioclase, (Q) quartz, (Hn) hornblende, (B) biotite. Crossed polars. B. Alkali granite of the Aconchi batholith, (m) microcline, (Q) quartz, (P) plagioclase, (Mu) muscovite. Crossed polars.

yses (Table 4) have the highest SiO_2 values of the area. In the silica variation diagram (Fig. 6), these rocks continue the trends of the alkali granites of the Aconchi batholith.

ISOTOPIC DATA

Only a few isotopic ages have been reported for the El Jaralito batholith; not all the lithologies have been dated. They range from 51.8 to 69.6 ± 2 Ma (Mead, 1982; Anderson and others, 1980; Gastil, written communication, 1986). Mead and Kesler (1984) reported an age of 46.6 Ma for skarn in Mina San Antonio, near El Jaralito. This younger age could be related to pegmatites or other granitic rocks within the Aconchi batholith in this area.

The only isotopic age for the Aconchi batholith is by Damon (*in* Roldán, 1979), who obtained a K/Ar date of 35.96 ± 0.70 Ma for a sample from Cajón de Las Casas in the northeastern portion of Sierra Aconchi. The Aconchi batholith corresponds to the peraluminous two-mica granites reported by Damon and others (1983a), who showed that the ages of these rocks in Sonora vary from 40 to 65 Ma.

For the San Felipe stock, Damon and others (1983b) reported a K/Ar age from orthoclase as 50.4 ± 1.6 Ma.

CONCLUSIONS

The intrusive rocks of the Aconchi–El Jaralito region are divided into three large units: the granitic-granodioritic batholith of El Jaralito; the alkali granite of the Aconchi batholith, and the porphyritic rhyolite stocks of San Felipe and Sierra del Tiznado. The batholiths have been barely unroofed, as evidenced by the numerous roof pendants exposed. Outcrop dimensions do not represent the true size of the batholiths because most of the exposed contacts are tectonic. Hence, these batholiths must be much larger.

TABLE 1. CHEMICAL ANALYSES AND CIPW NORMS FOR ROCKS OF THE EL JARALITO GRANITIC BATHOLITH*

Sample No.	29-84[†]	277-86[§]	54-78[†]	4-84[†]	50-78[†]	55-78[†]	121-85[§]	122-85[§]	211-86[**]	106-85[§]	12-77[§]
SiO_2	73.40	72.00	71.02	70.84	66.94	65.44	62.54	62.43	58.34	60.85	56.96
TiO_2	0.24	0.19	0.27	0.14	0.55	0.72	0.45	0.54	1.03	0.63	0.76
Al_2O_3	13.24	14.42	16.59	15.78	16.22	16.62	15.18	15.37	17.21	15.65	17.55
Fe_2O_3	0.83	1.60	1.54	0.78	3.07	2.06	1.78	2.25	5.36	2.70	3.28
FeO	0.95	....	1.14	0.76	1.52	2.31	3.27	2.92	....	3.39	4.04
MnO	0.05	0.04	0.04	0.02	0.04	0.04	0.13	0.10	0.08	0.11	0.15
MgO	0.76	0.58	0.79	0.55	1.74	1.91	2.78	2.47	3.09	2.68	3.14
CaO	1.52	1.12	0.97	2.03	3.01	3.97	4.23	4.48	5.72	4.88	5.56
Na_2O	4.20	3.46	2.35	4.50	2.44	2.35	3.99	3.87	4.40	3.59	4.36
K_2O	4.30	4.76	3.00	4.10	2.90	2.60	2.78	3.40	1.96	3.55	2.25
P_2O_5	0.10	0.07	0.16	0.10	0.16	0.24	0.23	0.23	0.25	0.40	0.39
H_2O^+	0.66	0.37	1.28	0.67	0.99	0.68	1.95	0.65	0.70	0.66	0.66
H_2O^-	0.05	....	0.16	0.07	0.05	0.50	0.09	0.08	....	0.05	0.08
Total	100.30	98.61	99.31	100.34	99.63	99.44	99.40	98.79	98.14	99.14	99.18
CIPW Norms											
Qz	28.56	30.28	42.85	23.83	33.05	30.49	15.20	14.33	8.13	12.80	6.52
C	0.00	0.00	8.10	0.49	3.98	3.30	0.00	0.00	0.00	0.00	0.00
Or	25.41	28.60	17.73	24.23	17.14	15.30	16.43	20.09	11.87	20.98	13.30
Ab	35.54	29.78	19.88	38.08	20.65	19.88	33.76	32.75	38.18	30.38	36.89
An	4.57	5.23	3.77	9.42	13.89	18.13	15.30	14.53	21.97	16.10	21.67
Hm	0.00	0.00	0.00	0.00	0.70	0.00	0.00	0.00	0.00	0.00	0.00
Hy	1.75	3.21	2.42	1.93	4.33	6.18	9.30	6.53	10.76	7.69	10.24
Mt	1.20	0.38	2.23	1.13	3.43	2.99	2.58	3.26	1.29	3.91	4.76
Il	0.46	0.37	0.51	0.27	1.04	1.37	0.85	1.03	2.01	1.20	1.44
Ap	0.23	0.17	0.37	0.23	0.37	0.56	0.54	0.54	0.61	0.93	0.91
Di	0.00	0.00	0.00	0.00	0.00	0.00	3.40	5.00	4.59	4.44	2.72
Total	97.72	99.74	97.86	99.61	98.58	98.20	97.36	98.06	99.40	98.43	98.45

*Location of samples is shown in Figure 2.
[†]Instituto de Geología, UNAM (Atomic absorption).
[§]Université D'Aix Marseille III, Lab. Petrol., Marseille, France (Atomic absorption).
[**]San Diego State University, Department of Geological Sciences (X-ray fluorescence).

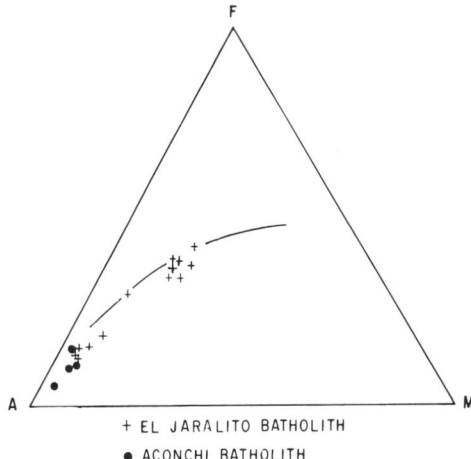

Figure 5. AMF diagram for rocks of the batholiths of Aconchi and El Jaralito, compared with a generalized trend for the Baja California batholith (Larsen, 1948). (A) $Na_2O + K_2O$; (F) $FeO + 0.9\ Fe_2O_3$; (M) MgO.

TABLE 2. CHEMICAL ANALYSES AND CIPW NORMS FOR ROCKS OF THE ACONCHI GRANITIC BATHOLITH

Sample No.	222-86*	223-86*	127-85†	214-86†
SiO_2	74.63	73.54	73.02	71.59
TiO_2	0.02	0.11	0.11	0.10
Al_2O_3	13.66	13.91	11.90	14.88
Fe_2O_3	0.53	1.06	0.53	1.01
FeO	....	....	0.87	....
MnO	0.06	0.05	0.06	0.19
MgO	0.30	0.48	0.62	0.40
CaO	0.63	1.21	1.63	0.88
Na_2O	3.67	3.96	4.30	3.77
K_2O	4.48	3.57	4.15	4.03
P_2O_5	0.04	0.06	0.13	0.24
H_2O	0.52	0.46	0.97	1.00
H_2O^-	....	....	0.05	....
Total	98.54	98.41	98.34	98.09
CIPW Norms				
Qz	34.75	33.78	28.94	32.62
C	1.75	1.49	0.00	3.33
Or	26.98	21.51	24.52	24.50
Ab	31.65	34.18	36.38	32.83
An	2.95	5.76	0.91	3.04
Hy	1.50	2.45	0.10	2.49
Mt	0.13	0.25	0.77	0.24
Il	0.04	0.21	0.21	0.20
Ap	0.10	0.15	0.30	0.59
Di	0.00	0.00	5.17	0.00
Total	99.85	99.78	97.30	99.84

*San Diego State University, Department of Geological Sciences (x-ray fluorescence).
†Université D'Aix Marseille III, Lab. Petrol., Marseille, France (Atomic absorption).

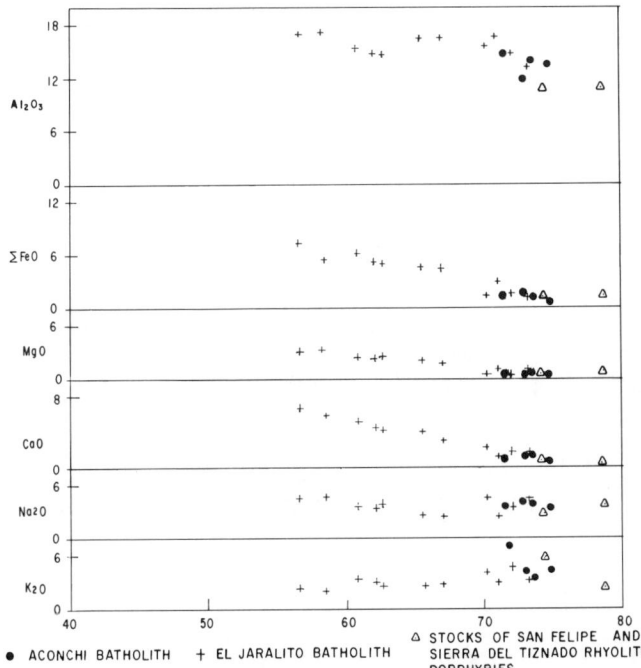

Figure 6. Harker silica-variation diagrams for analyzed samples of the Aconchi and El Jaralito batholiths, and the rhyolitic stocks of Sierra del Tiznado and San Felipe.

The El Jaralito batholith is a large plutonic body formed by multiple intrusive episodes, as shown by isotopic ages ranging from 51 to 69 Ma. However, chemical analyses from a variety of facies of the batholith suggest that the rocks follow the same trend. Damon and others (1983a) compared the modal composition of the "Sonoran Laramide Batholith" with the better known rocks of the Sierra Nevada batholith. They found a correspondence of the granitic and granodioritic facies of the El Jaralito batholith with those of the Sierra Nevada batholith; however, the latter is much older. The "Sonoran Laramide Batholith" is calc-alkaline and has been attributed to subduction processes by Damon and others (1983a). There is no clear separation in the chemistry of the batholiths of El Jaralito and Aconchi (Fig. 6). Instead, there is some overlap in major oxide trends above 71 percent SiO_2. From this information it is concluded that the rocks of El Jaralito and Aconchi batholiths, with only four analyses, fall along the same trend in Figure 6. Trace-element information would be useful in establishing genetic relations. This chemical overlap seems to be common in other batholiths, e.g., in granites of the Sierra Nevada and alkali granites of southern California (Rogers and Greenberg, 1981).

The muscovite-bearing alkali granite of the Aconchi batholith is of the type described by Miller and Bradfish (1980) in Arizona, eastern California, and Nevada. In northern Sonora, muscovite-bearing granites have been reported in Sierra Guaucomea by Anderson and others (1977), in Sierra de Magdalena, and possibly in the Sierra de la Madera, northeast of Magdalena, Sonora (Nourse, personal communication, 1988). In the western Cordillera, such rocks are considered as S-type granitoids, according to the classification of Chappell and White (1974).

Cataclastic deformation was observed within both batholiths, especially in the area of Puerta del Sol (Anderson and others, 1980), and in isolated localities such as Rancho Zunibiate, El Jaralito, and the northern portion of the Aconchi batholith. True mylonites were observed in only small areas where the granitic rocks show lineation. The structure of these deformed rocks was not studied in detail.

Chemical and isotopic data for these batholiths remain incomplete. Future studies should focus on areas where no information is available at present. It is highly recommended that future chemical data include trace-element concentrations.

Mineralization associated with the El Jaralito batholith consists of contact deposits of W and Fe. Some veins of Pb and Ag are hosted in granites of this batholith. Also, pegmatites containing beryl intrude the El Jaralito batholith (Fig. 2).

TABLE 3. COMPARATIVE CHEMICAL ANALYSES OF PERALUMINOUS GRANITES IN SOUTHWESTERN UNITED STATES AND CENTRAL SONORA

Location	SiO_2 (%)	Al_2O_3 (%)	CaO (%)	Na_2O (%)	K_2O (%)
Sta. Catalina Mts., Arizona*	72.9	14.6	1.6	4.1	3.7
Whipple Mts., California*	71.8	15.0	2.2	4.1	3.4
Ruby Mts., Nevada*	73.0	13.8	1.1	3.4	4.5
Aconchi batholith					
Rancho La Rosca[†]	71.59	14.88	0.88	3.77	4.03
1.5 km NW of Los Cuates[†]	74.63	13.66	0.63	3.67	4.48
5 km NW of El Torreón[†]	73.54	13.91	1.21	3.96	3.57
5.3 km NW of Agua Caliente[†]	73.02	11.90	1.63	4.30	4.15

*Data from Reynolds and Keith (1982)
[†]Chemical analyses shown in this report

TABLE 4. CHEMICAL ANALYSES AND CIPW NORMS FOR THE RHYOLITE STOCKS OF SIERRA EL TIZNADO AND SAN FELIPE

Sample No.	43-78*	126-85[†]
SiO_2	78.65	74.31
TiO_2	0.01	0.05
Al_2O_3	11.02	11.55
Fe_2O_3	1.36	0.91
FeO	0.24	0.65
MnO	0.01	0.10
MgO	0.71	0.52
CaO	0.53	1.03
Na_2O	3.50	2.63
K_2O	2.30	5.85
P_2O_5	0.20	0.02
H_2O^+	1.25	1.20
H_2O^-	0.26	0.10
Total	100.04	98.92
CIPW Norms		
Qz	47.85	33.99
C	2.29	0.00
Or	13.59	34.57
Ab	29.62	22.25
An	1.32	2.43
Hy	1.77	0.85
Mt	0.78	1.32
Hm	0.82	0.00
Il	0.02	0.09
Ap	0.47	0.05
Di	0.00	2.05
Total	98.53	97.60

*Instituto de Geología, UNAM (Atomic absorption).
[†]Université D'Aix Marseille III, Lab. Petrol., Marseille, France (Atomic absorption).

The most important mineralization related to the Aconchi batholith is associated with the pegmatites. It includes beryl and nonmetallic minerals such as muscovite, quartz, and microcline.

Veins and contact deposits of Pb, Ag, Zn, and Au are associated with the stocks of San Felipe and Sierra del Tiznado. In the area of Los Amoles there is U mineralization hosted in altered andesite near the contact with the El Jaralito batholith.

ACKNOWLEDGMENTS

This work was financed by the Institute of Geology of the National University of Mexico, as one of the research programs of its "Estación Regional del Noroeste" in Hermosillo, Sonora. I thank all of the people in support services of the regional office for the help they have provided. I would like to give special thanks to Paul E. Damon, University of Arizona, the pioneer of geochronological work in Sonora, who started his work more than 20 years ago. He invited me to cooperate with him on some field aspects of the study of the batholiths in Sonora. He also dated some rocks of the San Felipe District that are reported here. Alain Demant and Jean Jacques Cochemé of the "Université D'Aix-Marseille III," kindly did some of the chemical analysis and calculated the norms. They also provided good advice during one of their visits to Sonora. I thank the chemists, Angel Rodríguez M. and Irma Aguilera O. for some of the whole-rock analyses performed in the laboratories of the Institute in Mexico City. Gordon Gastil, San Diego State University, made the necessary arrangements for processing there some of the analyses by Joan Calhoun, analyst of the Department of Geological Sciences. Two radiometric dates were also obtained there; help of Gastil and Calhoun is very much appreciated. César Jacques A., from our regional office, kindly read the first draft of the manuscript and provided suggestions that improved its clarity. During different periods of field work, I was accompanied by Wilfrido Salazar, from the town of Baviácora. Juan Carlos García and Enrique Castillo also acted as field companions. To all of these persons, thanks. I would like to thank F. W. McDowell, J. Jorge Aranda-Gómez, and W. E. Elston for their helpful reviews of the manuscript.

REFERENCES CITED

Anderson, T. H., and Silver, L. T., 1974, Late Cretaceous plutonism in Sonora, Mexico, and its relationship to circum-Pacific magmatism: Geological Society of America Abstracts with Programs, v. 6, p. 484.

——, 1979, The role of the Mojave-Sonora megashear in the tectonic evolution of northern Sonora, in Anderson, T. H., and Roldán, J., eds., Geology of northern Sonora; Geological Society of America Guidebook, field trip 27: San Diego, California, San Diego State University.

Anderson, T. H., Silver, L. T. and Salas, G. A., 1977, Metamorphic complexes of the southern part of the North American Cordillera, northwestern Mexico: Geological Society of America Abstracts with Programs, v. 9, p. 881.

——, 1980, Distribution and U-Pb isotope ages of some lineated plutons, northwestern Mexico, in Crittenden, M. D., Jr., Coney, P. J., and Davis, G. H., Cordilleran metamorphic core complexes: Geological Society of America Memoir 153, p. 269-283.

Barker, S., 1983, Igneous rocks: Englewood Cliffs, New Jersey, Prentice-Hall Inc., p. 72-73.

Bateman, P. C., and Busacca, J., 1983, Millerton Lake Quadrangle, west-central Sierra Nevada California; Analytic data: U.S. Geological Survey Professional Paper 1261, p. 1-7.

Bateman, P. C., and Clark, L. D., 1974, Stratigraphic and structural setting of the Sierra Nevada batholith, California: Pacific Geology, v. 8, p. 79-89.

Chappell, B. W., and White, A.J.R., 1974, Two contrasting granite types: Pacific Geology, v. 8, p. 173-174.

Charoy, B., 1986, The genesis of the Cornubian batholith, southwest England; The example of the Carnmenellis pluton: Journal of Petrology, v. 27, part 3, p. 571-604.

Chávez-Aguirre, J. M., 1978, Geologie et metallogenie de la Sierra D'Aconchi (Sonora, Mexique) [These de Docteur-Ingenieur]: Paris, France, Universite Pierre et Marie Curie, Paris VI, 202 p.

Damon, P. E., Shafiqullah, M., and Clark, K. F., 1981, Age trends of igneous activity in relation to metallogenesis in the southern Cordillera: Arizona Geological Society Digest, v. 14, p. 137-154.

Damon, P. E., Shafiqullah, M., Roldán, Q. J., and Cochemé, J. J., 1983a, El Batolito Laramide (90-40 m.a.) de Sonora: Asociación de Ingenieros de Minas, Metalurgistas y Geólogos de México, A.C., Memoria de la XV Convención Nacional, Guadalajara, Jal., p. 63-95.

Damon, P. E., Shafiqullah, M., and Clark, K. F., 1983b, Geochronology of the porphyry copper deposits and related mineralization of Mexico: Canadian Journal of Earth Sciences, v. 20, p. 1052-1071.

Hamilton, W., and Bradley, M. W., 1967, The nature of batholiths: U.S. Geological Survey Professional Paper 554-C, p. 1-26.

Henry, C. D., 1975, Geology and geochronology of the granitic batholithic complex, Sinaloa, Mexico [Ph.D. thesis]: Austin, University of Texas, 159 p.

Larsen, E. S., Jr., 1948, Batholith and associated rocks of Corona, Elsinore, and San Luis Rey Quadrangles, southern California: Geological Society of America Memoir 29, 182 p.

Mead, R. D., 1982, Summary of K-Ar isotope ages determined at Ohio State University: Unpublished report, 12 p.

Mead, R. D., and Kesler, S. E., 1984, Relation of Sonoran tungsten mineralization to the metallogenic evolution of Mexico: Geological Society of America Abstracts with Programs, v. 16, p. 591-592.

Miller, F. C., and Bradfish, J. L., 1980, An inner Cordilleran belt of muscovite-bearing plutons: Geology, v. 8, p. 412-416.

Mills, A. R., and Hokuto, A. C., 1971, Geología y potencial mineral del tungsteno del área de Baviácora, Sonora: Asociación de Ingenieros de Minas, Geólogos y Metalurgistas de Mexico, IX Convención Nacional, Hermosillo, Sonora, Memoria, p. 491-503.

Peabody, C. E., 1979, Geology and petrology of a tungsten skarn; El Jaralito, Baviácora, Sonora, Mexico [M.S. thesis]: Stanford, California, Stanford University, 90 p.

Pedrazzini de Schlaepfer, C., 1961, Estudio geológico y petrológico de la pegmatita de El Batamote en la Sierra de Aconchi, Municipio de Baviácora, Estado de Sonora [B.Sc. thesis]: Mexico City, Mexico, Facultad de Ingeniería, U.N.A.M., 90 p.

Pérez-Segura, E., 1985, Carta Metalogenética de Sonora 1:250,000: Dirección de Minería, Geología y Energéticos del Gobierno del Estado y Depto. de Geología UNISON, Publicación No. 7, 65 p.

Raisz, E., 1964, Landforms of Mexico: Cambridge, Massachusetts, map with text, scale approximately 1:3,000,000.

Reynolds, S., and Keith, S. B., 1982, Chemistry and mineral potential of peraluminous granitoids: Arizona Bureau of Geology and Mineral Technology Fieldnotes, v. 12, no. 4, p. 4–6.

Rodríguez-C., J. L., 1984, Geology of the Tuape region, north-central Sonora, Mexico [M.S. thesis]: Pittsburgh, Pennsylvania, University of Pittsburgh, 157 p.

Rogers, J., Del, J. W., and Greenberg, J. K., 1981, Trace elements in continental-margin magmatism; Part 3, Alkali granites and their relationship to cratonization: Geological Society of America Bulletin, part 1, v. 92, p. 6–9.

Roldán-Q., J., 1979, Geología y yacimientos minerales del Distrito de San Felipe, Sonora: Universidad Nacional Autónoma de México, Instituto de Geología, Revista, v. 3, no. 2, p. 97–115.

——, 1989, Geología de la Hoja Baviácora en Sonora centro-oriental: Universidad Nacional Autónoma de México, Instituto de Geología, Revista, v. 8, no. 1, p. 1–14.

Sillitoe, R. H., 1976, A reconnaissance of the Mexican copper belt: Transactions, Section B of the Institution of Mining and Metallurgy, v. 85, p. 169–190.

Streckeisen, A., 1974, Classification and nomenclature of plutonic rocks: Geologische Rundschau, v. 63, no. 2, p. 773–786.

Vassallo, F., L., 1985, Sobre la evolución geológica de la parte central del Estado de Sonora-México y su relación con los depósitos de grafito: Instituto de Geología, U.N.A.M., Memoria Simposio sobre floras del Triásico Tardío, su fitogeografía y paleontología, p. 87–100.

MANUSCRIPT ACCEPTED BY THE SOCIETY APRIL 18, 1990

Geology of the Yécora area, northern Sierra Madre Occidental, Mexico

Jean-Jacques Cochemé and Alain Demant
Laboratoire de pétrologie magmatique, URA 1277, Faculté des Sciences de St. Jérôme, Université d'Aix-Marseille III, 13397 Marseille Cedex 13, France

ABSTRACT

This study presents new field, structural, petrographic, and geochemical data on the Cenozoic volcanic rocks of the northwestern Sierra Madre Occidental (SMO). Since the Mesozoic, this region has been the site of three successive magmatic events: (1) A calcalkaline sequence (90 to 40 Ma), which in Sonora is characterized by the emplacement of the granitoids of the Cordilleran coastal batholiths coeval with the subduction of the East Pacific plate under the North American continent. (2) An Eocene-Oligocene (around 35 Ma) volcanic sequence represented by calcalkaline andesites and high-K ignimbrites. The plagioclase-biotite phenocrystic assemblage is typical of the ignimbrites of this sequence in the northern SMO. Some of these rocks are caldera-related but they are much thinner and restricted in exposure than those in the southernmost areas. This magmatic activity probably accounts for much of the mineralization in the region. (3) A bimodal Oligocene-Miocene sequence (30 to 17 Ma) consisting of successive eruptions of basalt and basaltic andesite associated with acidic calcalkaline tuffs and lavas. Basic rocks form the predominant outcrops in the northern area of the SMO. The trap-forming flows were extruded from fissures and small calderas. These mafic rocks are transitional between the calcalkaline and tholeiitic series and show a progressive change from orogenic to anorogenic affinities with time. This volcanic event is related to pre–Basin and Range normal-faulting tectonism. Extensional tectonism at 17 Ma accounts for the modern basin-range morphology, which is well developed in the northern SMO. The volcanism progressively disappeared along the central axis of the SMO during this time. Quaternary volcanism is rare in Sonora; there are scattered basaltic fields such as the Quaternary lavas of Moctezuma, which crop out in central Sonora. Basalts are typically anorogenic and coeval with strike-slip faults, which affected the whole western margin of the North American continent, and with the opening of the Gulf of California. Based on the age and geochemical characteristics of the studied rocks, it is possible to conclude that, since Mesozoic time, the magmatism in the northern SMO probably reflects successive changes in the geodynamic regime.

INTRODUCTION

The Sierra Madre Occidental (SMO) is a 1,400-km-long and 300-km-wide physiographic province representing one-sixth of Mexico's surface and extending from the U.S.–Mexico border on the north to the Trans-Mexican volcanic belt (latitude 20°N) on the south. The SMO is a volcanic plateau mainly made up of Cenozoic ash-flow tuffs and with an average altitude of ca. 2,000 m above sea level. A Cenozoic crustal extensional phase has disrupted the plateau into a characteristic Basin and Range morphology both east- and westward. An impressive relief exists along the western slope of the SMO, with excellent exposures of the volcanic sequences.

Cochemé, J.-J., and Demant, A., 1991, Geology of the Yécora area, northern Sierra Madre Occidental, Mexico, *in* Pérez-Segura, E., and Jacques-Ayala, C., eds., Studies of Sonoran geology: Geological Society of America Special Paper 254.

The abundance of rhyolitic rocks in the SMO has been known since the beginning of this century (Ordoñez, 1900). Summaries of the recent work on the volcanic rocks can be found in the papers by McDowell and Clabaugh (1979), Cameron and others (1980a, b), Cameron and Hanson (1982), and Swanson and McDowell (1984). Recent studies in the Yécora-Maicoba-Mulatos area consist only of reconnaissance geology of the Yécora-Ocampo area by Bockoven (1980). These previous studies in the SMO have established that an Upper Volcanic Complex (UVC) composed mainly of Oligocene rhyodacitic ash-flow tuffs, locally more than 1.5 km thick, unconformably overlies a Late Cretaceous to Eocene complex, the Lower Volcanic Complex (LVC; McDowell and Keizer, 1977). The LVC is made up of plutonic and volcanic calcalkaline rocks, to which economic mineral deposits are related.

The mapped area (3,700 km^2 at a scale of 1:50,000) is located along the western slope of the SMO plateau in the state of Sonora, near its border with Chihuahua (Fig. 1).

MAGMATIC SEQUENCES

A generalized stratigraphic column of the area is given in Figure 2. The relations between the three magmatic sequences are as follows: the lower sequence, which partly correlates to the LVC, is mainly composed of Late Cretaceous to early Tertiary plutonic rocks, unconformably overlain by a predominantly ignimbritic Eocene-Oligocene sequence, covered in turn, by a basaltic Oligocene-Miocene sequence. In the basins, a continental molasse (the Báucarit Formation) of Miocene age overlies the basaltic sequence. After this depositional phase, volcanism resumed only along the borders of the SMO.

The lower sequence

This mainly plutonic sequence is part of the Western Mexican batholithic belt, which belongs to the Cordilleran coastal batholiths. In the studied area, granodiorites and quartz-diorites of the lower sequence crop out at the Santa Rosa–San Nicolás area, as well as east of Maicoba. The plutonic suite intruded and produced contact metamorphism on previously tectonized country rocks affected by low-grade regional metamorphism.

In Sonora, the granitoids crop out over a total area of ca. 19,000 km^2, corresponding to more than 10 percent of the state's area and to 25 percent of the total outcrops excluding the graben fillings. The age of this sequence ranges from 90 to 40 Ma (Damon and others, 1983a).

The Santa Rosa–San Nicolás area. East of Tepoca, a granodioritic body crops out along 11 km of the new Hermosillo–Ciudad Obregón road. At San Nicolás, the granodiorite is microgranular on its margin, and it has been dated at 49.6 Ma (Damon and others, 1983a). Its subvolcanic character is marked by its well-preserved dome-shaped top, which is cut by subhorizontal mafic dikes.

The main ore deposits in the area are porphyry copper, veins, and other mineralization related to contact metamorphism. At Los Verdes mine, 1.5 km north of Santa Ana, a Mo-Cu-Wo-U deposit occurs in a tourmaline breccia pipe within the granodioritic dome (K-Ar average age on biotite + amphibole = 66.3 Ma; Damon and others, 1983a). Several roof pendants of limestone have been mined for scheelite at San Nicholás. At Santa Rosa, mining is active today on the Mo-porphyry of Tres Piedras and at the Ag-veins of La Trinidad.

The Maicoba area. The Maicoba granodiorite is found at 1,600 m above sea level in a horst of the SMO where it has been dated at 63.6 Ma (K-Ar age on biotite; Bockoven, 1980). The marginal zone is microgranular, whereas the main body has hypidiomorphic granular texture composed of andesine, biotite, amphibole, and scarce clinopyroxene. Some quartz and K-feldspar are present as interstitial phases. The Maicoba granodiorite intrudes a strongly deformed and metamorphosed andesitic sequence, with Ag, Cu, and Au as main ore deposits. The pluton shows a characteristic morphology, with a central topographic low and a double system of dikes (N25 and N125).

The andesites appear as flows, breccias, and tuffs, quite often reworked into volcano-sedimentary complexes. Some minor dacites and rhyolites also occur in the sequence. The andesites, found in close spatial association with the plutons, contain labradorite, two pyroxenes, and hornblende and show regional, low-grade metamorphic assemblages characterized by the association chlorite-calcite-albite-epidote. As first pointed out by Hawkins (1970) and then by McDowell and Keizer (1977), field and age relations, as well as petrographic and geochemical characteristics, suggest that these rocks belong to the same calcalkaline magmatic complex.

The Eocene-Oligocene sequence

The Eocene-Oligocene sequence in the studied area can be divided into a chiefly andesitic lower member and an upper member consisting of ash-flow tuffs. A major unconformity separates this plateau-forming ignimbritic sequence from the LVC. In the mapped area, this unconformity can be observed near Maicoba where pebbles of granodiorite occur in a thin conglomerate stratigraphically correlatable with the Navosaigame conglomerate of the Ocampo-Moris area (Bockoven, 1980). The lower member begins with andesitic flows, tuffs, and breccias (Mulatos area, Santa Rosa area, and the volcano-sedimentary complex of La Iglesia) alternating with ignimbrites. Good exposures of this lower member are found in the section of the Rancho La Cueva (Fig. 3).

The typical phenocryst assemblage for the andesites is plagioclase (An 50) + amphibole. Some samples contain embayed quartz or clinopyroxene, and less commonly, orthopyroxene.

Although some authors have already mentioned the occurrence of post-Laramide andesites (Hisazumi, 1929), in the absence of detailed stratigraphic studies it is difficult to differentiate them from the Cretaceous ones. Moreover, their alteration precludes the application of the K-Ar dating method in most cases.

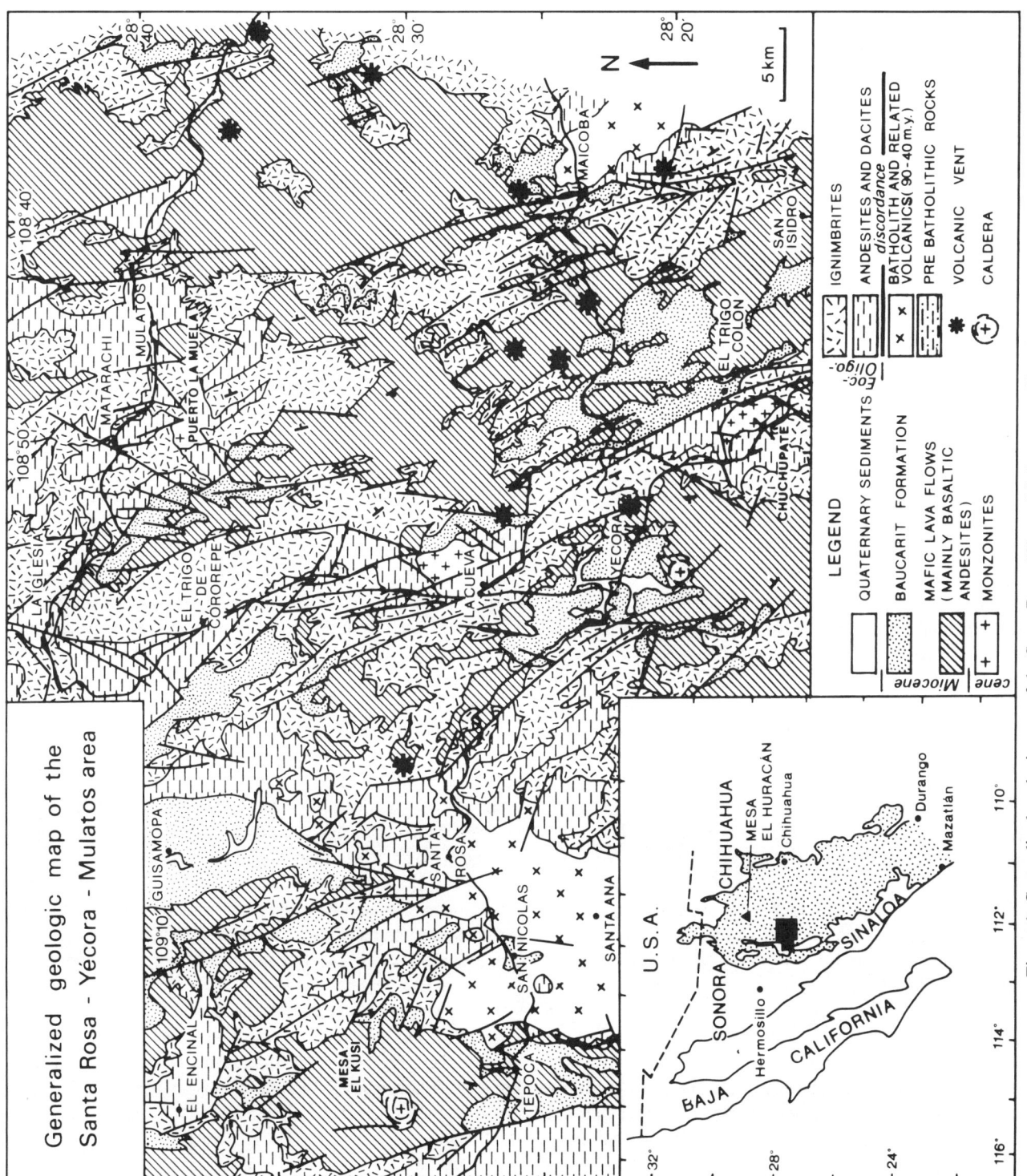

Figure 1. Generalized geologic map of the Santa Rosa–Yécora–Mulatos area. The stippled area in the location map (insert) represents Tertiary volcanic rocks of the SMO.

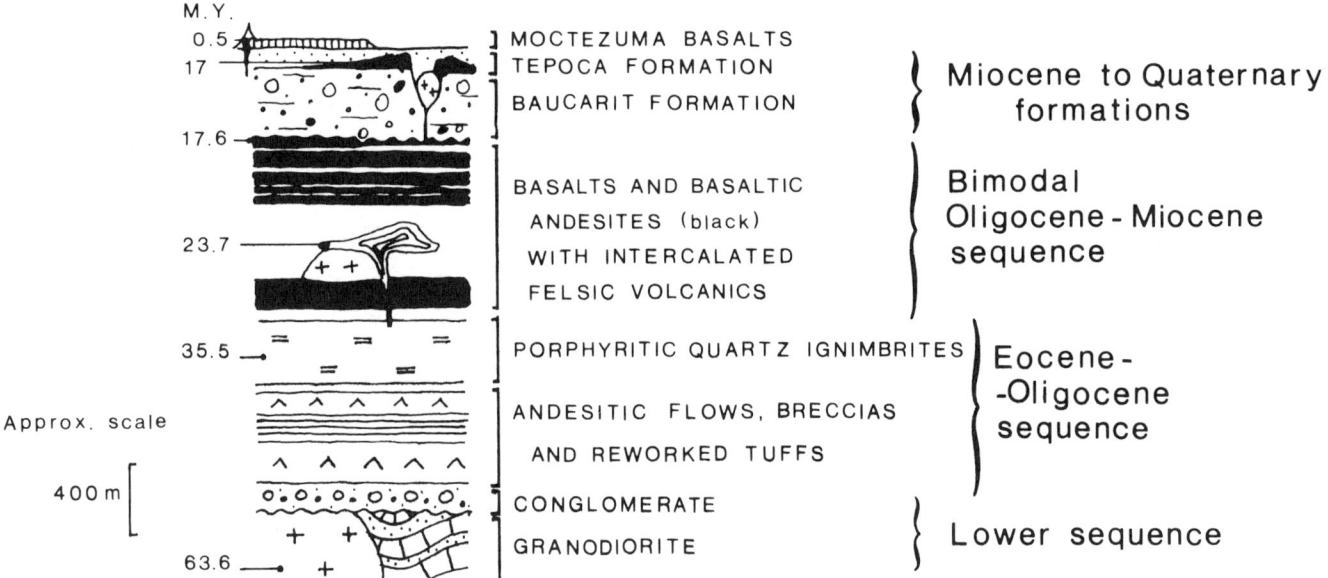

Figure 2. Generalized stratigraphic column of the area. K-Ar ages are from Bockoven (1980), Damon (unpublished data), and Montigny and others (1987).

However, Damon and others (1981) report an age of 31 Ma for a purple andesite west of Yécora.

The huge volume of rhyolitic tuffs in the central part of the SMO is commonly referred to as the Upper Volcanic Series and was erupted between 34 and 27 Ma (McDowell and Clabaugh, 1979). Ignimbritic flows followed the andesitic volcanism without a major discontinuity (Fig. 3). They show different petrographic facies and thicknesses, from rhyodacitic ignimbrites with Pl + Qtz + Bt + Hbl phenocrysts to rhyolitic ignimbrites with Qtz + Sa + Pl + Bt. Some of these flows are clearly related to monzonites (La Cueva, Puerto La Muela), forming possible cauldron structures with associated mineralization (gold mines of Mulatos; La Chipriona silver mine).

In the mapped area, one rhyolitic unit is best exposed and corresponds to the densely welded ash-flow tuff of Cerro La Cebadilla. The rock is typically reddish to light pink, xenolith-poor, and strongly porphyritic (30 to 40 percent phenocrysts) with quartz (14.5 percent), oligoclase (10 percent), biotite (2 percent), and scarce hornblende. Its maximum thickness is in excess of 200 m at Puerto La Muela, but no caldera structure could be outlined with certainty there.

Thus, the lavas of this Eocene-Oligocene episode form a continuous high-K suite from andesite to rhyolite (Fig. 11). Based on their mineralogic characteristics, chalcophile mineralization, and lack of Fe-enrichment, we regard the entire suite as calcalkaline.

The basaltic Oligo-Miocene sequence

Previous studies in the central and easternmost parts of the SMO have shown that almost everywhere, and particularly at the margin of the plateau, scattered outcrops of basaltic rocks overlie the ignimbrites of the Eocene-Oligocene sequence. In the mapped area, basic volcanic rocks are strikingly more widespread than the ignimbrites. On top of the Eocene-Oligocene sequence, a bimodal sequence formed by poorly welded ash-flow tuffs, flow-banded perlitic rhyolites, and dacites intercalated with basalts and basaltic andesite flows is present. Toward the upper part of this sequence, the rhyolites completely disappear, and flows of basaltic andesites appear instead. The sequence is morphologically analogous to the flood basalts of the continental tholeiitic provinces (Fig. 4). Basalts (SiO_2 <53 percent), basaltic andesites (53 <SiO_2 <57 percent; MgO <6 percent), and andesites (57 <SiO_2 <63 percent) have been distinguished by petrographic and chemical studies. However, all facies characteristically look like basalts (color, texture, mineralogy) and have similar phenocryst content; Plagioclase (dominates over Olivine in basaltic andesites) + Olivine ± clinopyroxene. In addition, some basaltic andesites contain phlogopite or pigeonite.

A sequence of basaltic rocks thicker than 300 m crops out at Rancho La Pinosa within the Rio Pilares–Rio Maicoba graben (Fig. 4). This NNW-SSE–trending graben is 15 km wide and 60 km long. At an elevation of 1,400 m it is bounded to the west by a normal fault of the Yécora basin and to the east by the steps of the Maicoba horst (Fig. 5). The flat-lying top of the sequence is contemporaneous with the Báucarit Formation. Some volcanic cones (Los Charcos–La Pinosa) are still visible north and west of Rancho La Pinosa. The basalts and basaltic andesites of this area have similar mineralogic (Ol, Pl, Cpx phenocrysts) and geochemical (low contents of MgO and CaO; high contents of Al_2O_3, K_2O, and P_2O_5) characteristics. However, a geochemical evolu-

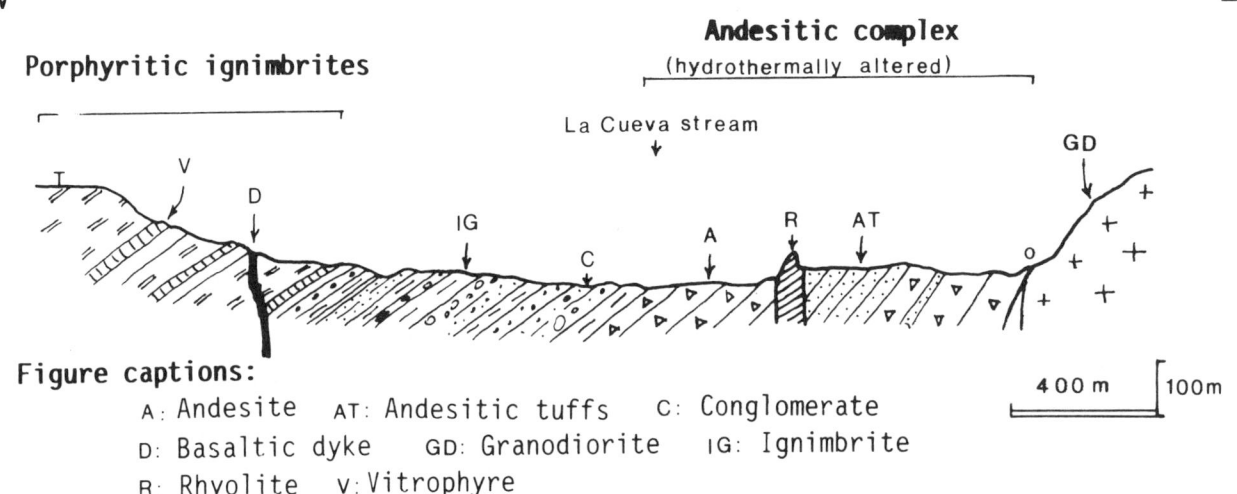

Figure 3. Geologic cross section of the La Cueva Formation.

tion appears within the sequence of Rancho La Pinosa according to the stratigraphic position of the flows. The Th/Ta ratio changes from 10 at the bottom to 1.4 at the top (Fig. 12). Th, transition elements, light REE, and alkali elements are more scarce at the bottom.

The acid rocks of the bimodal sequence are dacites, rhyodacites, and rhyolites (perlitic domes and rare ignimbrites). They are interbedded with basaltic flows. Most ash-flow tuffs seem to be related to the Sierra Chuchupate cauldron, as suggested by stratigraphic, petrographic, and geochemical relations.

The Sierra Chuchupate cauldron (Fig. 6) is located 12 km southeast of the town of Yécora. Along its northeastern margin, a micromonzonite plug intrudes an andesitic complex of unknown thickness. Au-mineralization occurs along the cauldron-related faults and within auto-brecciated and silicified ring and NNW-trending dikes. The micromonzonite is composed of andesine, orthoclase, hornblende, and biotite phenocrysts in a granophyric groundmass.

The bimodal volcanism is clearly related to extensional tectonics (pre–Basin and Range phase) as shown by a NNW alignment of volcanic feeders found in the area (dacitic or rhyolitic domes and basaltic cones) and by some marker flows that thicken toward the center of the Rio Pilares–Maicoba graben.

Mapping of the Nácori Chico–Mesa El Huracán area is in progress. Only a reconnaissance geologic cross section is presently available for comparison with the southernmost areas. Along this section (Fig. 10), basalts and basaltic andesites largely predominate, and practically no ignimbrite has been found. Some dacite and pumice tuffs are found within the sequence.

The basalts form a wide plateau disrupted by Basin and Range faults. Horizontal traps form Mesa Tres Rios and Mesa El Huracán, and the morphology presents striking similarities with many continental basaltic provinces of the world. About 13 km east of Nácori Chico, the thickness of the volcanic pile exceeds 800 m, and extends as far south as the town of Madera. Some basaltic cones are still preserved (Cueva Tres Rios). The oldest ages found on basalts or basaltic andesites of this episode, approximately 30 Ma (Swanson, 1977; Montigny and others, 1987), are those of Mesa El Huracán and Mesa Tres Rios in Chihuahua. This basaltic complex rests conformably upon apparently fresh andesites of probable Eocene age. Coarse diabasic textures are typical in the basalts. Magnesian olivine (Fo 70 to 75), labradorite, and clinopyroxene (Ca-rich augite) are the most common phenocrysts. Some phlogopite and orthopyroxene appear in a few samples as products of late crystallization. Characteristically, basalts and basaltic andesites are alkali-rich and have low MgO (<6 percent wt) and high Al_2O_3 (16 to 18 percent wt) contents. No iron enrichment has been established in the sequence.

Basalts and basaltic andesites with similar petrographic and geochemical characteristics have been found in Sierra del Nido, Chihuahua (Mauger, 1977, 1981; Zarate-Cruz, 1983), and possibly in Sierra del Gallego (Milagro basalt dated at 28.7 Ma; Keller,

Figure 4. Arroyo La Pinosa, 5 km east of Yécora town. The gorge is cut in horizontal basaltic andesite lava flows.

1977) and in the Cuauhtemoc area, where a basaltic flow has been dated at 27.3 Ma (Swanson, 1977). Cameron and others (1980b) have already mentioned the occurrence of similar basalts and basaltic andesites in Chihuahua, assigning them to a mildly alkalic basaltic suite.

In the southwestern United States, similar basaltic andesites are known at the Mogollon Plateau within the Bearwallow Mountain Formation (Elston and others, 1976) where they have been dated between 29 and 21 Ma (Elston and Northrop, 1976) and interpreted as products of a contamination of basalts by crustal material and differentiation processes.

Contrary to what occurs within the SMO, the acidic volcanic rocks associated with basaltic andesites of the Mogollon sequence show a typical anorogenic mineralogy.

The bimodal volcanism ended with the beginning of accumulation of the Báucarit Formation, i.e., with the start of the development of a Basin-and-Range structure. The basal portion of this formation consists of lahars between basaltic flows, as found in the Yécora basin, and changes progressively upward to a detrital formation consisting of debris from all the older units of the surrounding highlands. This upper part is a conglomerate with poorly sorted, subrounded pebbles and cobbles in an indurated sandy matrix.

The Báucarit Formation, first described by Dumble (1900), was deposited on top of the basaltic sequence. It is thickest on the western side of the SMO where it can be hundreds of meters thick in paleovalleys. There, erosion has produced characteristic ruiniform landscapes. The age of the Báucarit Formation can be tentatively bracketed, within the studied area, between 17.6 (age of one of the last basalts in the Yécora basin, Figs. 2 and 6; Bockoven, 1980) and 17 Ma (age of the lower group of the Tepoca Formation, Figs. 2 and 8; Damon, unpublished data). Moreover, 8 km west of Tepoca, the Báucarit Formation is capped by a 14 Ma basalt (Damon and others, 1981).

Post-Báucarit magmatism: the Tepoca caldera

Tepoca is a small village located 210 km east of Hermosillo, at the foot of the SMO plateau (Fig. 1) and traversed by the ring-road Ciudad Obregón–Hermosillo. The area around the village provides good exposures of a contact between andesites and the underlying Báucarit Formation. As estimated from areal photographs, the andesites cover an area of 100 km^2 and form a plateau with a maximum elevation of 1,200 m. On the western side of the Tepoca basin, the volcanic pile reaches 250 m in thickness and rests conformably on top of the Báucarit Formation.

The Tepoca Formation (Fig. 7) comprises three groups of andesitic flows, which are, from bottom to top: (1) amphibole-bearing andesites (6 percent hornblende, 16 percent [An 60] plagioclase, 6 percent clino and ortho- pyroxenes); a flow has been dated at 17 Ma by Damon (personal communication, 1983); (2) aphyric andesites; and (3) plagioclase-bearing andesites (29 percent plagioclase [An 64], 6 percent two pyroxenes), forming the Mesa el Kusi plateau. The two upper groups consist of a number of thin flows with an average thickness of 1 m.

The summit of the plateau near Rancho Mesa el Kusi, is cut by a circular caldera, 2 km in diameter. Ring dikes and faults

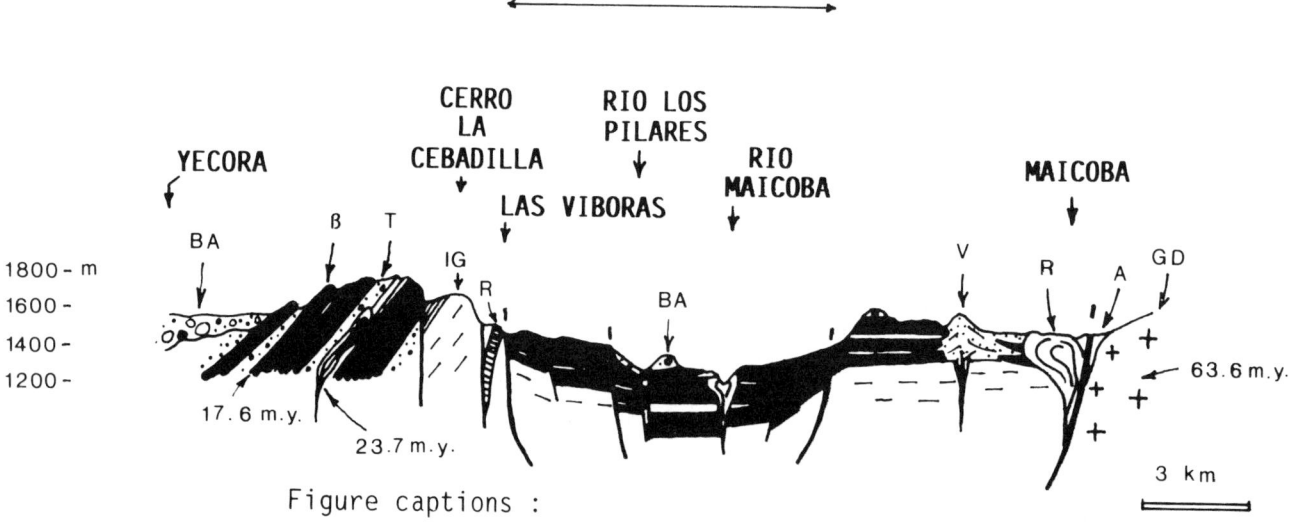

Figure 5. Generalized cross section between Yécora and Maicoba. (see Fig. 1 for location). Ages are from Bockoven (1980).

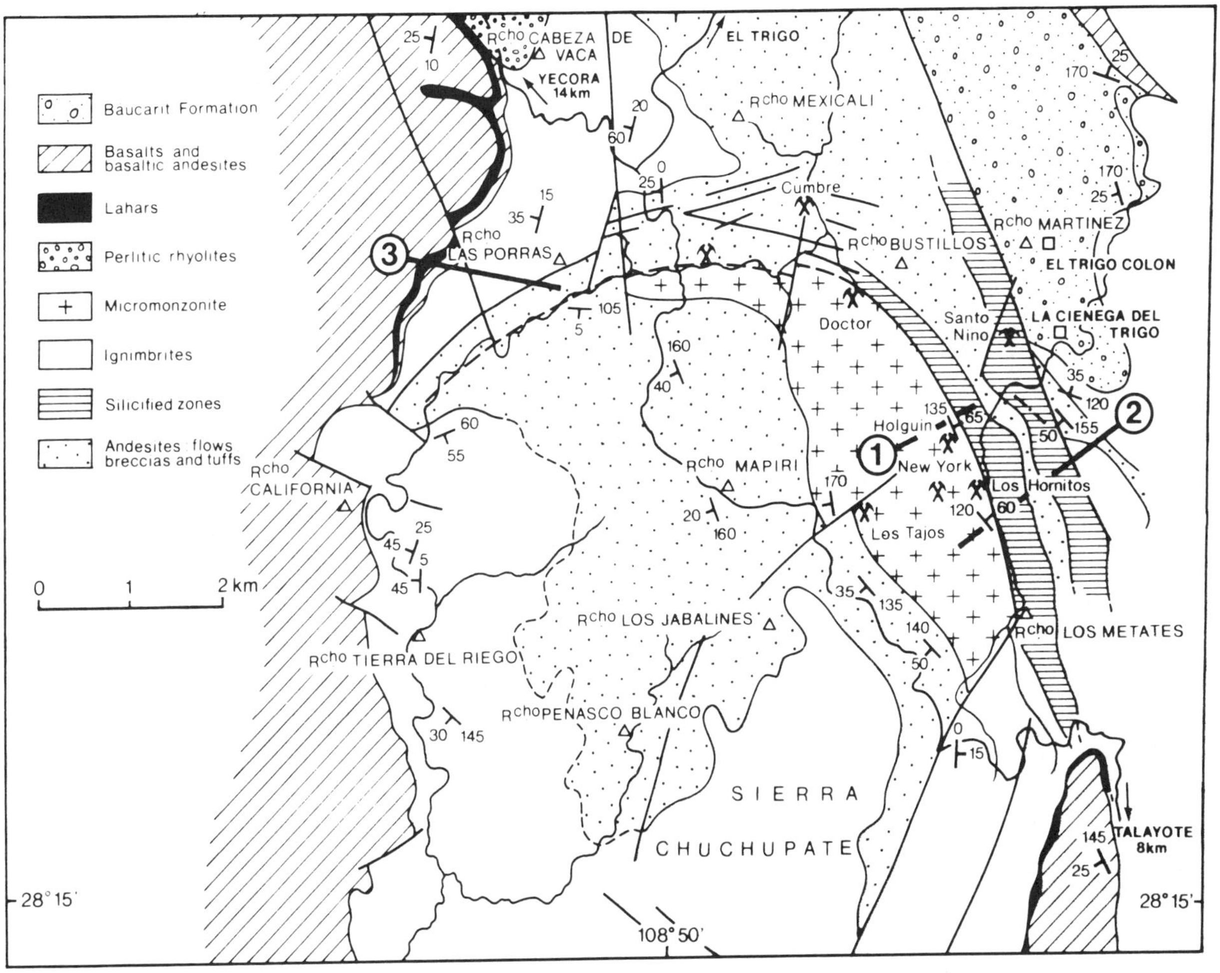

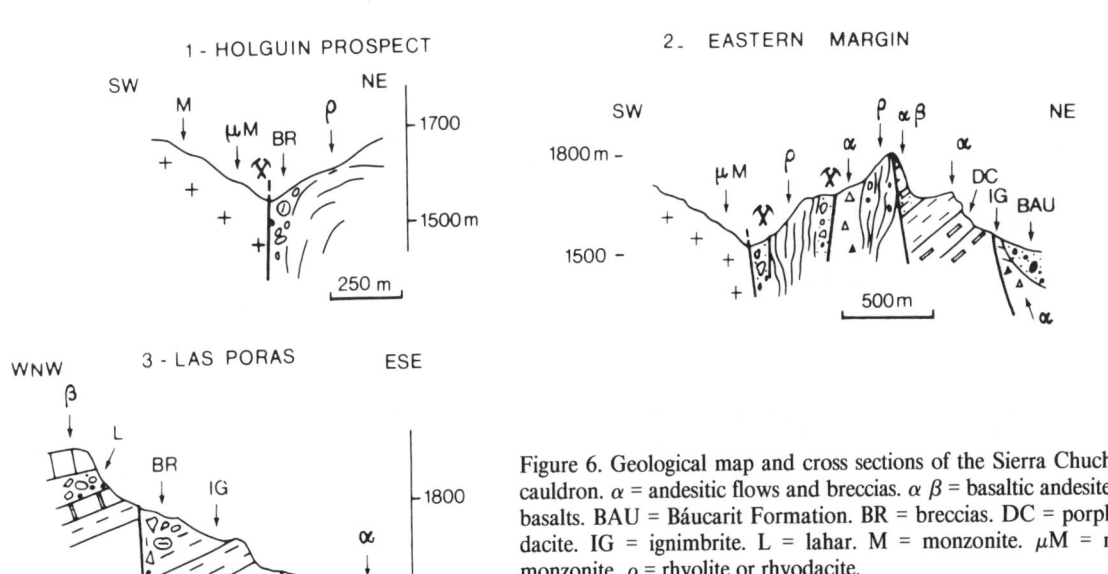

Figure 6. Geological map and cross sections of the Sierra Chuchupate cauldron. α = andesitic flows and breccias. $\alpha\beta$ = basaltic andesites. β = basalts. BAU = Báucarit Formation. BR = breccias. DC = porphyritic dacite. IG = ignimbrite. L = lahar. M = monzonite. μM = micromonzonite. ρ = rhyolite or rhyodacite.

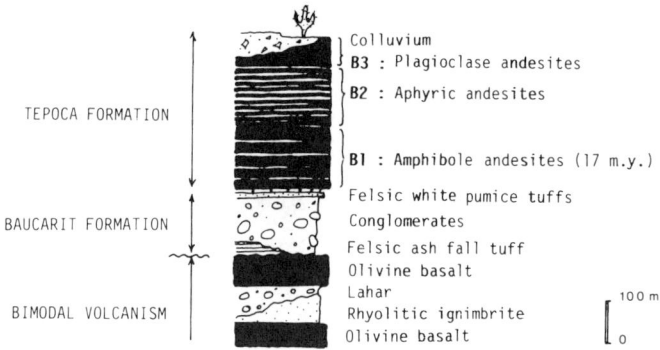

Figure 7. Generalized stratigraphic column of the Tepoca Formation.

within reddish tephra (where bombs are commonly found) delimit the walls of the caldera (Fig. 8). A fine-grained monzonite, consisting of plagioclase (An32, K-feldspar rim), clinopyroxene, orthopyroxene, and rare biotite, crops out at the center of this structure. Toward the margin, it changes to a microgranular facies with development of columnar jointing. Trace-element data show a homogeneous composition and a good correlation between the hygromagmaphile (HGM) elements (Th, Ta, U, La, Rb) for both andesites and monzonite. Thus, field observations and geochemical data show that the plutonic and volcanic facies belong to the same magmatic complex.

STRUCTURAL ANALYSIS

In northwestern Mexico, the younger compressional structures are of late Paleocene–early Eocene age. They are best developed in the Sierra Madre Oriental (Tardy, 1980). In the state of Chihuahua, the last episode of compressional tectonics is of late Eocene age (Chaulot-Talmon, 1984). In eastern Sonora, the major Tertiary tectonic structures correspond to an extensional regime that originated a basin-and-range morphology. The normal faulting trend is remarkably constant throughout northwestern Mexico, as evidenced by satellite imagery and field observations. The same holds true for the SMO, although grabens are less numerous there. In Sonora, N10°W- to N40°W-trending normal faults delimit grabens that represent 59 percent of the surface of the whole state.

In eastern Sonora, grabens are essentially asymmetric because of a 20 to 30° rotation of blocks along listric faults, giving a mean extension of about 15 percent. A generalized cross section between Yécora and Maicoba shows the extensional deformation (Fig. 5).

From microstructural analysis, Chaulot-Talmon (1984) distinguished three tectonic phases: (1) The first extension direction was N40°E to N64°E (Φ1, Fig. 9) and is represented by regional N10°W to N40°W fault trends resulting in basins with molassic deposits (Báucarit Formation). (2) The next extensional direction was oriented N56°W to N88°W (Φ2, Fig. 9) and corresponds to normal faults trending N10°W to N20°E that cut the basins. The extensional tectonics affected all the geologic formations of the area. Nevertheless, because rhyolitic domes (23 to 25 Ma) of the bimodal sequence follow the regional N10°W to N40°W fault trend, this tectonic event might have begun earlier, as in the southwestern United States where its inception seems to be contemporaneous with the emplacement of the first ignimbrites around 44 to 45 Ma (Zoback and others, 1981). (3) Along the Yécora-Maicoba road, widespread microstructural strike-slip faults (Φ3, Fig. 9) with horizontal displacement are found within both basaltic andesites and Báucarit Formation. These structures

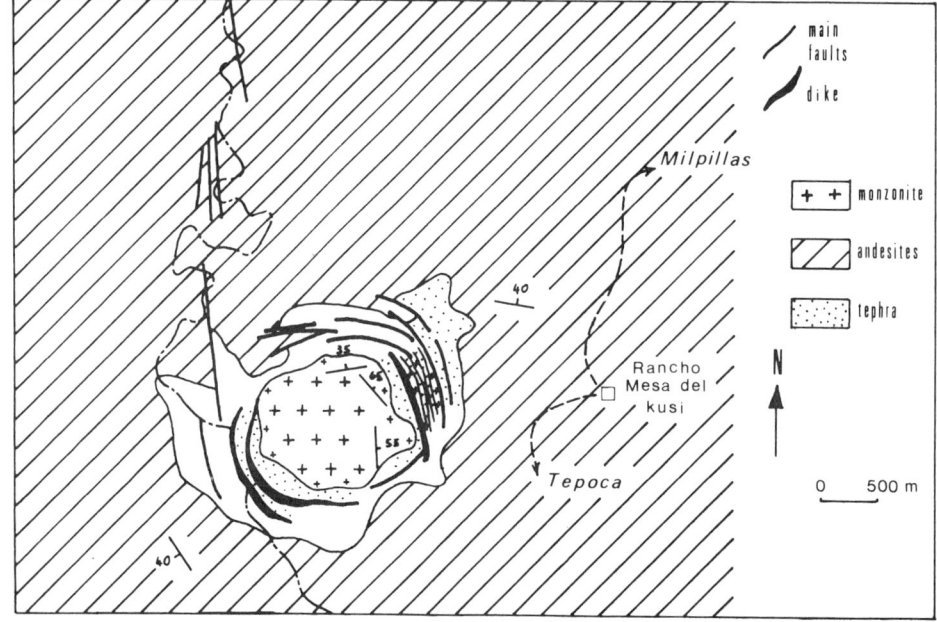

Figure 8. Geologic map of the Tepoca caldera.

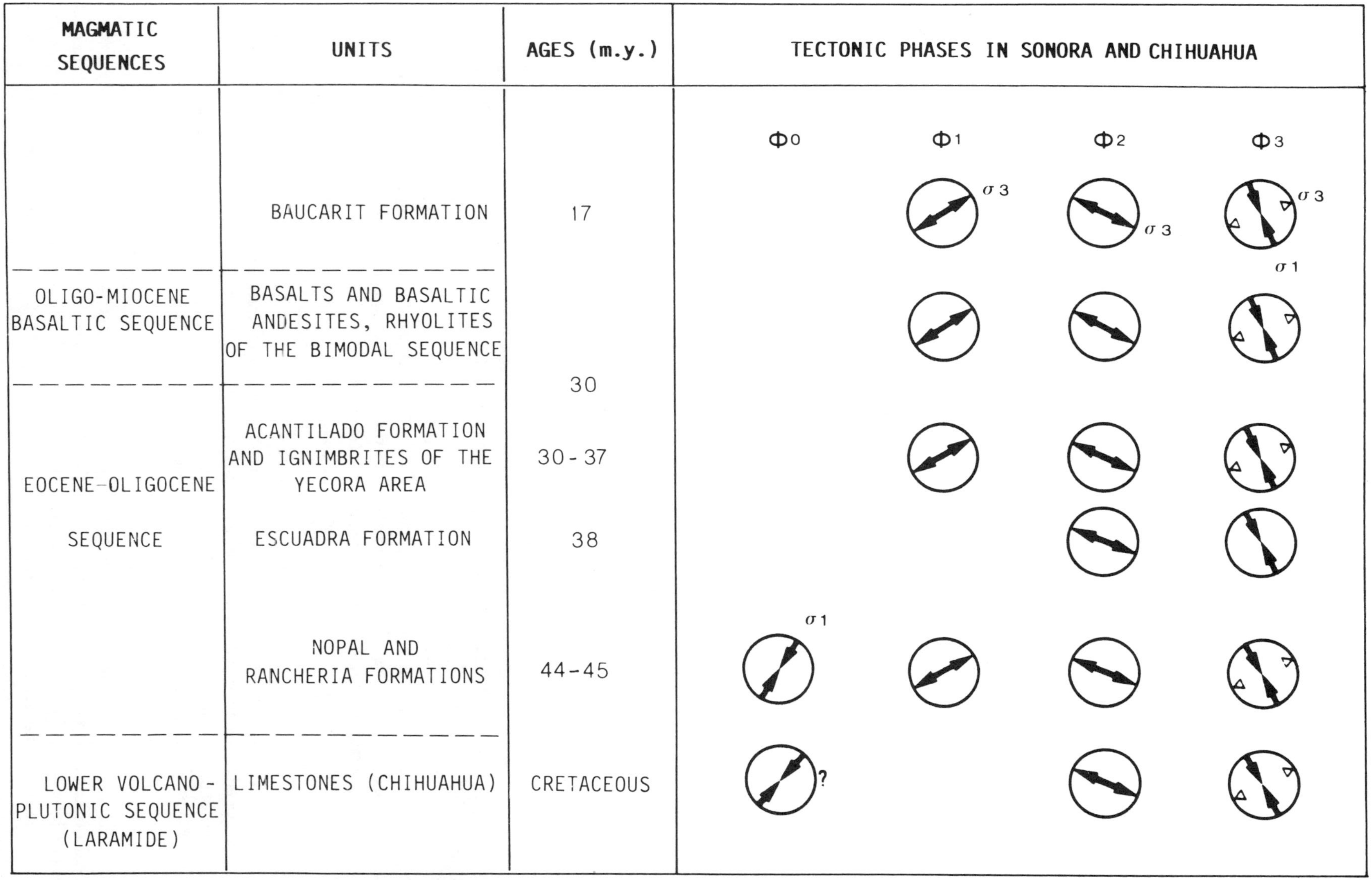

MAGMATIC SEQUENCES	UNITS	AGES (m.y.)	TECTONIC PHASES IN SONORA AND CHIHUAHUA
OLIGO-MIOCENE BASALTIC SEQUENCE	BAUCARIT FORMATION	17	φ0, φ1, φ2, φ3 phases shown
	BASALTS AND BASALTIC ANDESITES, RHYOLITES OF THE BIMODAL SEQUENCE	30	
EOCENE-OLIGOCENE SEQUENCE	ACANTILADO FORMATION AND IGNIMBRITES OF THE YECORA AREA	30–37	
	ESCUADRA FORMATION	38	
	NOPAL AND RANCHERIA FORMATIONS	44–45	
LOWER VOLCANO-PLUTONIC SEQUENCE (LARAMIDE)	LIMESTONES (CHIHUAHUA)	CRETACEOUS	

Figure 9. Tectonic phases in northwest Mexico. Principle stress orientations are from Chaulot-Talmon (1984) for Sonora and Chihuahua. Acantilado, Escuadra, Nopal, and Rancheria Formations are from Sierra del Nido, Chihuahua. $\phi 0$ is a N20°E-striking compressional phase represented by reverse faults (only found in Chihuahua). $\phi 1$ is a N40°E to N64°E extensional phase corresponding to the N10°W-N40°W-trending normal faulting in Sonora. $\phi 2$: N56°W to N88°W extensional phase corresponding to N10°W-N20°E normal faulting. $\phi 3$: strikeslip faulting.

have an almost north-south compressional component and an ENE-WSW tensional component and appear to occur along some preexisting normal faults. These dextral or sinistral (depending on fault strike) strike-slip faults are apparently related to the opening of the Gulf of California and have been dated at around 5 to 6 Ma in Baja California (Colletta and others, 1981). Figure 9 shows the correlations between these tectonic phases and the magmatic sequences in northwestern Mexico.

DISCUSSION AND CONCLUSIONS

Detailed geological mapping and generalized cross sections provide insight into the structure of the northern SMO and allow some comparisons with the section drawn by McDowell and Clabaugh (1979) for the Durango-Mazatlán transect. The mapped area displays the characteristic Basin and Range structure with normal faults trending north-south to NNW-SSE. Sections through the SMO plateau (Fig. 10) and satellite imagery interpretation show structural and lithologic changes from north to south. In the northern section, all of the SMO plateau has been under the influence of Basin and Range tectonics, whereas in the two southern sections, only the margins have been affected by such tectonics.

The occurrence of basalts and basaltic andesites is coeval with extensional tectonics. Basaltic vents and rhyolitic domes of the bimodal episode are located in the vicinity of NW-SE–trending faults. Furthermore, the basalts thicken toward the north, whereas ignimbrites totally disappear. From their petrochemical characteristics, andesites, dacites, and ignimbrites of the Eocene-Oligocene sequence, as well as the acid rocks of the bimodal sequence, represent a single calcalkaline magmatic complex. Their high Th/Ta ratios (>10; Fig. 12; Magonthier, 1984) suggest that they were derived from a subduction-related parental source and that the different rock types are linked by fractional crystallization processes, as shown by Cameron and others (1980a) and Cameron and Hanson (1982) for the rocks of the Upper Volcanic Complex in Chihuahua.

Petrographically (Pl + Ol ± Cpx phenocrysts) as well as geochemically (Al_2O_3, K_2O, P_2O_5 rich and MgO, CaO poor), the basalts, basaltic andesites, and andesites of the Oligocene-Miocene sequence form a separate group, although all of them display strong affinities with the orogenic suites except for their high Ti content (Table 1). Between Yécora and Maicoba, the Th/Ta ratio in basaltic andesites changes stratigraphically upward from 10 to 1.4 (Fig. 12) and suggests, for this particular area, a change in the genetic conditions within the sequence characterized by an early affinity with a magmatic subduction regime and by a late affinity with an extensional one.

According to Gill's classification (1981), all facies from the Tepoca caldera are high-K and Si-rich orogenic andesites related to the calcalkaline series (low FeO*/MgO ratio). Nevertheless, in contrast with the older calcalkaline andesites, these andesites are not related to mineralization. The behavior of the HGM elements analyzed by neutron activation is interpreted according to Treuil's method (Treuil and Varet, 1973; Joron et Treuil, 1977; Treuil and others, 1979). It shows that most basalts and basaltic andesites have a Th/Ta ratio (4.2) intermediate between those commonly found in volcanic rocks from typical extensional settings, i.e., implying a direct mantle origin, and those of compressional tectonic settings, i.e., derived from a subduction-modified mantle (Joron and Treuil, 1984).

The occurrence and characteristics of Cenozoic magmatism in the northern SMO are in good agreement with the geodynamic events recorded in the northeastern Pacific. During the classical Laramide orogeny (80 to 40 Ma; Coney, 1976), subduction occurred at a rapid convergence rate (Coney, 1970, 1972; Rea and Dixon, 1983). Because of the opening of the Atlantic, the North American plate was moving southwest. At this time, a plutonic belt (mainly granodiorites and quartz-monzonites) was emplaced in Sonora, with cogenetic volcanism generally older. A histogram of published ages (Damon and others, 1983a, b) shows that in Sonora, the paroxysmal phase occurred near 60 Ma and that magmatism progressively migrated eastward. Following Snyder and others (1976), we consider that the subduction of the Farallon plate under the continental margin is responsible for this calcalkaline magmatism.

At about 42 Ma, a plate reorganization occurred (Clague and Jarrard, 1973; Dalrymple and Clague, 1976), reducing the convergence rate from 14 to 8 cm/yr (Coney, 1978) until 29 Ma (Heirtzler and others, 1968; Atwater, 1970; Handschumacher, 1976) when the Pacific and North American plates collided in the vicinity of the Mendocino fracture zone.

TABLE 1. MAJOR-ELEMENT COMPOSITION OF BASALTIC ANDESITES FROM DIFFERENT TECTONIC SETTINGS

	Basalts and basaltic andesites from the Yécora area*		Basaltic andesites from different tectonic environments†			Mogollon plateau§
	$\bar{x}$(21)	σ	I	II	III	
SiO_2	51.81	(2.79)	53.07	53.65	52.05	51.55
TiO_2	1.61	(0.38)	2.06	1.01	1.31	1.95
Al_2O_3	17.01	(0.80)	14.17	17.44	15.69	16.73
FeO**	8.66	(1.75)	12.09	8.91	9.20	9.56
MnO	0.14	(0.04)	0.21	0.15	0.16	0.15
MgO	4.43	(1.01)	5.10	5.13	7.47	4.24
CaO	7.70	(0.86)	8.73	9.22	10.76	6.72
Na_2O	3.63	(0.29)	2.75	3.01	2.77	3.57
K_2O	1.83	(0.30)	1.14	0.91	0.30	2.18
P_2O_5	0.71	(0.27)	0.35	0.21	0.10	0.84

*σ = standard error; $\bar{x}$ = average.
†I = Continental; II = Orogenic; III = Ocean ridge and floor (Pearce and others, 1977).
§Fodor, 1976.
**Total iron as FeO.

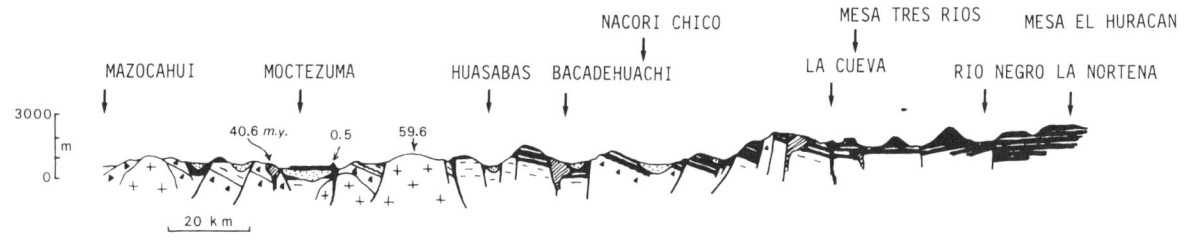

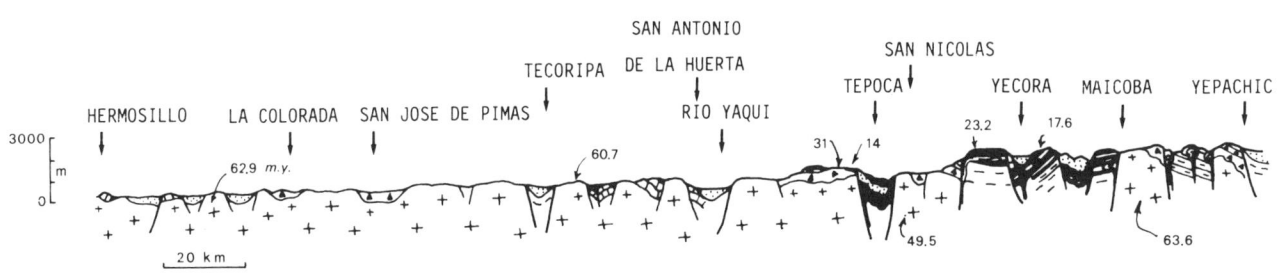

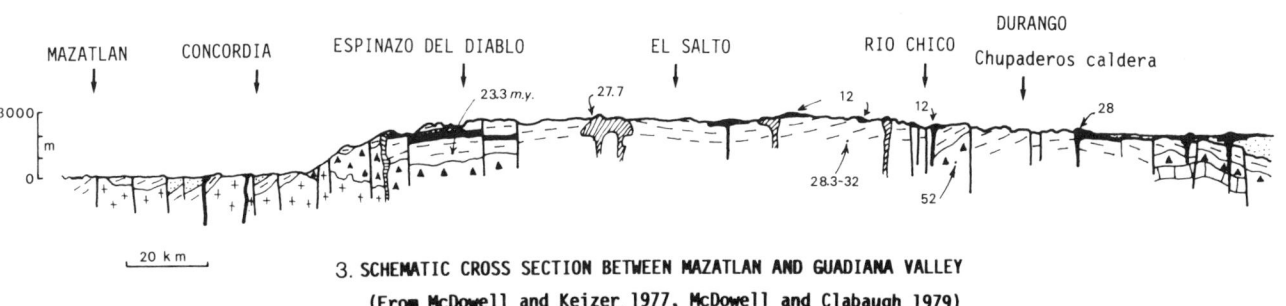

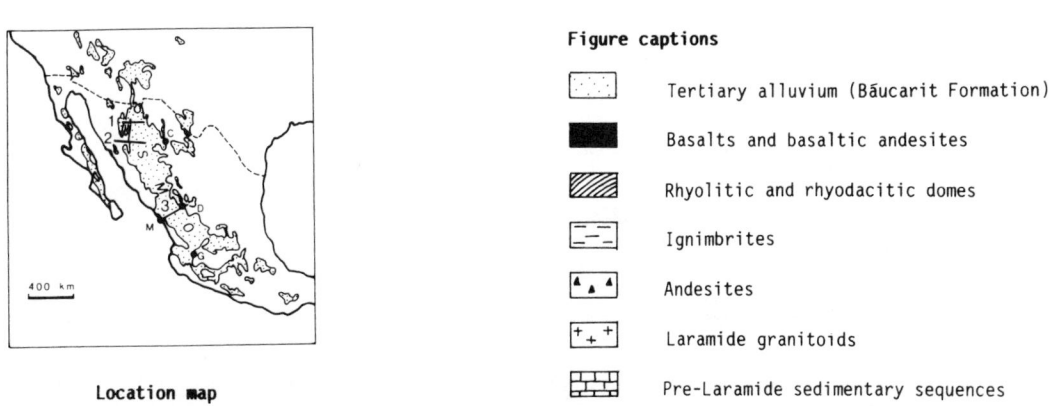

Figure 10. Geologic cross sections and location map for the Sierra Madre Occidental. Ages are from Bockoven (1980), Damon and others (1981, 1983a), McDowell and Clabaugh (1979), McDowell and Keizer (1977), and Montigny and others (1987).

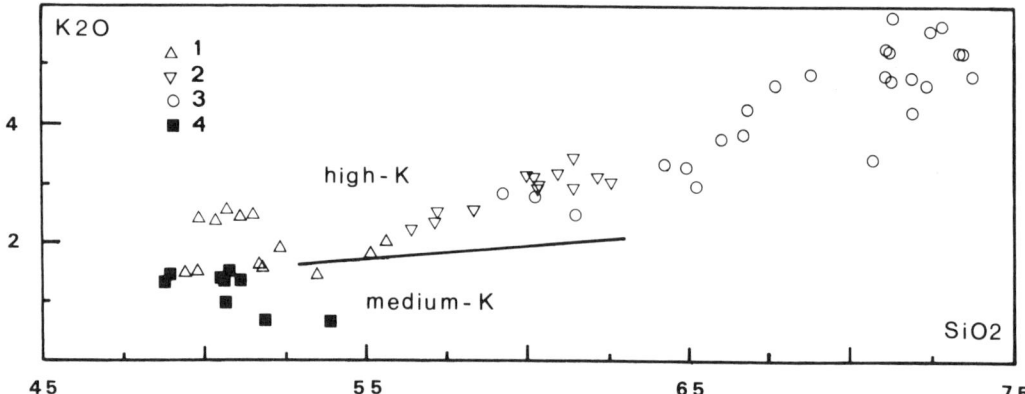

Figure 11. Variation diagram of K_2O plotted against SiO_2 content for analysis of mid-Tertiary volcanic rocks of northwest SMO. 1 = basalts and basaltic andesites of the Yécora area; 2 = volcanic and plutonic rocks of the caldera of Tecopa; 3 = intermediate and acidic rocks of the Yécora area; 4 = basalts of Moctezuma.

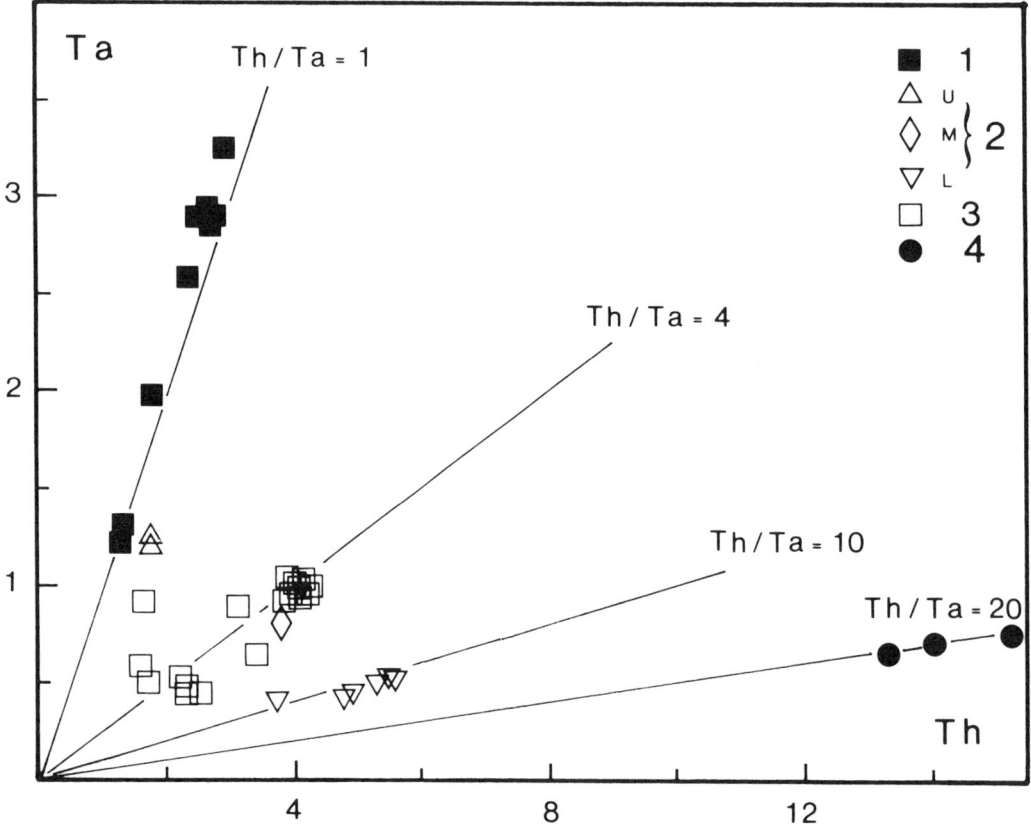

Figure 12. Plot of Ta content (ppm) against Th (ppm). 1 = basalts of Moctezuma; 2 = basalts from Rancho La Pinosa (upper, middle, lower part of the stratigraphic column); 3 = basalts and basaltic andesites of the Yécora area including volcanic and plutonic rocks of the caldera of Tepoca; 4 = andesites of Mulatos.

Magmas formed during rapid convergence under the continental plate moved up as the convergence rate slowed. Calcalkaline magmatism resumed and ignimbritic volcanism occurred in a very short time span. This time of ignimbritic flare-up in the SMO is also the time of the beginning of extensional tectonics (pre–Basin and Range phase). In this changing tectonic regime, and by analogy to what has been described in the Andes (Aguirre and others, 1989), the subduction geometry possibly became more of a "Mariana type" (Uyeda and Kanamori, 1979) and resulted in the generation of new magmas, producing basaltic andesites. Emplacement of late felsic calcalkaline magmas and rising of newly formed basic magmas resulted in a bimodal volcanism that progressively became more basaltic and more extensional in affinity.

After collision with the East Pacific rise, transform motion along the western continental margin resulted in a direct coupling of a continental wedge with the Pacific plate, and Quaternary volcanism occurred along pull-apart basins in the Gulf of California (Colletta and Angelier, 1983). In Sonora, the Moctezuma basaltic field shows that the Quaternary basaltic volcanism occurred according to the regional strain pattern, i.e., along Basin and Range faults (fissural volcanism) and possible northeast-southwest normal faults (Cerro Blanco, Villalobos and El Barríl volcanoes). The Cerro Blanco, south of the town of Moctezuma, has been dated at 0.52 Ma (Montigny and others, 1987). This volcanism differs in composition from the basaltic andesites and has typical extensional characteristics (Th/Ta = 0.92; Fig. 12).

In the northwestern SMO, the petrochemical evolution reflects the geodynamic evolution. A subduction-related calcalkaline magmatism, with fractional crystallization as the main evolutionary process prior to 30 Ma, progressively changed to a basic volcanism controlled by extensional tectonics. This change at 30 Ma marks the decreasing influence of the subduction component in the magma source.

ACKNOWLEDGMENTS

This study has been carried out through a cooperative program between the Instituto de Geología, Universidad Nacionál Autónoma de México (UNAM), and the University of Aix-Marseille III. The authors benefitted from the facilities of the Estacion Regional del Noroeste (ERNO–UNAM) which provided important logistical support for the field work. A large part of both field and laboratory activities was supported by the Laboratory of Geochemistry of M. Treuil (Paris VI) and CEA–DAMN (grant MC/14986). Neutron activation analyses were performed by J. L. Joron and H. Jacquemin in P. Süe laboratory at Saclay. We acknowledge the help of several Mexican geologists, particularly D. Córdoba, J. Guerrero, J. Roldán, Efrén Pérez S. and F. Paz Moreno. We are also grateful to our colleagues of the ERNO and the students of the Universidad de Sonora. Discussions with C. Coulon and L. Radelli greatly improved earlier versions of the manuscript. We thank F. McDowell, K. Cameron, J. Aranda and L. Aguirre for helpful reviews.

REFERENCES CITED

Aguirre, L., Levi, B., and Nyström, J. O., 1989, The link between metamorphism volcanism, and geotectonic setting during the evolution of the Andes, *in* Daly, J. S., Cliff, R. A., Yardley, B.W.D., eds., Evolution of metamorphic belts: Geological Society of London Special Publication 43, p. 223–232.

Atwater, T., 1970, Implications of plate tectonics for the Cenozoic tectonic evolution of western North America: Geological Society of America Bulletin, v. 81, p. 3513–3536.

Bockoven, N. T., 1980, Reconnaissance geology of the Yécora-Ocampo area, Sonora and Chihuahua, Mexico [Ph.D. thesis]: Austin, University of Texas, 197 p.

Cameron, K. L., and Hanson, G., 1982, Rare earth elements evidence concerning the origin of voluminous mid-Tertiary rhyolitic ignimbrites and related volcanic rocks, Sierra Madre Occidental, Chihuahua, Mexico: Geochimica et Cosmochimica Acta, v. 46, p. 1489–1503.

Cameron, M., Bagby, W. C., and Cameron, K. L., 1980a, Petrogenesis of voluminous mid-Tertiary ignimbrites of the Sierra Madre Occidental, Chihuahua, Mexico: Contributions to Mineralogy and Petrology, v. 74, p. 271–284.

Cameron, K. L., Cameron, M., Bagby, W. C., Moll, E. J., and Drake, R. E., 1980b, Petrologic characteristics of mid-Tertiary volcanic suites, Chihuahua, Mexico: Geology, v. 8, p. 87–91.

Chaulot-Talmon, J. F., 1984, Etude géologique et structurale des ignimbrites du Tertiaire de la Sierra Madre Occidental, entre Hermosillo et Chihuahua, Mexique [Thèse 3ème cycle]: Orsay, University of Paris-Sud, 259 p.

Clague, D. A., and Jarrard, R. D., 1973, Tertiary Pacific plate motion deduced from the Hawaian-Emperor chain: Geological Society of America Bulletin, v. 84, p. 1135–1154.

Colletta, B., and Angelier, J., 1983, Fault tectonics in northwestern Mexico and opening of the Gulf of California: Bull. Centres Rech. Exlor.-Prod. Elf-Aquitaine, 7, 1, p. 433–446.

Colletta, B., Angelier, J., Chorowicz, J., Ortlieb, L., and Rangin, C., 1981, Fracturation et évolution néotectonique de la péninsule de Basse-Californie (Mexique): Comptes Rendus de l'Académie des Sciences Paris, t. 292, série II, p. 1043–1048.

Coney, P. J., 1970, The geotectonic cycle and the new global tectonics: Geological Society of America Bulletin, v. 81, p. 739–748.

—— , 1972, Cordilleran tectonics and North American plate motion: American Journal of Science, v. 272, p. 603–628.

—— , 1976, Plate tectonics and the Laramide orogeny: New Mexico Geological Society Special Publication 6, p. 5–10.

—— , 1978, Mesozoic–Cenozoic Cordilleran plate tectonics, *in* Smith, R. B., and Eaton, G. P., eds., Cenozoic tectonics and regional geophysics of the Western Cordillera: Geological Society of America Memoir 152, p. 33–50.

Dalrymple, G. B., and Clague, D. A., 1976, Age of the Hawaiian–Emperor bend: Earth and Planetary Science Letters, v. 31, p. 313–329.

Damon, P. E., Clark, K. F., Shaffiqullah, M., Roldan-Q., J., and Islas, L. J., 1981, Geology and mineral deposits of southern Sonora and the Sonoran Sierra Madre Occidental, *in* Ortlieb, L., and Roldan, J., eds., Geology of northwestern Mexico and southern Arizona; Geological Society of America Cordilleran Section Fieldguide and Papers: Hermosillo, Sonora, Instituto de Geologia, p. 369–426.

Damon, P. E., Shafiqullah, M., Roldan-Q., J., and Cochemé, J. J., 1983a, El Batolito Larámide (90–40 m.a.) de Sonora: 15th Convención Naciónal, Asociación de Ingenieros de Minas, Metalurgistas y Géologos de México, Memoria, Guadalajara, p. 63–95.

Damon, P. E., Shafiqullah, M., and Clark, K. F., 1983b, Geochronology of the porphyry-copper deposits and related mineralization of Mexico: Canadian Journal of Earth Sciences, v. 20, p. 1052–1071.

Dumble, E. T., 1900, Notes on the geology of Sonora, Mexico: American Institute of Mining Engineering Transactions, v. 29, p. 122–152.

Elston, W. E., and Northrop, S. A., 1976, Cenozoic volcanism in southwestern New Mexico: New Mexico Geological Society Special Publication 5, 151 p.

Elston, W. E., Rhodes, R. C., Coney, P. J., and Deal, E. G., 1976, Progress report on the Mogollon Plateau volcanic field, southwestern New Mexico; No. 3, Surface expression of a pluton, in Elston, W. E., and Northrop, S. A., eds., Cenozoic volcanism in southwestern New Mexico: New Mexico Geological Society Special Publication 5, p. 3–28.

Fodor, R. V., 1976, Volcanic geology of the northern Black Range, New Mexico, in Elston, W. E., and Northrop, S. A., eds., Cenozoic volcanism in southwestern New Mexico: New Mexico Geological Society Special Publication 5, p. 68–70.

Gill, J. B., 1981, Orogenic andesites and plate tectonics: Berlin, Springer-Verlag, 390 p.

Handschumacher, D. W., 1976, Post-Eocene plate tectonics of the eastern Pacific, in The geophysics of the Pacific Ocean Basin and its margin; The Woollard volume: American Geophysical Union Monograph 19, p. 177–202.

Hawkins, J. W., 1970, Petrology and possible tectonic significance of late Cenozoic volcanic rocks, southern California and Baja California: Geological Society of America Bulletin, v. 83, p. 3323–3396.

Heirtzler, J. R., Dickson, G. O., Herron, E. M., Pitman, W. C., and Le Pichon, X., 1968, Marine magnetic anomalies, geomagnetic field reversals, and motions of the ocean floor and continents: Journal of Geophysical Research, v. 73, p. 2119–2136.

Hisazumi, H., 1929, Informe geológico preliminar de la parte norte del Estado de Sinaloa: Anales del Instituto de Geología de México, v. 3, p. 95–109.

Joron, J. L., and Treuil, M., 1977, Utilisation des propriétés des éléments fortement hygromagmatophiles pour l'étude de la composition chimique et de l'hétérogénéité du manteau: Bulletin de la Société Géologique de France, t. 19, v. 6, p. 1197–1205.

—— , 1984, Etude géochimique et pétrogenèse des laves de l'Etna, Sicile, Italie: Bulletin Volcanologique, v. 47-4, no. 2, p. 1125–1144.

Keller, P. C., 1977, Geology of the Sierra Gallego area, Chihuahua, Mexico [Ph.D. thesis]: Austin, University of Texas, 124 p.

Magonthier, M. C., 1984, Les ignimbrites de la Sierra Madre Occidentale et de la province uranifère de la Sierra Peña Blanca, Mexique [Thèse d'Etat]: Paris, Univ. P. et M. Curie. Mem. Sci. Terre, 84-17, 283 p.

Mauger, R. L., 1977, A progress report on the geology of the Sierra de la Calera–Sierra Del Nido block, Chihuahua, Mexico: Geological Society of America Abstracts with Program, v. 9, p. 62.

—— , 1981, Geology and petrology of the central part of the Calera del Nido block, Chihuahua, Mexico, in Goodell, P. C., and Waters, A. C., eds., Uranium in volcanic and volcaniclastic rocks: American Association of Petroleum Geologists Studies in Geology, v. 13, p. 205–242.

McDowell, F. W., and Clabaugh, S. E., 1979, Ignimbrites of the Sierra Madre Occidental and their relation to the tectonic history of western Mexico, in Chapin, C. E., and Elston, W. E., eds., Ash-flow tuffs: Geological Society of America Special Paper 180, p. 113–124.

McDowell, F. W., and Keizer, R. P., 1977, Timing of mid-Tertiary volcanism in the Sierra Madre Occidental between Durango City and Mazatlán, Mexico: Geological Society of America Bulletin, v. 88, p. 1479–1487.

Montigny, R., Demant, A., Delpretti, P., Piguet, P., and Cochemé, J. J., 1987, Chronologie K/A des séquences volcaniques du Nord de la Sierra Madre Occidental (Mexique): Comptes Rendus de l'Académie des Sciences Paris, t. 304, série II, no. 16, p. 987–992.

Ordoñez, E., 1900, Las rhyolitas de México: Boletín del Instituto de Geología de México, v. 15, 76 p.

Pearce, T. H., Gorman, B. E., and Birkett, T. C., 1977, The relationship between major element chemistry and tectonic environment of basic and intermediate volcanic rocks: Earth and Planetary Science Letters, v. 36, p. 121–132.

Rea, D. K., and Dixon, J. M., 1983, Late Cretaceous and Paleogene tectonic evolution of the North Pacific Ocean: Earth and Planetary Science Letters, v. 65, p. 145–166.

Snyder, W. S., Dickinson, W. R., and Silberman, M. L., 1976, Tectonic imlications of space-time patterns of Cenozoic magmatism in the western United States: Earth and Planetary Science Letters, v. 32, p. 91–106.

Swanson, E. R., 1977, Reconnaissance geology of the Tomochic–Ocampo area, Sierra Madre Occidental, Chihuahua, Mexico [Ph.D. thesis]: Austin, University of Texas, 123 p.

Swanson, E. R., and McDowell, F. W., 1984, Calderas of the Sierra Madre Occidental volcanic field, western Mexico: Journal of Geophysical Research, v. 89, no. B10, p. 8787–8799.

Tardy, M., 1980, Contribution à l'étude géologique de la Sierra Madre Orientale du Mexique [Thèse Doctorat d'Etat]: University of Paris VI, 445 p.

Treuil, M., and Varet, J., 1973, Critères volcanologiques, pétrologiques et géochimiques de la genèse et de la différenciation des magmas basaltiques: exemple de l'Afar: Bulletin de la Société Géologique de France, t. XV, no. 5-6, p. 506–540.

Treuil, M., Joron, J. L., Jaffrezic, H., Villemant, B., and Calas, G., 1979, Géochimie des éléments hygromagmatophiles, coefficients de partage minéraux/liquides et propriétés structurales de ces éléments dans les liquides magmatiques: Bulletin Minéralogique, v. 102, p. 402–409.

Uyeda, S., and Kanamori, H., 1979, Back-arc opening and the mode of subduction: Journal of Geophysical Research, v. 84, p. 1049–1061.

Zarate-Cruz, G., 1983, Etude pétrologique et géochimique des roches volcaniques de la Sierra Del Nido, Chihuahua, Mexique [Thèse 3ème cycle]: University of Paris VI, 68 p.

Zoback, M. L., Anderson, R. E., and Thompson, G. A., 1981, Cainozoic evolution of the state of stress and style of tectonism of the Basin and Range province of the western United States: Philosophical Transactions of the Royal Society of London, ser. A, v. 300, p. 407–434.

MANUSCRIPT ACCEPTED BY THE SOCIETY APRIL 18, 1990

Quaternary shorelines along the northeastern Gulf of California; Geochronological data and neotectonic implications

Luc Ortlieb*
Institut Français de Recherche Scientifique pour le Développement en Coopération (ORSTOM), 213 rue La Fayette, 75010 Paris, France

ABSTRACT

A general reconnaissance of the remnants of Pleistocene high sea-level stands was conducted along the Sonora coast, between the Rio Colorado delta and the Guaymas area, to determine the neotectonic comportment of the mainland margin of the Gulf of California plate boundary.

The generally low relief of this coastal region and the extensive cover of the late Quaternary eolian sands (in the northern gulf) and alluvium (in the eastern gulf) somewhat limit the study of Pleistocene shorelines, but coastal deposits corresponding to several episodes of high sea level are preserved at low elevations all along the coastline. The morphostratigraphic interpretations and lateral correlations of these deposits support the conclusion that, unlike the Baja California peninsula and coastal California, the northeastern Gulf of California remained vertically stable, at least during the late Quaternary. Pleistocene vertical motions have been insignificant on the edge of the North American plate, except in the Rio Colorado delta area and along the Cerro Prieto fault zone, which links the San Andreas fault system and the northernmost Gulf of California.

This chapter emphasizes the problems of correlation and age determination of the terrace remnants and particularly those met in the identification of the last interglacial maximum (isotopic substage 5e, about 125,000 yr ago). Uranium-series, radiocarbon, and amino-acid stereochemistry data from fossil pelecypod shells provide useful geochronologic information, but in some cases proved to be inaccurate (large spread of Th/U ages from a single locality), or unreliable (radiocarbon), because of contamination and diagenetic alteration.

The Holocene coastal deposits are generally well developed along the Sonoran shores; several lines of evidence indicate that the sea level nearly reached its present position about 4,000 yr ago.

MARINE TERRACES, QUATERNARY CLIMATOSTRATIGRAPHY, AND RECENT VERTICAL MOTIONS

A marine terrace is a coastal landform resulting from littoral erosion produced when sea level remains long enough at a given position relative to the land. In the "stable" and slowly uplifted regions, emerged marine terraces generally correspond to the highest sea stands coeval with the Quaternary warmest episodes (global glacial minima).

It is now widely accepted that the Quaternary was characterized by glacial/interglacial climatic cycles, identified in the deep-sea-core oxygen-isotope records. The ^{18}O isotopic variations measured in foraminifers of deep oceanic cores register the basis of the Quaternary chronostratigraphy (Shackleton and Opdyke, 1973, 1976). The episodes of eustatic high sea-level stands,

---

*Present address: Misión ORSTOM en el Perú, Apartado 18-1209, Lima 18, Peru.

Ortlieb, L., 1991, Quaternary shorelines along the northeastern Gulf of California; Geochronological data and neotectonic implications, *in* Pérez-Segura, E., and Jacques-Ayala, C., eds., Studies of Sonoran geology: Geological Society of America Special Paper 254.

and thus of marine terrace formation, are correlatable with the main interglacial stages (IS), as defined in the oceanic isotopic curves: IS 5, 7, 9, 11, etc. During these interglaciations, which occurred, respectively, at ~125 ka (125,000 yr ago), ~200 ka, ~320 ka, etc., the sea level was close (within a few meters) to the present mean sea level (MSL).

On rapidly uplifted coastal areas of the world, three marine terraces were formed during the last interglacial period (IS 5) by high sea-level stands identified as substages 5e, 5c, and 5a, and radiometrically dated as ~125, ~105, and ~85 ka (Broecker and others, 1968; Bloom and others, 1974; Chappell, 1974; Aharon, 1983; Chappell and Shackleton, 1986). In so called "stable" areas and in coastal regions that suffered slow uplift motions in the late Quaternary (less than 100 mm/10^3 yr), only the earliest substage IS 5e marine terrace is present, which means that the last time the sea level reached a eustatically higher position than present was at 125 ka. Worldwide comparisons of last-interglacial marine-terrace elevations (involving reconstructions of uplift rates) led to the consideration that the IS 5e eustatic high sea-level stand reached a +6-m elevation above the present datum.

In this chapter, the following divisions of the Quaternary were adopted: early Pleistocene (1.8 to 0.7 Ma), middle Pleistocene (0.7 to 0.15 Ma), late Pleistocene (150 to 10 ka), Holocene (10 ka to present). The term "late Quaternary" is here meant to include the late Pleistocene and Holocene.

VERTICAL MOTIONS ALONG THE PACIFIC/NORTH AMERICAN PLATE BOUNDARY

Along the Pacific coast of the United States and northwestern Mexico, the marine terrace coeval with the IS 5e high sea-level stand is generally well preserved and identified through faunal analyses, radiometric dating, and aminostratigraphic measurements (Wehmiller and others, 1977; Kennedy and others, 1982; Kern, 1977; Ortlieb, 1984c, 1987, 1990). In this wide region, only one recent marine terrace is locally identified below the IS 5e; according to one or the other of the models of sea-level fluctuations (Bloom and others, 1974; Stearns, 1976; Aharon, 1983; Chappell and Shackleton, 1986), this terrace may be correlated with the IS 5c high sea-level stand.

Marine-terrace studies in coastal California provided estimates of middle and late Quaternary uplift rates, which generally range between 100 and 300 mm/10^3 yr (Palmer, 1967; Bradley and Griggs, 1976; Lajoie and others, 1979; McLaughlin and others, 1983).

The distribution of Pleistocene marine terraces around Baja California peninsula indicates that, since Pliocene time, the mean regional uplift rate has been of the order of 100 mm/10^3 yr, but that it has much diminished in the last few hundred thousand years in most of the area (Ortlieb, 1978, 1979, 1980, 1982b, 1984b, 1987, 1990). There are only three coastal areas that experienced recent vertical motions with rates above 100 mm/10^3 yr: east-central (Santa Rosalía), west-central (Vizcaino peninsula), and northwestern Baja California (Punta Banda). In these tectonically active areas, the IS 5e shoreline is preserved at more than +15 m, and an IS 5c (?) high sea-level stand is registered well above the present MSL (Ortlieb, 1982a, 1984a, 1984c, 1987).

Marine terraces along the eastern Gulf of California have not been studied as late as the last decade; the generally accepted idea, based on limited work in the northernmost Gulf of California (Ives, 1951, 1959, 1964; Merriam, 1965), was that the margin of the North American continent also registered strong uplift motions, particularly during the late Quaternary (Richards, 1973).

PREVIOUS STUDIES ON THE QUATERNARY COASTAL DEPOSITS IN SONORA

One of the first mentions of emerged Quaternary marine deposits on the east coast of the Gulf of California concerned the finding of marine fossils on the Sonora side of the Infiernillo Straits (McGee and Johnson, 1896; Fig. 1). Later, some brief descriptions of Pleistocene marine terraces on several islands of the eastern Gulf of California were given by Beal (1948) and Anderson (1950).

Ives, the first author to publish several papers (1951, 1959, 1964) on Quaternary shorelines, worked along the northern coast of the Gulf of California, in the vicinity of Puerto Peñasco. He supposedly identified several Pleistocene shorelines, and inferred that the area had undergone recent uplift. In contrast, Hertlein and Emerson (1956) described a relatively low-lying (+7 m maximum elevation), late Pleistocene shoreline in the same area, which would indicate that the area had not suffered vertical motions in the late Quaternary.

On the Sonoran side of the Rio Colorado delta, Merriam (1965) observed uplifted marine sediments that he supposed were fairly recent, and that thus would have documented strong Holocene uplift motions associated with the "San Jacinto" fault activity. This fault, later renamed Cerro Prieto fault, lies along the eastern side of the Rio Colorado delta and is the principal structural link between the San Andreas and Gulf of California systems (Elders and others, 1972; Fuis and others, 1982).

At a short distance from the Baja California/Sonora state boundary, Pleistocene marine deposits of the southwestern Rio Colorado delta area were studied by Thompson (1968), Walker and Thompson (1968), Ortlieb and Malpica (1978), and Ortlieb (1982b, 1987); these workers concluded that a late Pleistocene (most probably the IS 5e) shoreline is preserved at +7 to +10 m and that it shows little vertical deformation. Holocene deltaic sediments have been investigated by Gorsline (1967), Meckel (1975), and Thompson (1968); a few radiocarbon dates indicate that sea level reached its present position at least 3,000 yr ago (Thompson, 1968). It should be noted that, immediately north of the Mexico/U.S. border, the Salton trough has had a long history of marine, lagoonal, and lacustrine flooding since the end of the Miocene (Blake, 1854; Tarbet, 1941; Dibblee, 1954; Arnal, 1961; Thomas, 1963; Stanley, 1962; Van den Kamp, 1973; Waters, 1983; Johnson and others, 1983). During the late Quater-

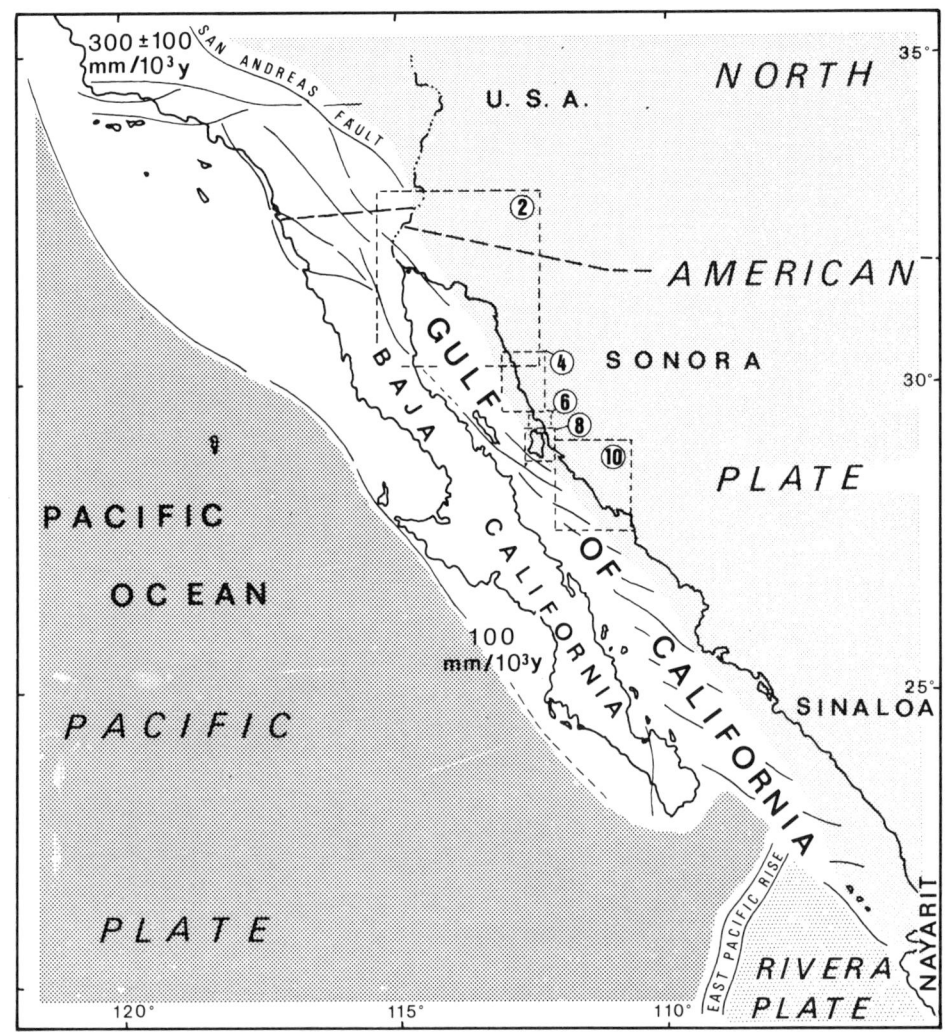

Figure 1. Tectonic setting of the Sonoran coast, northeastern Gulf of California, on the edge of the North American plate. The order of magnitude of middle and late Quaternary mean uplift rates, deduced from marine-terrace studies along southwestern California and Baja California, is indicated. Map locations for this chapter are indicated by circled numbers, which refer to Figures 2, 4, 6, 8, and 10.

nary, the episodic flooding of the Salton basin (−85 m deep at its lowest point) by northerly diversions of the Rio Colorado, has been controlled by the building up of the deltaic fan and by the neotectonic comportment of this actively faulted area (Gilmore and Castle, 1983; Gilmore, 1985, 1986; Sharp, 1982, 1986).

Marine-terrace studies along the eastern coast of the Gulf of California prior to 1975 have been very limited. The reconnaissance geologic map covering the central Sonoran coast and Tiburon Island by Gastil and Krummenacher (1974, 1977) shows a few localities of Quaternary marine deposits, which were not specifically studied.

In the last decade, detailed studies on Quaternary littoral sediments along the Sonoran coast were carried out by a Franco-Mexican group within the framework of a general reconnaissance of Quaternary marine terraces in the Gulf of California region (Geocortez program, between ORSTOM and the Instituto de Geología, Universidad Nacional Autónoma de México). This multidisciplinary work included several unpublished theses (Chavez, 1975; Celis-Gutierrez, 1975, 1979; Luna-Guerin, 1981; Gonzalez-Gonzalez, 1982), as well as published reports (Malpica and others, 1978; Ortlieb and Malpica, 1978; Coletta and Ortlieb, 1979, 1981; Bernat and others, 1980; Celis-Gutierrez, 1980; Gastil and Ortlieb, 1981), and has been updated by more recent geochemical and geochronological analyses (Ortlieb and Triclot, 1984; Ortlieb, 1987, 1990). This chapter presents a more complete panorama of the distribution of the Quaternary marine remnants, synthesizing the available information on recent vertical motions along the Sonoran coast.

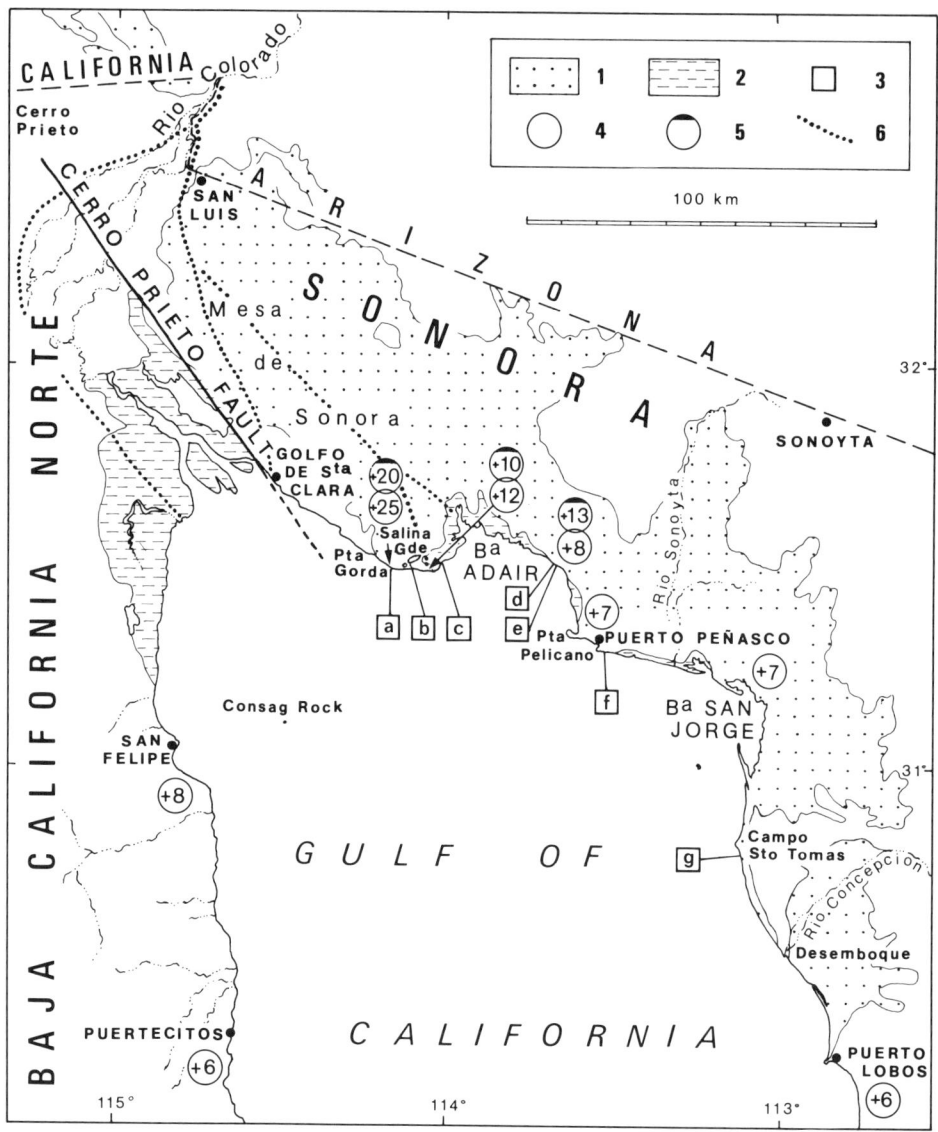

Figure 2. The northernmost Gulf of California: distribution and height of Pleistocene shorelines. 1. Late Quaternary eolian sands. 2. Lagoonal and deltaic areas. 3. Location of the sections of Quaternary coastal deposits (a to g) shown in Figure 3. 4. Maximum elevation (in meters above present MSL) of the early late Pleistocene (IS 5e) shoreline. 5. Elevation of middle Pleistocene shorelines (meters above present MSL). 6. Former courses of the Rio Colorado estuaries during the IS 5e maximum sea-level stand.

THE NORTHERNMOST GULF OF CALIFORNIA AND THE RIO COLORADO DELTA

Pleistocene deltaic deposits. The mouth of the Rio Colorado delta is bordered on its eastern margin by a dissected plateau called Mesa de Sonora (Fig. 2), which is essentially formed by fluvio-deltaic sediments and covered by active desert sands. The relief of the Mesa de Sonora results from an uplift motion linked to the Cerro Prieto fault activity (Colletta and Ortlieb, 1979, 1981, 1984). The substrate of the plateau had been assigned several ages: undifferentiated Tertiary (Kniffen, 1932), Pliocene (Beal, 1948), and Quaternary (Merriam, 1965). It is now well established that the sediments overlying the 1- to 3-km-thick Pleistocene deltaic sequence, revealed by recent drilling in the northernmost Gulf of California region (Eberly and Stanley, 1978; Trinidad-Reyes and Rueda-Gaxiola, 1982; Viñas-Gómez, 1982, 1984), are of middle to late Quaternary age (Ortlieb, 1987).

The coastal cliffs of the Mesa de Sonora cut fluvial continental sediments between Golfo de Santa Clara and Punta Gorda, and nearshore and marine deltaic sediments between Punta Gorda and the western shores of Bahia Adair (Fig. 2). The last-

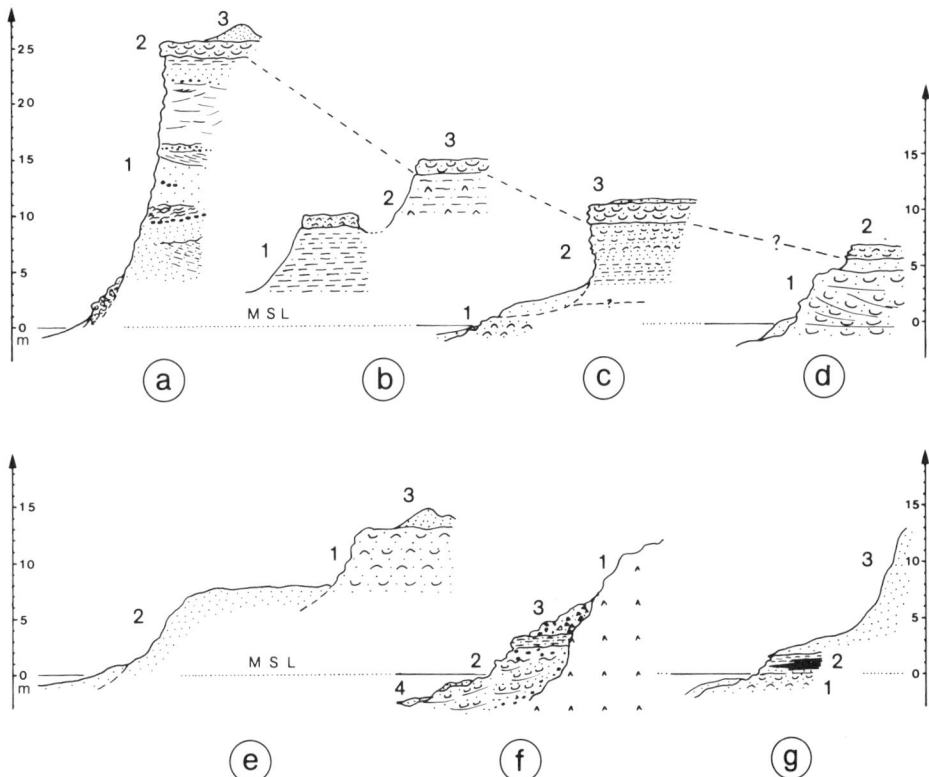

Figure 3. Sections of Quaternary coastal deposits along the northern Gulf of California (see location in Fig. 2). Section a: 1. Composite sequence of fluvio-deltaic deposits (earliest late Pleistocene, or latest middle Pleistocene?). 2. Early late Pleistocene *Chione* coquina bed. 3. Holocene sand dunes, and late Quaternary terrestrial reddish silts. Section b: 1. Lagoonal silty sands capped with a fossiliferous calcarenite layer (middle Pleistocene?). 2. Silty and clayey sands, with abundant gypsum crystals (latest middle Pleistocene, or earliest late Pleistocene). 3. Early late Pleistocene *Chione* coquina bed. Section c: 1. Sublittoral calcarenitic sandstone, rich in *Encope* sp. shells (middle Pleistocene). 2. Series of intertidal sandy and silty layers, including weathered shell fragments (late middle Pleistocene?). 3. Early late Pleistocene *Chione* coquina bed (LU 981) covered with late Quaternary eolian sands and terrestrial silts. Section d: 1. Calcarenitic sandstones rich in *Chione californiensis* shells (middle Pleistocene). 2. Fossiliferous bioclastic sands with *Chione cortezi* and *C. californiensis* (late Pleistocene). Section e: 1. Calcarenitic sandstones with weathered shells (middle Pleistocene?). 2. Lower terrace covered with eolian sands (late Pleistocene?). 3. Holocene sand dunes. Section f: 1. Volcanic substratum. 2. Early late Pleistocene beachrock, capped with fossiliferous sands and organic-matter-rich sediments. 3. Cemented scree deposits (late Quaternary). 4. Holocene beachrock forming atop the unit (2). Section g: 1. Cemented sands including *Chione fluctifraga* and *C. cortezi* (LT 933) (mid-Holocene). 2. Fine sands and clayey silts with peat layers (LT 936), plant remains, and lagoonal fauna capped with alluvial clays (mid-Holocene). 3. Sandy coastal ridge (second half of the Holocene).

mentioned coastal deposits may be correlated with two, and possibly three, episodes of high sea level: a younger sequence largely cropping out in the recent sea cliffs of the southeasternmost Mesa de Sonora and along the western shores of Bahia Adair, and a few isolated remnants of older Pleistocene transgression.

The younger sequence includes a great variety of interfingering deltaic units capped by a coquina (Figs. 3a, b, c). The deltaic units, composed of sands, sandy silts, and silty sands, correspond to fluvial, estuarine, tidal flats, shoreface, and shallow sublittoral environments. The marine coquina is formed by slightly rounded shells of *Chione cortezi* and *Chione gnidia,* and laterally grades to fossiliferous sublittoral sandy units containing a more varied fauna (predominantly *Trachycardium panamense, Chione californiensis, Polinices recluzianus, Anomia adamas*). The coquina is generally covered by a thin sheet of oxidized sandy silts (including reworked material), and by eolian sands of late Quaternary age. This unit, which is relatively resistant to erosion and which has a wide lateral extent, is interpreted to register the maximum of a transgressive event. It is assigned an IS 5e age, on the triple basis of its morphostratigraphic position (highest elevated and most recent marine remnant of the area), of a uranium-series date on *Chione* shell of 129 ± 13 ka (LU 981, Table 1), and of three

TABLE 1. RADIOMETRIC ^{230}Th/^{234}U DATA FROM PLEISTOCENE PELECYPOD SHELLS COLLECTED ALONG THE SONORAN COAST*

Sample Number	Genus†	Calcite§ %	Lab Number**	^{238}U (ppm)‡	^{234}U/^{238}U (activity)§§	^{230}Th/^{232}Th (activity)***	^{234}U/^{232}Th (activity)	Measured Age ^{230}Th/^{234}U (ka) †††	Individual Mean	Morphostratigraphic age (ka)§§§
LU 981	Chi	0	UQT 36	0.524 ± 0.012	1.226 ± 0.034	142.823 ± 30.298	199.455 ± 42.705	129 ± 13		125
LQ 202	Dos	<1	NICC 45	1.00 ± 0.03	1.21 ± 0.08			42 ± 8		125
LP 58	Dos	3	NIMB a	1.35 ± 0.04	1.16 ± 0.10			118 ± 17		125
LQ 181	Dos	17	NIMB b	0.89 ± 0.02	1.38 ± 0.10◊			180 ± 25		125
LP 49	Dos	<1	NIMB c	0.60 ± 0.02	1.26 ± 0.06			93 ± 13		125
LP 134 a	Dos	<1	NIMB 37	0.85 ± 0.02	1.22 ± 0.08			83 ± 12		
LP 134 b	Dos	0	NIMB 38	0.27 ± 0.01◊	1.24 ± 0.08			102 ± 14	98.5 ± 18	
LP 134 b			NICC 38	0.32 ± 0.02◊	1.14 ± 0.10	31.80 ± 7.500	49.43 ± 15.10	95 ± 18		
LP 134 c	Dos	<1	NICC 39	0.38 ± 0.02	1.39 ± 0.12◊	58.06 ± 16.02◊	95.898 ± 32.82	96 ± 14	108.0 ± 16	125?
LP 134 c			NIMB 39	0.43 ± 0.01◊	1.14 ± 0.08	48.60 ± 5.50◊	71.4 ± 12.5	120 ± 16		
LP 134 d	Dos	1	NICC 40a	0.55 ± 0.02	1.16 ± 0.05	115.30 ± 42.50	253.14 ±114.64	64 ± 10	64.5 ± 10	(or
LP 134 d			NICC 40b	0.57 ± 0.02	1.23 ± 0.08	74.50 ± 32.30◊	252.06 ±120.10	65 ± 10		105)
LP 134 e	Dos	<1	NIMB 41a	0.92 ± 0.02	1.15 ± 0.08		>100	88 ± 12		
LP 134 e			NIMB 41b	1.02 ± 0.03	1.17 ± 0.10	80.70 ± 15.00◊		84 ± 12	85 ± 13	
LP 134 e			NICC 41	0.96 ± 0.02	1.17 ± 0.08	31.00 ± 7.80◊	57.37 ± 17.96	83 ± 13		
LP 134 f	Dos	0	NIMB 42	0.48 ± 0.01◊	1.21 ± 0.10	5.8 ± 0.05◊	10.8 ± 1.1	82 ± 11		
LQ 149 a	Dos	<3	NIMB d	0.40 ± 0.01◊	1.26 ± 0.10			84 ± 12		125
LQ 149 b	Dos	<3	NIMB e	0.46 ± 0.01◊	1.18 ± 0.08			100 ± 14		
LP 1075e	Dos	0	UQT 42	1.147 ± 0.029	1.229 ± 0.031			66.6 ± 5		125?
LQ 141	Dos	<1	NIMB f	0.71 ± 0.02	1.21 ± 0.08			86 ± 12		125
LQ 175	Dos	<1	NIMB g	0.92 ± 0.02	1.26 ± 0.08			83 ± 12		125
LP 28	Dos	<3	NIMB h	2.66 ± 0.07◊	1.25 ± 0.04			22.8 ± 3	20.9 ± 3	125
LP 28			NIMB i	1.53 ± 0.4	1.28 ± 0.08			19 ± 2.5		

*See location of the samples in the maps of Figures 2, 4, 6, and 10, and in the sections shown in Figures 3, 5, 7, and 9. The Th/U geochronological results are poorly reliable, and generally provide only minimum ages (see text)
†Dos = *Dosinia ponderosa*; Chi = *Chione* sp.
§Calcite/(aragonite + calcite) x 100.
**UQT = Lab. Géotop (UQAM, Montréal); NICC = Lab. Géochronol., Univ. Nice (C. Causse); NIMB = idem (M. Bernat).
‡ ◊ = low (<0.5 ppm), or high (>2 ppm), uranium content.
§§ ◊ = Activity >1.30.
*** ◊ = ^{232}Th content <1 percent.
†††Calculated error corresponding to ± 1σ.
§§§See discussion in text.

out of four aminostratigraphic analyses (LS 647, LS 649, and LS 660, Table 2).

Between Punta Gorda and Bahia Adair, the younger deltaic and marine sequence unconformably overlies one or two series of marine sediments (Figs. 3b, c). The older units are distinguished from the IS 5e deposits by the stronger cementation of the sediments and by the fact that the invertebrate skeletal remains are much more altered than are the fossils of the younger unit. In the older unit(s), the remaining fossils are those that were originally composed of low-magnesium calcite, like *Encope* sp. (echinoderms), *Anomia* sp., *Ostrea palmula,* and *Pecten* sp. (pelecypods); the originally aragonitic mollusk shells (like *Chione* sp., *Trachycardium* sp., etc.) are found mostly in casts, many of which are now filled with sparry calcite. Petrographic studies clearly indicate that the older sediments suffered a long diagenetic history, involving several cycles of calcite dissolution/precipitation and neomorphic transformations, while the late Pleistocene sediments show only one main cycle of partial dissolution of bioclasts and intergranular cementation (Ortlieb, 1987).

The older sediments, which generally crop out underneath (or at least morphostratigraphically below) the late Pleistocene sequence, are interpreted to be coeval with one of the latest middle Pleistocene high sea-level stands (IS 7 and/or 9 ?). It has not been determined whether distinct outcrops of pre-late Pleistocene marine and nearshore units correspond to a single, or to several episode(s) of high sea level. From the present position of these sediments, and by extrapolating the uplift evidenced in the late Quaternary, it is inferred that during the last interglaciation(s) of the middle Pleistocene, sea level reached an elevation close to its present position.

The *Chione* coquina, and all previous nearshore sediments, are totally lacking in the sea cliffs from Punta Gorda northwestward. The thick fluvio-deltaic sands, which crop out in the cliff along the shore from Punta Gorda to the Rio Colorado mouth, are stratigraphically older than the late Pleistocene marine-deltaic sequence, and thus are assigned a late middle Pleistocene age. North of Golfo de Santa Clara, these relatively old fluvio-deltaic sediments are locally eroded and overlain by a narrow fluvial channel deposit that has been tentatively correlated with the late Pleistocene marine-deltaic sequence (Colletta and Ortlieb, 1981, 1984). The stratigraphic and geographic distribution of late Quaternary sediments (Ortlieb, 1987) suggests that during the last high sea-level stand (IS 5e) the Rio Colorado delta was forked, with another arm running southeastward from Yuma to the area of Salina Grande (Fig. 2).

Tectonic deformation. The *Chione* coquina, which is considered an index bed and is believed to have been originally horizontal (intertidal unit), shows that the southeastern part of the Mesa de Sonora underwent recent vertical motion. This unit crops out at a maximum elevation of +23 m at Punta Gorda, then slopes down toward the east, and is observed at +12 to +10 m between Punta Gorda and Bahia Adair. The altimetric position of the *Chione* coquina thus depicts both a local relative upwarping of more than 10 m at Punta Gorda and a wider regional uplift, which only amounts to a few meters (with respect to an assumed "eustatic" +6-m position of the sea during IS 5e).

At the top of the sea cliff at Punta Gorda, the *Chione* coquina is the only unit that is not displaced by the Punta Gorda fault. This WNW-ESE–oriented fault is the southeasternmost feature associated with the Cerro Prieto fault system, along the Mesa de Sonora (see location in Fig. 11; Colletta and Ortlieb, 1984). As a major structural feature, this fault system has probably been active during the entire Quaternary, and possibly since the Pliocene. From the deformation evident on the southwestern edge of the Mesa de Sonora, it is deduced that a major faulting event, accompanied by a strong uplift (or more than 100 m), occurred at the end of the middle Pleistocene. The uplift may be the result of a transpressive folding, like the one described at present on Durmid Hill along the Banning–Mission Creek (San Andreas) fault (Hudnut and others, 1985; Maloney, 1986). During the late Quaternary, the locus of the main faulting activity apparently migrated toward the southwest, along the present-day coastline, and the uplift continued, probably at a lesser rate than at the end of the middle Pleistocene. The vertical motion experienced since 125 ka by the southeastern rim of the Mesa de Sonora, rapidly decreased from a maximum (mean) rate of 200 mm/10^3 yr (at Punta Gorda) to a low rate of less than 30 mm/10^3 yr.

THE NORTHEASTERN EXTREMITY OF THE GULF OF CALIFORNIA

Pleistocene nearshore deposits. The coastal area and a large part of the hinterland of the northeastern extremity of the Gulf of California are mantled with thick dunes and eolian sand sheets, which were deposited during the major part of the Quaternary. This wide sand cover hampers the recognition of emerged remnants of high sea-level stands. Actually, the only well-identified Pleistocene marine deposits crop out along the present coastline. These partially cemented fossiliferous sands and gravels, which reach a maximum elevation of +7 to +8 m, have been described in the northeasternmost Bahia Adair (Ortlieb, 1987), in the Punta Pelícano–Puerto Peñasco area (Hertlein and Emerson, 1956; Ortlieb and Malpica, 1978; Malpica and others, 1978; Ortlieb, 1987), and in Bahia San Jorge (Estero San Jorge, Isla San Jorge near the railroad station at Almejas; Beal, 1948; Ortlieb, 1987; Fig. 2); no Pleistocene marine deposits were observed in southeastern Bahia San Jorge or along the Rio Concepcion delta (Fig. 2).

The outcrops from the Puerto Peñasco area provided ^{14}C ages of 35,200 and 43,000 B.P. on mollusk shells (Inman, *in* Sandusky, 1969), which are not considered reliable, and did not yield Th/U dates. Nevertheless, morphostratigraphic, petrographic, and paleontologic criteria support a chronologic correlation of these marine sediments with the last interglacial maximum (IS 5e).

In the vicinity of Puerto Peñasco, Ives (1951, 1959, 1964, 1971) reported possible remains of several distinct shorelines: at a

TABLE 2. AMINOSTRATIGRAPHIC DATA, AND COMPARISON WITH OTHER GEOCHRONOLOGICAL RESULTS OBTAINED ON PELECYPOD SHELLS FROM EMERGED PLEISTOCENE COASTAL DEPOSITS OF SONORA*

Sample Number	Genus[†]	Morphostratigraphic Age (IS)[§]	Radiometric Apparent Age (ka)**	alle/Ile Ratios[‡]	Racemized State[§§]	Aminostratigraphic Interpretation (IS)
LP 28	Dos	5e	32(◊)/21(¤)	0.90(·)		5e
LS 642 a	Chi	5e? (or 5c)	33.5(◊)	0.81		5e
LS 681 a	Chi		31(◊)	1.06	R	
LT 700 i	Pro			0.862		
LT 700 o	Chi	5e? (or 5c)	31–33(◊)	0.71		5e or 5c?
LT 700 o				0.69		
LQ 202	Dos	5e	42(¤)	0.86(·)		5e
LP 49 Dp	Dos			1.09(·)		
LP 49 d	Dos	5e	93(¤)	0.82		5e
LP 49 d				0.81		
LP 58 Dp	Dos	5e	118(¤)	0.89(·)		5e
LP 63	Chi			0.86(·)		
LP 134 Dp	Dos			0.88(·)		
LP 134 a	Dos	5e	82(¤)/100(¤)	1.33	R	5e
LP 134 a				1.24	R	
LP 134 d	Chi			0.84		
LQ 141 a	Chi	5e	86(¤)	1.11	R	5e
LQ 141 Dp	Dos			1.05(·)		
LQ 149 a	Dos	5e	84(¤)/100(¤)	0.91(·)		5e
LQ 149 b	Dos			0.96(·)		
LQ 181 Dp	Dos	5e	(180)(¤)	0.98(·)		5e
LQ 181 a	Dos			1.12		
LQ 175 Dp	Dos			0.89(·)		
LQ 175 a	Chi	5e	83(¤)	0.89		5e
LQ 175 d	Dos			1.20		
LQ 159 c	Dos	5e	....	0.95		5e
LS 647	Chi			0.86		
LS 649	Chi	5e	129(¤)	0.77		5e
LS 660	Chi			0.93		
LT 755	Chi			1.18	R	

*See location of the samples in Figures 2 to 10. The accuracy of the aminostratigraphic interpretation is limited by high late Quaternary temperatures in the Gulf of California, but it may be considered that localities that contain at least one unracemic sample are not older than early late Pleistocene (IS 5e. Ortlieb, 1987).
[†]Dos = *Dosinia ponderosa*; Chi = *Chione* sp.
[§]See discussion in text; IS = Isotopic stage.
**¤ = apparent Th/U age (see Table 1); ◊ = apparent ^{14}C age (see Table 3).
[‡]Analyses by G. Miller (INSTAAR, University of Colorado), except for (·) by R. Mitterer (University of Texas).
[§§]R = racemic samples.

few meters above MSL, at +23 m ("*Chione cancellata* shoreline"); at +35 m; and at +60 to 90 m ("*Turritella* shoreline"). Later studies in the area (Hertlein and Emerson, 1956; J. F. Schreiber, Jr., personal communication, 1977; Ortlieb and Nalpica, 1978) established that the only well-preserved coastal remnants are those related to the ~+7-m (IS 5e) high sea-level stand (Figs. 3d, f), and that the more elevated "shorelines" were either misinterpreted eolian sands or artificial accumulations of marine shells (by prehistoric Indians or by animals) (Ortlieb, 1987). In one locality in northeasternmost Bahia Adair (Fig. 2), not mentioned by Ives, two Quaternary marine deposits can be distinguished at maximum elevations of +7 and +13 m (Figs. 3d, e). The strong alteration of the sediments and poor exposure conditions hindered a definitive chronostratigraphic interpretation of these deposits; the lowest-lying unit shows many sedimentologic and faunal similarities with the *Chione* coquina of southwesternmost Bahia Adair and, on this basis, is tentatively assigned an IS 5e age.

Thus, in the northeasternmost Gulf of California, the remnants of a late Pleistocene transgression indicate a +6- to +7-m sea-level maximum, while middle Pleistocene emerged marine sediments have been preserved in only one locality (northeastern Bahia Adair) at as much as +13 m.

Beachrock and Holocene coastal deposits. Wide beachrock units crop out in the intertidal and upper sublittoral zones, between Bahia Adair and Bahia San Jorge (particularly at Playa Hermosa and Playa de Oro, respectively west and east of Puerto Peñasco), and were interpreted to be of possible Holocene age (Sandusky, 1969; Jones, 1975; Rose, 1975). Petrographic studies show that this extensive beachrock unit is primarily of late Pleistocene age (Ortlieb, 1987); evidence of dissolution by meteoric and vadose water of the aragonitic bioclasts and of the primary cement indicate that the unit was originally formed in the intertidal zone and subsequently emerged during a relatively long period. Recent typical beachrock cementation is now in process in the intertidal zone and affects Holocene sands and submerged outcrops of the IS 5e sandstone and beachrock (Fig. 3f). The succession of dissolution and cementation phases evidenced by petrographic analyses are diagnostic to distinguish between presently submerged Pleistocene and Holocene nearshore sediments.

In the northern Gulf of California, the high tidal range (about 4 m), the seasonally elevated temperature (above 30°C) and the abundance of sands and biogenic carbonates in the coastal area have been favorable conditions for beachrock formation during both the last (IS 5e) and the present interglacial periods.

A Holocene coastal sequence, including consolidated fossiliferous beach sands, lagoonal silty clays, and marsh peat beds, crops out below the high-tide level at Campo Santo Tomas, 30 km north of Rio Concepcion mouth (Fig. 2). This sequence pre-dates the formation of the Holocene dune ridge that rims the coastline all along the Rio Concepcion delta (Fig. 3g). Two ^{14}C ages of 4,000 ± 170 and 5,570 ± 270 B.P. were obtained on shell and peat samples (LT 933 and LT 936, Table 3). A stratigraphic study of these deposits suggests that Holocene sea level reached its present position about 5,000 yr ago; the regressive part of the sequence is more probably due to a progradation of the shoreline rather than to a fall of the sea level in mid-Holocene time.

This model of Holocene sea-level evolution, according to which the sea would have been close to its present position for several thousand years, is supported by many observations along the northern Gulf of California, and especially in the southwestern Rio Colorado delta (Thompson, 1968), Bahia Adair, and Bahia San Jorge (Ortlieb, 1987).

THE CENTRAL SONORAN COAST

Between Puerto Lobos and Guaymas (Fig. 4), the eastern coast of the Gulf of California is characterized by an alternation of large bays, protected by hilly headlands and steep rocky stretches, where the coastline cuts the N-S–oriented ranges. Three typical coastal landscapes may be distinguished in central Sonora: rocky cliffs, sandy beaches backed by eolian dunes, and piedmont sequences (bajadas) cut in vertical cliffs by the Holocene sea. Pleistocene high sea-level stands are registered as abraded littoral platforms or as sequences of nearshore sediments capped by late Quaternary eolianites or thick bajada deposits. The erosional features register the position of former sea-level maxima better than the depositional sequences do.

Puerto Lobos area

The coastal cliffs of Bahia Lobos, south of Puerto Lobos, contain remains of two Pleistocene marine sequences (Fig. 4).

The younger marine unit crops out on the abraded littoral platform, which constitutes the peninsula of Puerto Lobos (Fig. 5a), in the low sea cliffs located immediately east of Puerto Lobos village (Fig. 5b), and in the southeastern Bahia de Lobos (Campo Julio) (Figs. 5d, e). These beach deposits, which reach a maximum elevation of +7 m, are characterized by a weak cementation and a well-preserved fauna. They are overlain by a few-meter-thick, poorly consolidated alluvium. A *Dosinia ponderosa* shell from the marine unit (LQ 202; 1 km east of Puerto Lobos) provided a Th/U age of 42 ± 8 ka (Bernat and others, 1980; Table 1) and an aminostratigraphic result that suggests an IS 5e age (Table 2). Because of their morphostratigraphic position (below a late Quaternary continental sequence) and petrographic characteristics (similar to that of early late Pleistocene sediments of the northernmost gulf), these beach deposits are interpreted to be coeval with the IS 5e.

The older marine sequence crops out in the middle part of Bahia Lobos, at the base of the vertical sea cliffs, under 10 to 30 m of consolidated alluvium (Fig. 5c). The nearshore sediments, which include marine conglomerate and fossiliferous sandstones, as well as tidal-flat silts and estuarine gravels, are more indurated and altered than the younger marine unit. In these sediments, the aragonitic bioclasts have been dissolved and/or replaced by sparry calcite; the only preserved fossil shells are oysters (*Ostrea californica* cf. *osunai, O. columbiensis, O. corteziensis*) and cirripeds (*Tetraclita squamosa*). This composite nearshore sequence is seen from below sea level up to a maximum elevation of +12 m. Its age has not been determined, but is most probably middle Pleistocene.

Puerto Libertad area

The surroundings of Puerto Libertad (Fig. 4) contain the most numerous, and highest-elevated Quaternary marine remnants that have been identified along the Sonoran coast.

The youngest Pleistocene marine deposits crop out extensively to the north and south of Punta Bola, the headland that limits Bahia Libertad to the north (Fig. 4). They typically consist of poorly consolidated fossiliferous sands and gravels that overlie littoral platforms cut in substrates of various lithologies. These coastal marine sediments, in turn, are mantled by alluvium and/or several-meter-thick eolianites, especially at Playa Santa Maria (Fig. 5f), Playa Lobitos (Figs. 5h, i), and southeastern Bahia Libertad (Figs. 5j, k, l). The sea-level maximum coeval with these deposits reached an elevation of about +6 m. The faunal content of the unit, which is particularly fossiliferous northwest of the Puerto Libertad village, has been described by Stump (1975), Celis-Gutierrez (1979, 1980), Celis-Gutierrez and Malpica-Cruz (1981), and Ortlieb (1987).

A *Dosina ponderosa* shell, collected 1 km west of Puerto Libertad (LP 58), yielded a Th/U age of 118 ± 17 ka (Table 1). The same *Dosina ponderosa* specimen and a *Chione* sp. shell (LP

TABLE 3. RADIOCARBON DATA FROM LATE QUATERNARY COASTAL DEPOSITS ALONG THE SONORAN COAST*

Locality	Sample Number	Species	Calculated ^{14}C Age (uncorrected)	$\delta^{13}C/_{PDB}$	$\delta^{18}O/_{PDB}$	Analysis Laboratory
Estero Las Lisas	LS 662	*Chione* sp.	21,000 ± 1,600	+1.14	-1.14	University Paris Sud UPS 2435
Bahia San Jorge	LT 726	*Chione* sp.	26,500 ± 200	+1.13	-1.20	University Paris Sud UPS 3779
Santo Tomás	LT 933	Various mollusks	4,000 ± 170	-4.65	-1.50	University Paris Sud UPS 2442
	LT 936	Peat	5,570 ± 270			University Paris Sud UPS 2465
Rio San Ignacio Estuary	HGR 1	*Lyropecten* sp.	29,550 ± 1,115			Results from H. G. Richards, 1973
	HGR 2	*Muricanthus* sp.	26,770 ± 525			
	HGR 3	*Anomia peruviana*	≥42,000			
Tepopa Region	HGR 4	(*Chione* sp. ?)	26,770 ± 525			
	NPM	*Chione fluctifraga*	20,150 ± 230			University Paris VI
	LTF 1	*Chione fluctifraga*	27,900 ± 1,000			University Paris VI
	LTF 3	*Chione fluctifraga*	28,500 ± 950			University Paris VI
	LR 642	*Chione fluctifraga*	33,500 +2,945 / -2,140			Queens College (N.Y.) QC 472
	LS 666a	*Chione fluctifraga*	35,300 +2,500 / -1,900			Queens College (N.Y.) QC 556
	LS 666b	*Chione fluctifraga*	32,520 +1,725 / -1,420	+0.61		Queens College (N.Y.) QC 556/2
	LS 669	*Chione undatella*	30,580 ± 1,200	+0.88	-1.63	University Paris VI
	LS 680	*Chione fluctifraga*	33,800 ± 1,500	+0.48	-1.41	University Paris VI
	LS 681a	*Chione fluctifraga*	31,200 ± 850	+0.95	-1.09	University Paris VI
	LT 681b	*Chione fluctifraga*	24,150 ± 150	+0.61	-1.44	University Paris Sud UPS 2578
	LT 700a	*Ostrea palmula*	33,000 ± 1,500	+1.06	-1.22	University Paris VI
	LT 700b	*Trachycardium panamense*	31,530 ± 1,500	+1.06	-0.79	University Paris VI
Punta Kino	LP 28Dp	*Dosinia ponderosa*	32,500 ± 1,500			University Paris Sud

*Beside the results from Santo Tomas, which correspond to mid-Holocene samples (LT 933 and LT 936), all the apparent ^{14}C ages are interpreted to be minimum ages; emerged pelecypod shells, which provided dates in the range 20,000 to 35,000 B.P., are most probably of early late Pleistocene (IS 5) age. The measured activity is attributed to contaminations by modern organic matter (1 to 3 percent).

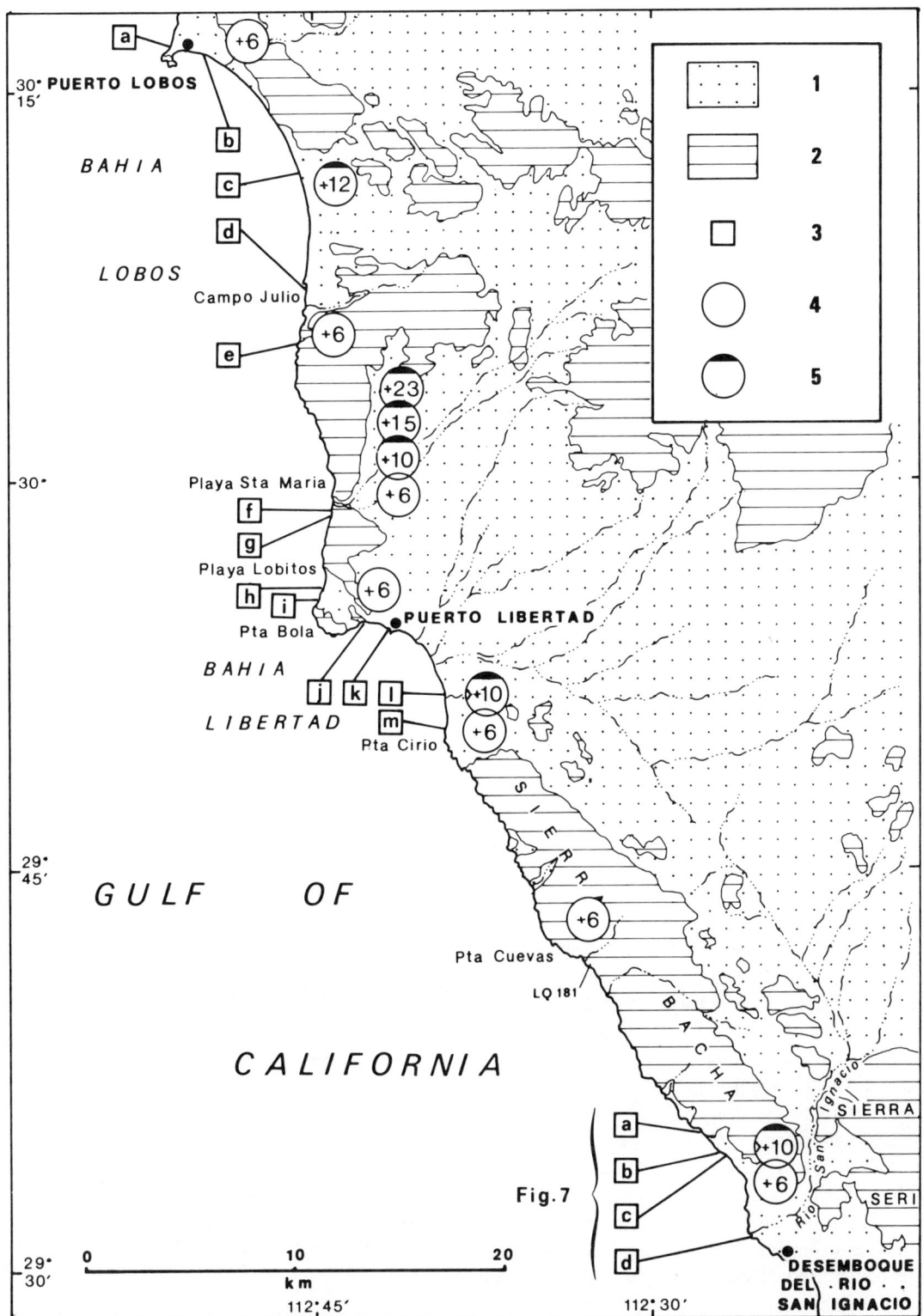

Figure 4. Puerto Lobos–Punta Cuevas region: distribution and height of Pleistocene shorelines. 1. Quaternary alluvium and eolian sands. 2. Pre-Quaternary units (geologic contours simplified from Gastil and Krummenacher, 1974). 3. Location of the sections of Quaternary coastal deposits shown in Figure 5 (a to m, upper half), and in Figure 7 (a to d, bottom). 4. Maximum elevation (meters above present MSL) of the early late Pleistocene (IS 5e) shoreline. 5. Elevation of middle Pleistocene shorelines (meters above present MSL).

63) gave aminostratigraphic data (Table 2) that further confirm the IS 5e age previously inferred solely on a morphostratigraphic basis (Malpica and others, 1978; Ortlieb and Malpica, 1978; Ortlieb, 1980).

Along southeastern Bahia Libertad, marine and fluvial sediments have been preserved below the early late Pleistocene marine unit and the more recent eolianite (Figs. 5l, m). These sediments are mainly infralittoral coarse sandstones with abundant echinoids (*Encope micropora, E. grandis*), oysters shells (*Ostrea fisheri, O. columbiensis*), and badly altered shells, or molds of aragonitic mollusks (*Dosinia ponderosa, Oliva* sp., *Strombus* sp., *Polinices* sp., *Cypraea* sp.); the matrix is well cemented with calcite and shows evidence of a complex diagenetic history. This marine unit has been abraded, at an elevation of a few meters above present MSL, during two successive episodes of high sea-level stands. Before the last interglacial maximum, identified by the most recent unconsolidated coastal sediments, another transgression left a few remnants that consist of a condensed fluvio-marine sequence capped by a layer of very thick oysters (*Ostrea fisheri* and *O. corteziensis*) (Fig. 5l).

Thus, two middle(?) Pleistocene transgressions have been registered in southeastern Bahia Libertad, before the deposition of the IS 5e sediments. The maximum sea level coeval with the oldest marine unit was probably at more than +10 m, while the reconstructed paleo-sea level responsible for the intermediate fluvio-marine unit probably reached a lower elevation, between +5 and +10 m.

In the embayment of Playa Lobitos, north of Puerto Libertad (Fig. 4), the main outcrops of Quaternary coastal deposits are those left by the late Pleistocene highest sea-level (IS 5e) episode, which point to a +6-m sea level (Figs. 5h, i). In the center of the bay, a single outcrop of cemented gravel may be observed below high-tide level; this small remnant of an old unit, with recrystallized bioclasts, is tentatively correlated with the oldest middle Pleistocene sublittoral sequence described in southeastern Bahia Libertad.

At Playa Santa Maria (named Playa Santa Margarita by Gastil and Krummenacher, 1974; Fig. 4), coastal deposits, which have many similarities with those described at Playa Lobitos and Bahia Libertad and which reach the same +5 to +6-m elevation, are also assigned an IS 5e age (Fig. 5f). Three other littoral platforms, veneered with marine pebbles, are preserved at elevations of +10 to +12, +15, and +23 m (Fig. 5g). These remains of Pleistocene high sea-level stands are devoid of any fossil material and therefore could not be dated; as the shore platforms do not support consolidated sediments, no correlation based on petrographic analyses could be established with the middle Pleistocene marine remnants of Bahia Libertad. Nevertheless, the geometric relations existing between these marine benches and the late Quaternary coastal dunes overlying the IS 5e marine unit indicate that the former must have been carved out before the late Pleistocene. The +23-m marine bench of Playa Santa Maria is the highest-elevated Pleistocene shoreline known in the eastern Gulf of California.

South of Punta Cirio, along the Sierra Bacha coast, a low-lying marine terrace, correlatable with the IS 5e sea-level maximum, is relatively well preserved. A conspicuous marine bench, which reaches a maximum +5-m elevation and supports associated nearshore deposits, is visible north and south of the Punta Cuevas fishermen camp. A Th/U age of 180 ± 25 ka, obtained on a *Dosinia ponderosa* shell from Punta Cuevas (LQ 181, Table 1), is not significant because 17 percent of the original aragonite has recrystallized to calcite. Aminostratigraphic data from the same shell and another one from the same locality (LQ 181, Table 2) rule out the possibility that this marine terrace is older than IS 5e (or that the shells had been reworked from a pre–late Pleistocene deposit). In the same bay of Punta Cuevas, there are some faint morphologic indications of an older (middle Pleistocene) shoreline several hundred meters inland from the coastline and under a relatively thick unit of late Quaternary alluvium (Ortlieb, 1987).

Tepopa area

Morphostratigraphy of Pleistocene nearshore units. The coastal region extending from the southern extremity of Sierra Bacha to the northern end of the Canal del Infiernillo (Figs. 4 and 6) contains two sequences of Pleistocene marine sediments.

The two sequences are observed in stratigraphic order at the

Figure 5. Sections of Quaternary coastal deposits in the Puerto Lobos (a to e)–Puerto Libertad (f to m) region (see locations in Fig. 4). Section a: 1. Marine platform cut in Mesozoic volcanic substrate. 2. Late Quaternary terrestrial sediments. Section b: 1. Consolidated alluvium (latest middle Pleistocene). 2. Early late Pleistocene nearshore sequence (LQ 202). 3. Unconsolidated bajada sediments (late Quaternary). Section c: 1. Cemented alluvium (middle Pleistocene). 2. Middle Pleistocene nearshore sequence. 3. Consolidated bajada sediments (latest middle Pleistocene) (see b1). Section d: 1. Tidalites and bioclastic littoral sands (middle Pleistocene). 2. Late Pleistocene nearshore sequence. 3. See b3. Section e: 1. See c1. 2. See d2. 3. See d3. Section f: 1. Marine platform cut in Mesozoic substrate. 2. Early late Pleistocene nearshore sequence. 3. Late Quaternary sand dune. Section g: 1. Mesozoic substrate. 2. Littoral platforms veneered with marine pebbles, at +10 to +12 m and +15 m (middle Pleistocene). 3. Highest observed marine platform (+23 m), covered with poorly cemented gravels (middle Pleistocene). 3. See f3. Section h: 1. Consolidated conglomerate including shell fragments (middle Pleistocene). 2. Early late Pleistocene nearshore sequence. 3. Late Pleistocene lagoonal fine sands and alluvium. 4. Late Quaternary sand dune. Section i: 1. Granodioritic substrate. 2. Early late Pleistocene nearshore sequence. 3. Late Quaternary slope deposits and alluvium. 4. See h4. Section j: 1. Early late Pleistocene nearshore sequence, ending with a *Tagelus californianus–Chione californiensis* coquina. 2. Late Quaternary sand dune. Section k: 1. Wave-abraded Miocene volcanoclastic substrate. 2. Early late Pleistocene nearshore sequence (LP 58 and LP 63). 3. Late Quaternary alluvial cover. Section l: 1. Fossiliferous sublittoral sand, rich in echinoids and oysters, locally covered by unconformable bioherms (middle Pleistocene). 2. Early late Pleistocene upper-beach deposits. 3. Late Quaternary sand dunes. Section m: 1. Mesozoic granitic substrate. 2. Middle Pleistocene nearshore sequence. 3. See l3.

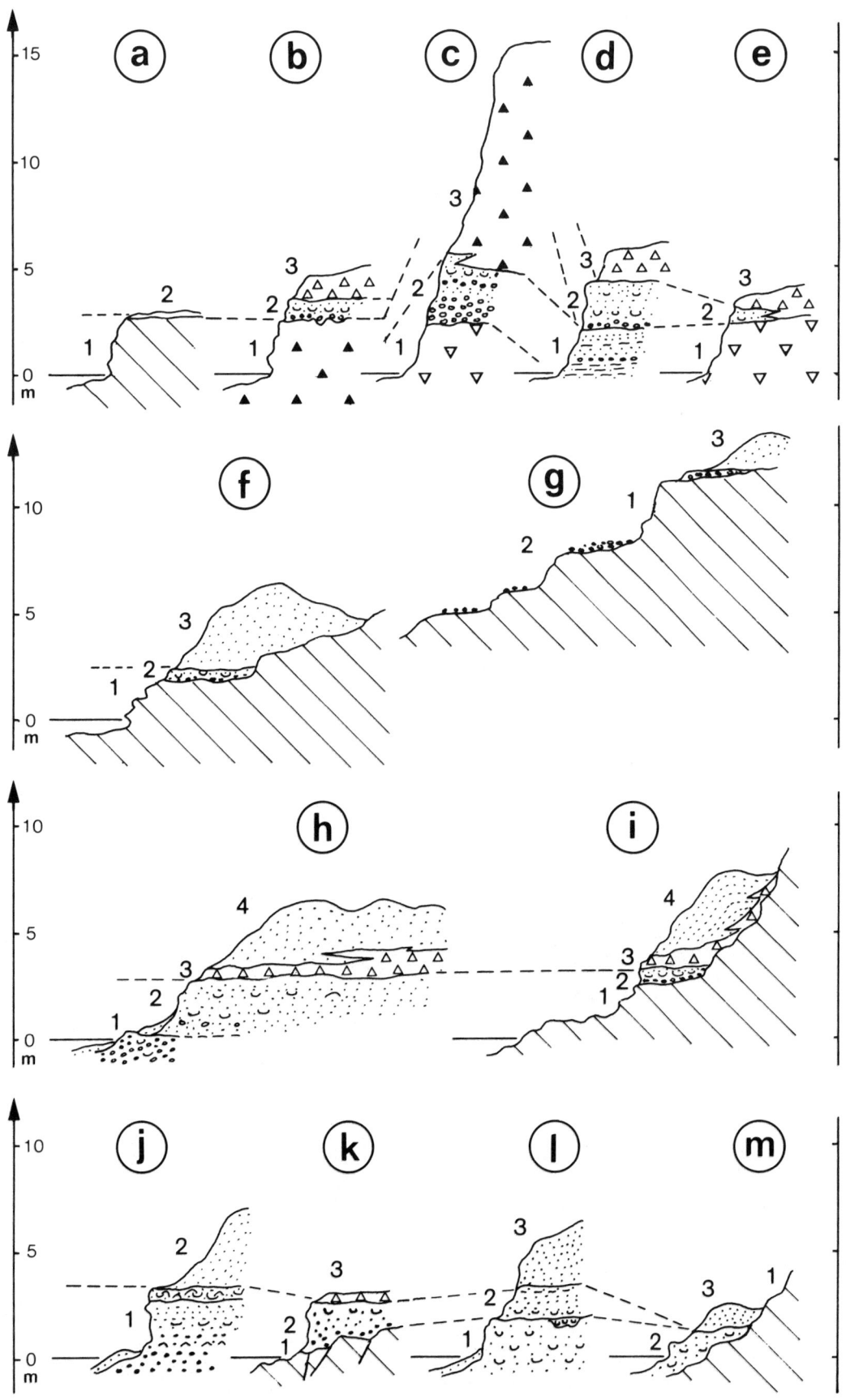

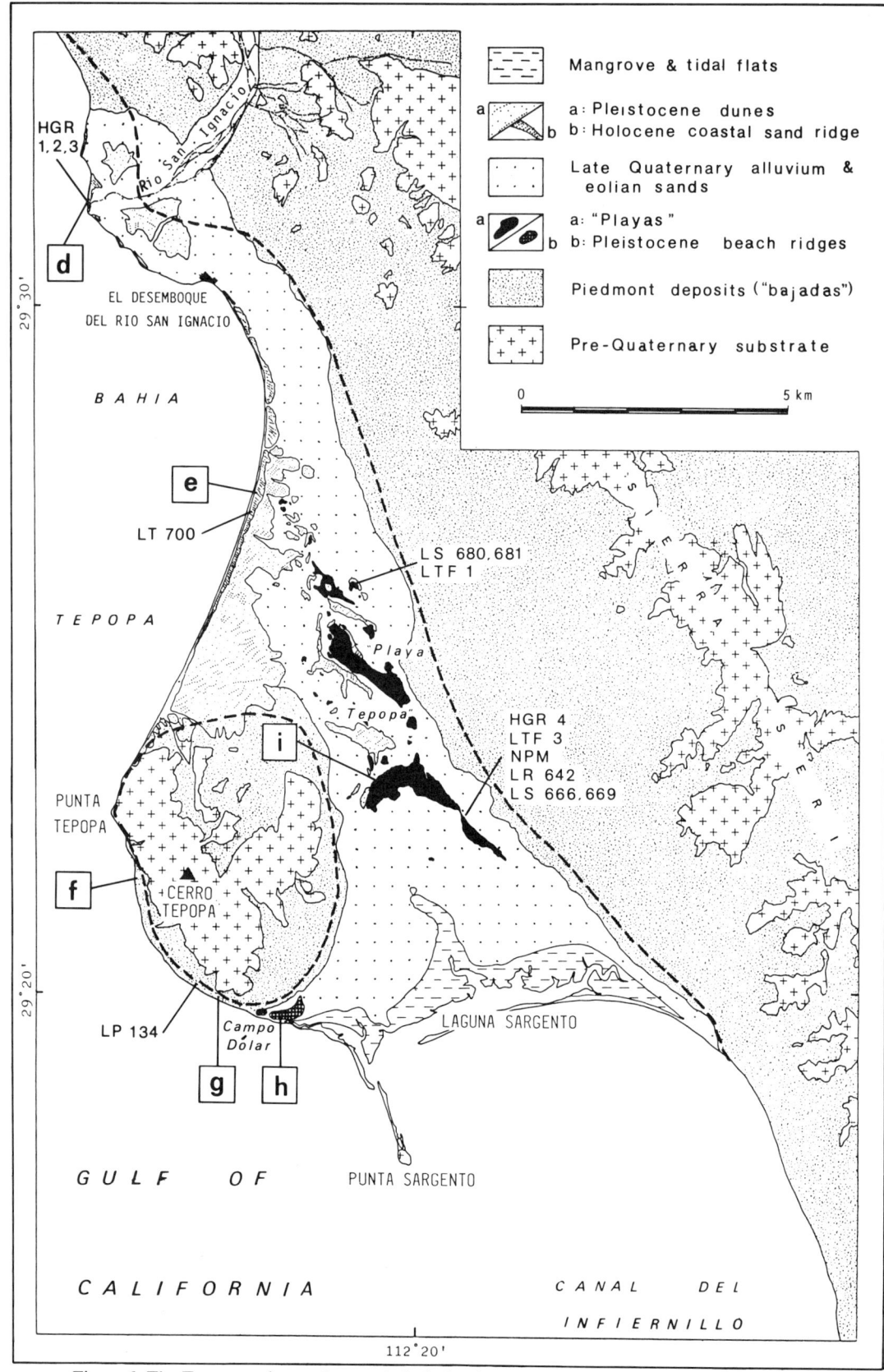

Figure 6. The Tepopa region: main morphological units, and locations of the sections of Quaternary coastal deposits shown in Figure 7 (d to h) and geochronological samples of Tables 1, 2, and 3.

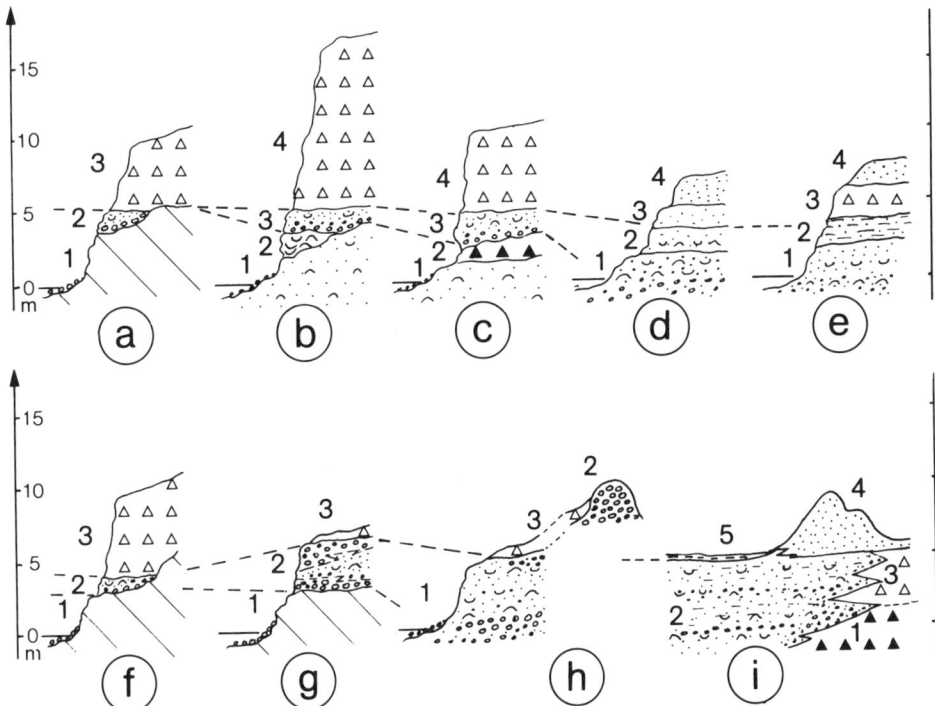

Figure 7. Sections of Quaternary coastal deposits in the Rio San Ignacio delta, Tepopa region (see location in Figs. 4 [a to d] and 6 [d to i]. Section a: 1. Mesozoic granodioritic substrate. 2. Early late Pleistocene nearshore sequence. 3. Late Quaternary bajada deposits. Section b: 1. Fossiliferous cemented gravels and sandstones (middle Pleistocene). 2. Bioherm of *Ostrea fisheri* (middle Pleistocene?). 3. Early late Pleistocene nearshore sequence (LP 49). 4. See a3. Section c: 1. See a1. 2. Consolidated alluvium (latest[?] middle Pleistocene). 3. See b3. 4. See b4. Section d: 1. Poorly consolidated and very fossiliferous conglomerate (early late Pleistocene). 2. Upper-beach deposits (early late Pleistocene). 3. Late Pleistocene eolianite. 4. Holocene sand dune. Section e: 1. Early late Pleistocene beach deposits (LT 700). 2. Late Pleistocene lagoonal sediments. 3. Late Quaternary alluvium. 4. Late Quaternary eolian sands. Section f: 1. Mesozoic substrate. 2. Early late Pleistocene nearshore sequence. 3. Late Quaternary slope deposits. Section g: 1. See f1. 2. Early late Pleistocene nearshore sequence, including thick units of boulders (LP 134). 3. See f3. Section h: 1. Fossiliferous early late Pleistocene nearshore sequence. 2. Early late Pleistocene shingle ridge deposits. 3. Late Quaternary alluvium cover. Section i: 1. Latest middle Pleistocene alluvium. 2. Early late Pleistocene nearshore and lagoonal sequence. 3. Late Quaternary bajada deposits. 4. Late Quaternary eolian sands. 5. Playa reworked sediments (Holocene).

southwestern end of Sierra Bacha (S. Tordilla), 4 km north of the mouth of Rio San Ignacio (Figs. 7a, b, c). The older sequence crops out at the base of the coastal cliff, up to an elevation of +5 to +10(?) m, and mainly consists of well-cemented gravels and sandstones, which include pectens (*Pecten vogdesi, Argopecten circularis*), oysters (*Ostrea corteziensis*), and molds of other mollusk shells (*Trachycardium* sp., *Codakia distinguenda, Glycymeris* sp.). This several-meter-thick unit is locally covered by an unconformable bioherm of *Ostrea fisheri* and *Encope grandis*, which postdates it (distinct interglacial high sea-level stand, or separate high sea-level episode during the same interglaciation?).

Stratigraphically above all these deposits (Figs. 7b, c), or directly above a marine bench carved out in the granodioritic substrate (Fig. 7a), lies a younger marine unit formed by poorly consolidated beach sands and gravels. The sea level coeval with this deposit may be reconstructed at about +6 m above MSL. *Dosinia ponderosa* shells, collected in this marine unit (LP 49), yielded an apparent Th/U age of 93 ± 13 ka (Table 1) and two aminostratigraphical results compatible with an IS 5e age (Table 2).

The youngest marine unit is covered by several meters of unconsolidated bajada deposits that accumulated during the late Quaternary. The older marine sequence (Figs. 7b, c), which is also locally overlain by bajada deposits (twice as thick and a little more cemented than the late Quaternary alluvium), is interpreted to be of latest(?) middle Pleistocene age.

Around the headland of Punta Tepopa (Fig. 6), the IS 5e high-sea-level stand reached a maximum elevation of about +6 m, evidenced by almost continuous coastal deposits. During this early late Pleistocene high sea-level stand, the steep, rocky coastline was actively eroded by wave action and was bordered by narrow cobble beaches. Southeast of Cerro Tepopa, this late

Pleistocene nearshore conglomerate grades to more fossiliferous and finer sediments that were deposited in a protected environment.

Combined morphostratigraphic, sedimentologic, and paleontologic studies in the Tepopa region (Lecolle and Ortlieb, 1978; Lecolle, 1980; Ortlieb, 1981, 1984a, 1987; Luna-Guerin, 1981; González-González, 1982; González and others, 1984) indicate that all the marine sediments cropping out between the southwesternmost Sierra Bacha and Punta Tepopa are most probably penecontemporaneous with the last interglacial high stand(s) of sea level (IS 5). These deposits correspond to various nearshore paleoenvironments and include: fluvio-marine conglomerates at the Rio San Ignacio mouth, open-shore fossiliferous sandstones in Bahia Tepopa, lagoonal and shallow-bar fine sands in the center of the Tepopa bolson ("Playa Tepopa" of various authors), and shingle and cobble beach ridges to the southeast of Cerro Tepopa (Fig. 6).

A paleogeographic reconstruction of the Tepopa area suggests that at the beginning of the last interglaciation (IS 5e) the Cerro Tepopa was an island, which later became attached to the mainland by a tombolo that grew southward from the area of El Desemboque village (Fig. 6). After the building-up of this land bridge and the encroachment of coastal dunes, the "Playa Tepopa" area constituted a protected embayment that rapidly infilled with fine-grained and fossiliferous (*Chione californiensis, C. fluctifraga, Rhynocoryne humboldti, Cerithium stercusmuscarum, Certhidea mazatlanica*) sediments. Subsequently, sea level dropped, and the paleoembayment of Tepopa was covered with various types of eolian dunes: the sand source was provided by the nearby exposed infralittoral area. During the glacial period (encompassing IS 4, 3, and 2), the foot of the bajadas flanking the Sierra Seri and the Cerro Tepopa prograded toward the center of the bolson and totally buried the former shorelines. In the Holocene, the post-glacial sea invaded the southern part of the bolson and formed the El Sargento lagoon, which was promptly bordered by a thick mangrove (Sherwin, 1971; Fig. 6). In the last few thousand years and up to the present, the "Playa Tepopa" lows (altitude +4 m) have been periodically inundated by runoff waters.

In such reconstitution of the late Quaternary paleogeographic evolution, which takes into account all the available data from all the outcrops of marine and nearshore sediments in the Tepopa area, and from a series of auger perforations (maximum depth from the surface: 6 m) in the "Playa Tepopa" zone (Lecolle, 1980; Ortlieb, 1984a, 1987), it could not be unequivocally determined whether the early late Pleistocene high sea-level stand was a single one (IS 5e), or if the IS 5e was followed by another high stand (IS 5c?) that would have left its deposits above present MSL.

Geochronological analyses of mollusks from this region were of limited help in unravelling this uncertainty, and illustrated the limits of reliability of the chronological methods.

Geochronological problems and neotectonic implications. A series of six *Dosinia ponderosa* shells was collected in situ (with their articulated valves) in a single locality, 3 km west of Campo Dólar (Figs. 6 and 7g), and submitted for U-series analyses (LP 134, Table 1). The results, including repeated measurements on the same individuals, show a wide range of apparent ages (64 to 120 ka); such a scatter of results necessarily reflects an anomaly, since the conditions of occurrence of the fossils leave no doubt that the sampled shells lived at the same time (within a few centuries?) and none appear to have been reworked.

It is known that, unlike corals which are at equilibrium with sea water, mollusk shell carbonates incorporate most of their uranium after their depth (Broecker, 1963). This particular characteristic, which for many authors (Kaufman and others, 1971; Ku, 1976; and others) discredited mollusks as suitable material for U-series geochronology, explains that "molluscan" calculated ages are generally younger than "coral" ages and should be taken as minimum ages (Veeh and Valentine, 1967; Stearns, 1980). For the same reason it may be considered that in a series of dates from a collection of coexisting mollusk shells, the oldest apparent ages have more probability of being close to the true age than the statistical mean value. The set of data from the Campo Dólar locality, which is most probably 125,000 yr old, brings another confirmation of such interpretation.

Another geochronological problem met in the Tepopa area deals with a series of radiocarbon ages that suggest the main nearshore units cropping out above MSL are about 30,000 yr old.

After a preliminary ^{14}C dating of four shells by the late H. G. Richards (1973), a dozen additional radiocarbon dates were obtained on shells of the El Desemboque–Tepopa area (Table 3). The 16 dated shells were sampled in several places and various paleogeographic units: at the Rio San Ignacio mouth (three samples from a deltaic unit), 5 km south of El Desemboque (two samples from an intertidal unit on the former tombolo), and in the "Playa Tepopa" lows (11 samples from laguno-marine beds) (Fig. 6). The apparent ^{14}C ages of 15 of these samples range from 20,150 to 35,300 B.P. (most of the ages between 27,000 and 33,000 B.P.), while one last sample (from Richards, 1973) is apparently ≥42,000 B.P. (Table 3). Such a series of relatively consistent radiometric ages (except for the two extreme values of ~20 ka and ≥42 ka) suggests that the three main outcrops of nearshore sediments from the Tepopa–El Desemboque area are coeval with an ~30-ka episode of high sea level (last warm interstadial IS 3).

The significance of ^{14}C-dated 30-ka emergent shorelines in regions of the world that have not been notably uplifted has been much debated in the last three decades (see discussion by Morner, 1971; Thom, 1973; Giresse and Davies, 1980); in many cases (particularly in areas where the well-identified IS 5e shoreline is preserved at a low elevation above present MSL), it has been concluded that the radiocarbon age determination was inaccurate, and that the 30-ka age calculations probably resulted from small contamination by modern carbon.

Aminostratigraphic measurements on some coexisting shells (*Chione* sp. and *Protothaca* sp.) from the radiocarbon-dated "Playa Tepopa" beds and the Bahia Tepopa locality (Table 2) do not support an age much younger than 125 ka. As only one sample (LT 700o; *Chione* sp.) provided a relatively low alle/Ile ratio, the aminostratigraphic data are inconclusive as to whether these deposits are strictly coeval (IS 5e; 125 ka) with, or slightly younger (IS 5c?; ~105 ka?) than, the Punta Tepopa–Campo Dólar samples.

In conclusion, the morphostratigraphic evidence that the Tepopa area was only submerged once in the late Quaternary and that this high sea level almost certainly corresponds to the IS 5e are considered more reliable than most apparent ages (obtained by one method or the other). The idea that the Tepopa beds may be accurately dated by the radiocarbon method is rejected, and it is interpreted that the apparently consistent radiocarbon ages result from a generalized contamination by 1 or 2 percent modern carbon. Such contamination was probably facilitated by the pervious nature of the sandy sediments and endoreic character of the area.

It has been emphasized that the Th/U-dated shells from the Campo Dólar locality should be considered coeval with the IS 5e, in spite of the spread of individual geochronologic results. The aminostratigraphic results seem to confirm the IS 5e age assignment to two localities, although one sample may be interpreted as possibly younger (IS 5c?) than the IS 5e episode.

The proposed chronostratigraphic interpretation implies that no recent vertical displacements have been recorded in the area. Actually, the NW-SE faults (Gastil and Krummenacher, 1974) that controlled the typical Basin-and-Range morphology of the region do not seem to have been active during the late Quaternary.

Canal del Infiernillo and Bahia Kunkaak

On the sides of the Canal del Infiernillo, between Tiburon Island and the continent, late Quaternary marine sediments have been widely covered by an overwhelming accumulation of colluvium derived from the Sierra Seri (mainland side) and Sierra Kunkaak (on Tiburon Island) (Fig. 8). Furthermore, Holocene constructional coastal features (sand spits, marginal lagoons, and mangroves; see Sherwin, 1971; Lancin, 1979, 1985) contributed to restrict the exposures of the Pleistocene marine deposits. Thus, Pleistocene nearshore sediments, which are correlatable with the last interglaciation, are observed in only four localities: Palo Fierro (Fig. 9b), Punta Tormenta (Fig. 9c), Punta Santa Rosa, and 2 km north of Punta Chueca (Fig. 9j). In the first three localities, fossiliferous conglomerates, deposited in intertidal and uppermost sublittoral zones, point to a maximum sea-level stand at ~+5-m elevation. The sequence located between Punta Chueca and Punta Onah (Fig. 9j) is different from the deposits of the other outcrops: the top of the marine beds reaches a maximum +9-m elevation, and the base of the sequence is made of infralittoral fossiliferous sands. The latter unit contains micro- and macrofauna that suggest that the coeval sea level was more than 10(?) m above the mentioned bed (Celis-Gutiérrez, 1975, 1979; Stump, 1981). A *Dosinia ponderosa* shell (LU 1075e, Table 1) from this infralittoral unit yielded a Th/U age of 67 ± 5 ka. The fact that this sequence crops out in the only cliffed coastal segment of the Infiernillo Straits supports the hypothesis that this locality registered a limited fault-controlled uplift motion.

Without accepting at face value the single Th/U apparent age of the LU 1075 sample, it is conceivable that the sequence could be younger than the IS 5e episode, and that it was uplifted more than a few meters in the second half of the late Quaternary. Recent vertical deformation observed between Punta Chueca and Punta Onah has a limited lateral extent, although it should be noted that the prolongation of the same fault trace (Gastil and Krummenacher, 1974) crosses the Tepopa area. This fault activity is interpreted to result from a small structural reaccommodation at the margin between the Basin and Range province and the Gulf of California system (Ortlieb and others, 1989).

In the Palo Fierro locality (Fig. 9c), two *Dosinia ponderosa* shells submitted for Th/U dating gave apparent ages of 84 ± 12 and 100 ± 14 ka (Table 1), which are interpreted as minimum ages. Aminostratigraphic data (Table 2) support assignment of this locality to the IS 5e.

In Bahia Kunkaak, the remnants of a late Pleistocene high sea-level stand consist of gravelly sandstones deposited on wave-cut benches (maximum elevation +6 m) and in low-lying nearshore sequences that upgrade into alluvial and colluvial deposits (Fig. 9k). The reconstructed high sea-level stand is interpreted to be of IS 5e age, on the basis of the low degree of alteration of the sediments and the fauna (when the deposits are located above the Holocene high-tide level) and because of the assignment of the few-meter-thick bajada unit to the last glacial cycle. No evidence of older Pleistocene high sea-level stands has been found in the area; any emerged middle Pleistocene marine terrace would be deeply buried under the thick alluvium, inland from the present coastline, and/or would have been eroded by the early late Pleistocene transgression.

Isla Tiburon

Tiburon Island registers the IS 5 high sea-level stand relatively well, at least along the rocky stretches of the coast (Fig. 8). At its northern end, wave-cut benches covered with nearshore sediments and slope deposits clearly indicate that the sea level reached a ~+6-m elevation (Punta Ast Hoe Ben Oh Galp, Fig. 9a, southwestern Agua Dulce, Tecomate). Similar features were observed along the southeastern coast (embayments of El Perro, Fig. 9d, and La Cruz, Fig. 9g) and the southwestern part of the island (Punta Sauzal, Fig. 9h, Bahia Blanca, Bahia Vaporeta).

Two *Dosinia ponderosa* shells collected at Ensenada El Perro (LQ 141) and Ensenada La Cruz (LQ 175) provided Th/U ages of 86 ± 12 and 83 ± 12 ka, respectively (Table 1); aminostratigraphic results on the same individuals and other shells (Table 2) favor an IS 5e age. Like in the above mentioned Palo Fierro locality (northeastern coast of the island), it is interpreted that the radiometric dates should be taken as minimum ages, and that all these deposits are more probably coeval with the early late Pleistocene (IS 5e) high sea-level stand.

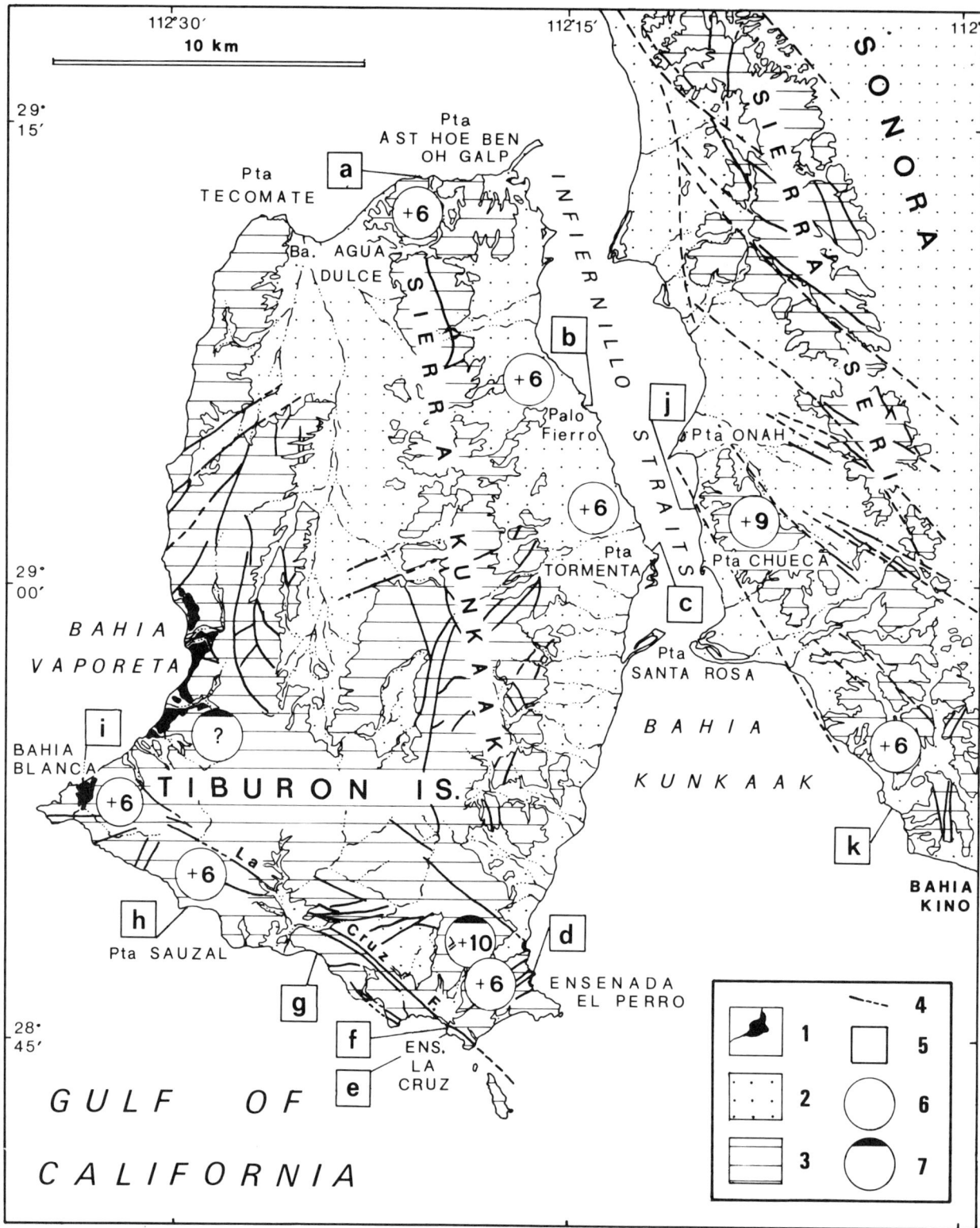

Figure 8. Tiburon Island and the Infiernillo Straits: location of the sections of Quaternary coastal deposits (a to k) shown in Figure 9. Simplified geology from Gastil and Krummenacher (1974). 1. Quaternary marine terraces, according to Gastil and Krummenacher (1974). 2. Quaternary alluvium. 3. Pre-Quaternary units. 4. Observed and inferred faults. 5. Location of the sections of Quaternary coastal deposits shown in Figure 9 (a to k). 6. Maximum elevation (meters above present MSL) of the early Late Pleistocene (IS 5e) shoreline. 7. Elevation of middle Pleistocene shorelines (meters above present MSL).

It should be added that the southern coast of Tiburon Island shows numerous evidences of post-Miocene faulting activity ("La Cruz fault" system of Gastil and Krummenacher, 1974, 1977: "Southern Tiburon fault" of Sanchez and others, 1985). If some of these faults had been recently active (Ortlieb and others, 1989), they do not appear to have induced important vertical offsets, at least during the late Quaternary, since the IS 5e shoreline is identified in several localities at its common elevation of +5 to +6 m (very close to its assumed "eustatic" position).

A few middle Pleistocene marine deposits are preserved in the southwesternmost and southeasternmost parts of Tiburon Island. At Ensenada La Cruz (Fig. 9e), indurated and poorly fossiliferous silts, unconformably overlain by IS 5e sediments, which are probably the localities previously studied by Anderson (1950) and Durham (1950), suggest that during their deposition the sea level was at least 10 m above the present datum.

In Bahia Blanca (Fig. 9i), a marine platform covered by a thick unit of consolidated eolianite is interpreted as probably pre–Late Pleistocene. In northern Bahia Vaporeta, elevated (middle and early?) Pleistocene marine terraces have been mentioned by Gastil and Krummenacher (1974, 1977) but have not yet been specifically studied. It would not be surprising if this region of Tiburon Island, which is thought to show exposures of the earliest Miocene marine rocks of the whole Gulf of California (Smith and others, 1985), would still be experiencing slow uplift.

Bahia Kino–Guaymas area

The Bahia Kino–Laguna La Cruz area (Fig. 10) is dominated by latest Quaternary (mostly Holocene) sediments; late Pleistocene marine sediments have only been observed around Cerro San Nicolas. Littoral embankments on the northern and western exposed coasts of this headland show that late Quaternary nearshore beds reach a maximum +5-m elevation, and that they are covered with several meters of eolianites and alluvium (Fig. 9l). Various morphostratigraphic arguments suggest that these deposits are of early late Pleistocene age (IS 5e), even if one of these localities (Punta Kino) yielded discordant radiochronologic data: coexisting shells provided a relatively young Th/U age (20.9 ± 3 ka, mean of two measurements on the same individual of *Dosinia ponderosa,* Table 1), an older ^{14}C age (32,500 ± 1,500 B.P., Table 3), and an aIle/Ile ratio that supports an IS 5e age (Table 2). On one hand, the high content of ^{238}U measured in the LP 28 sample reveals an anomalous intake of uranium, which invalidates the Th/U geochronologic determination; on the other hand, the radiocarbon age, which is close to the theoretical range limit, is not reliable either.

The wide coastal plain that spreads between Cerro San Nicolas and Tastiota ("Costa de Hermosillo", or "Llanos de San Juan Bautista," Fig. 8) is formed by late Quaternary alluvial and fluvio-deltaic fine sediments from the paleo-Rios Sonora and La Poza; its shore is lined with an uninterrupted dune ridge that has been built in the second half of the Holocene. Preliminary stratigraphic and sedimentologic studies in the coastal lagoons of the area (La Cruz, El Cardonal, Tastiota) suggest that the postglacial sea level practically reached its present position about 5,000 yr ago (Nichols, 1965).

Remnants of late Pleistocene high sea-level stands reappear as narrow abraded terraces along the rocky coastline south of Tastiota (Fig. 9m). These benches, which have generally been stripped of nearshore sediments (except near Tastiota), are observed in several places at about +5 m along the coast of Sierra Algodones southward to San Carlos (Fig. 10).

The hilly Guaymas area differs from the other rocky coastal regions of Sonora by the fact that its steep cliffs and screes do not present any wave-cut remnants above the Holocene shoreline. The harbor of Guaymas, located in a faulted Miocene caldera, has been recognized as a typical, recently submerged, drowned area (Anderson, 1950). The nearby Empalme lagoon probably constitutes another evidence of such recent local submergence. This late Quaternary subsidence, apparently related to some unusual microseismic activity (G. Ness, personal communication, 1986) and to local gravimetric anomalies (Harrison and Mathur, 1964), is interpreted as a possible consequence of the termination of a transform fault of the gulf system, close to the Guaymas basin "spreading center" (Moore, 1973; Bischoff and Henyey, 1974).

THE SOUTHWESTERN GULF OF CALIFORNIA

South of Guaymas, the coast of southern Sonora and Sinaloa (Fig. 1) is largely dominated by Holocene progradational features (coastal dunes, lagoonal complexes, deltaic sediments, alluvial plains, etc.) and by the hydrologic regime of several permanent rivers coming from the Sierra Madre Occidental; these conditions are particularly unfavorable for the study of Pleistocene shorelines (and explain why this region has not been more studied). The rarity of emerged remnants of Pleistocene high sea levels in the southwestern Gulf of California seems to be due to poor conditions of exposure rather than to a regional subsidence. In a few localities, such as Mazatlán (Gutiérrez-Estrada and Castro del Río, 1984) or northwestern Nayarit (Curray and others, 1969), late or middle Pleistocene marine sediments have been observed near present MSL.

In this part of the Gulf coast, between Mazatlán and San Blás, controversial ideas have been published concerning the late Quaternary vertical movements in relation to postglacial sea-level-rise evidences. Curray and others (1969) elaborated a local model of relative sea-level rise, according to which the coastal area could be considered stable, and in which sea level would have slowly risen in the last few thousand years (∼–10 m at around 7,000 B.P., and ∼-2 m at around 3,600 B.P.). More recently, new stratigraphic and paleosol evidence led Connally (1984) and Sirkin (1984) to suggest that this region registers an alternation of submersions and emersions (with amplitudes of relative vertical motions reaching several tens of meters) in the last 15,000 yr. The latter interpretation seems to be based on unreliable and insufficient data, which led to inaccurate reconstructions of successive sea-level positions.

GEOCHRONOLOGY OF THE PLEISTOCENE SHORELINES AND NEOTECTONIC IMPLICATIONS

Late Pleistocene shorelines

Along the Sonoran coast, the identification of the IS 5e shoreline is primarily based on morphostratigraphic evidence and lateral correlation. As found in numerous coastal regions of the world, the last Pleistocene transgression, during which the sea appears to have stood several meters above the present level, may be correlated with the last interglaciation maximum (IS 5e). The fact that in Sonora all the marine deposits that predate the most recent Pleistocene coastal sequence are significantly more altered, indicates that the latter sequence cannot be correlated with a late, last interglacial, high sea-level stand (IS 5a or 5c), and thus suggests that it corresponds to the IS 5e. Besides, the assignment of the well-preserved and most recent (Pleistocene) deposits to the IS 5e is supported, to some extent, by aminostratigraphic and radiometric data.

Measurements of aIle/Ile ratios of *Chione* and *Dosinia* shells of distinct Quaternary ages, collected in the Gulf of California (Ortlieb, 1982a, 1984c, 1987), established that the late Quaternary temperatures have been so high that only late Pleistocene fossils yield ratios below racemic equilibrium. In other words, unracemized *Dosinia* and *Chione* shells may be assigned a younger age than middle Pleistocene. It should be noted that this does not preclude the possibility that the late Pleistocene fossils be racemized because of local (warmer) diagenetic conditions. Aminostratigraphic analyses performed on Sonoran material (Table 2) thus proved useful to separate late Pleistocene from middle Pleistocene samples.

The Th/U dating method applied to mollusk shells is known to usually provide minimum ages. The series of 11 measurements made on six shells from a single locality (Campo Dólar, LP 134, Table 1) illustrates that the true age is most probably the oldest apparent age, and that the scatter of numerical ages may reach as wide as 60,000 yr. The fact that coexisting shells from a given locality show the same scatter of apparent ages of the whole set of (radiochemically reliable) data from the northeastern coast of the Gulf of California supports the interpretation that the individual radiometric results should not be taken at face value, and, on the contrary, all the analyzed shells may well be coeval with the single IS 5e high sea-level stand.

As a corollary, it may be added that the U-series geochronological method applied to mollusks would not be an efficient means to distinguish IS 5e from 5c (or 5a) samples. The only places where the neotectonic setting would have made possible the emersion of IS 5c (or 5a) deposits are the Mesa de Sonora and Tepopa areas.

Middle Pleistocene shorelines

When middle Pleistocene shorelines are not preserved as emerged features along a coastal region (where late Pleistocene marine sediments do crop out above MSL), it may be due either to the fact that the coast is subsident or the erosion has destroyed (or sedimentation has hidden) the middle Pleistocene interglacial shorelines. In Sonora, several outcrops of pre–late Pleistocene marine sediments have been observed from the northernmost gulf southward to Tiburon Island (and possibly to the southeasternmost gulf).

In most cases, the middle Pleistocene deposits crop out below the late Pleistocene units, thus at a very low elevation; reconstructions of paleo–sea level based on paleobathymetric interpretations indicate that the corresponding high sea-level stands were at elevations of the order of +10 to +20 m. The locality of Playa Santa Maria, where the highest elevated shoreline reaches +23 m, may have been uplifted by a few meters at the end of the middle Pleistocene and before the late Quaternary. The scarce middle Pleistocene remains are taken as indications that the eastern coast of the Gulf of California, as a whole, has not been subsiding in the last few hundred thousand years. If, during the middle Pleistocene interglaciations, sea level reached a position close to the present MSL, it is inferred that some uplift may have occurred in several places (if not all the Sonoran coastline) prior to the late Pleistocene.

Neotectonic conclusions

In spite of the poorly preserved condition of early and middle Pleistocene shorelines, the vertical distribution of Quaternary marine deposits indicates that the recent vertical displacements have been of limited amplitude along most of the eastern coast of the Gulf of California. In this chapter, it has been stressed that the

Figure 9. Sections of Quaternary coastal deposits around Tiburon Island (a to i) and in the Infiernillo Straits–Bahia Kino–Tastiota area (j to m). See locations on Figure 8 (a to k) and Figure 10 (k to m). Section a: 1. Mesozoic granitic substrate. 2. Early late Pleistocene nearshore sequence (LQ 159). 3. Late Quaternary slope deposits. Section b: 1. Early late Pleistocene very fossiliferous sublittoral sands (LQ 149). 2. Late Quaternary bajada deposits. Section c: 1. Early late Pleistocene nearshore sequence. 2. Late Quaternary sand dune. Section d: 1. Tertiary volcanic substrate. 2. Early late Pleistocene nearshore sequence (LQ 141). Section e: 1. See d1. 2. Poorly fossiliferous, lagoonal, fine-grained sandstone (middle Pleistocene). 3. Early late Pleistocene nearshore sequence. 4. Late Quaternary alluvium. Section f: 1. See d1. 2. Early Pleistocene fossiliferous nearshore sequence (LQ 175). 3. See e4. Section g: 1. See d1. 2. Early late Pleistocene condensed nearshore sequence. 3. Late Quaternary slope deposits. Section h: 1. See d1. 2. Lagoonal siltstones and sandstones (middle Pleistocene). 3. Middle Quaternary slope deposits. 4. Early late Pleistocene shingle ridge. Section i: 1. See d1. 2. Middle(?) Pleistocene nearshore sequence. 3. Consolidated eolianite (middle Pleistocene?). Section j: 1. Early late Pleistocene fossiliferous sublittoral and nearshore sequence (LU 1075). 2. Late Quaternary slope deposits. Section k: 1. Mesozoic granitic substrate. 2. Early late Pleistocene nearshore sequence. 3. Thick late Quaternary bajada sequence. Section l: 1. Mesozoic granitic substrate. 2. Early late Pleistocene nearshore sequence (LP 28). 3. Late Pleistocene eolianites. 4. Late Quaternary alluvium. Section m: 1. Miocene volcanoclastic substrate. 2. Early late Pleistocene nearshore sequence. 3. Late Pleistocene alluvium. 4. Late Quaternary eolian sands.

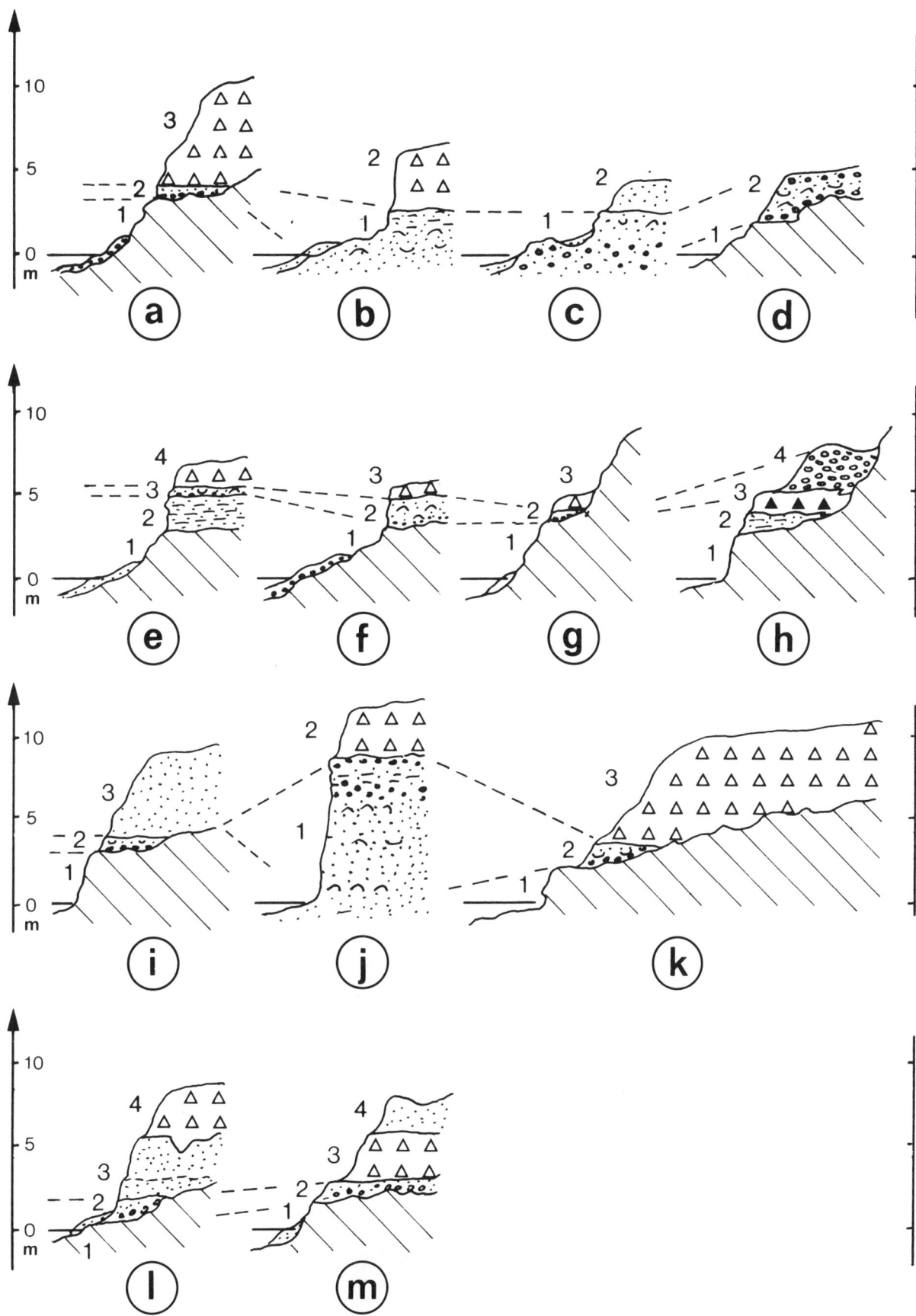

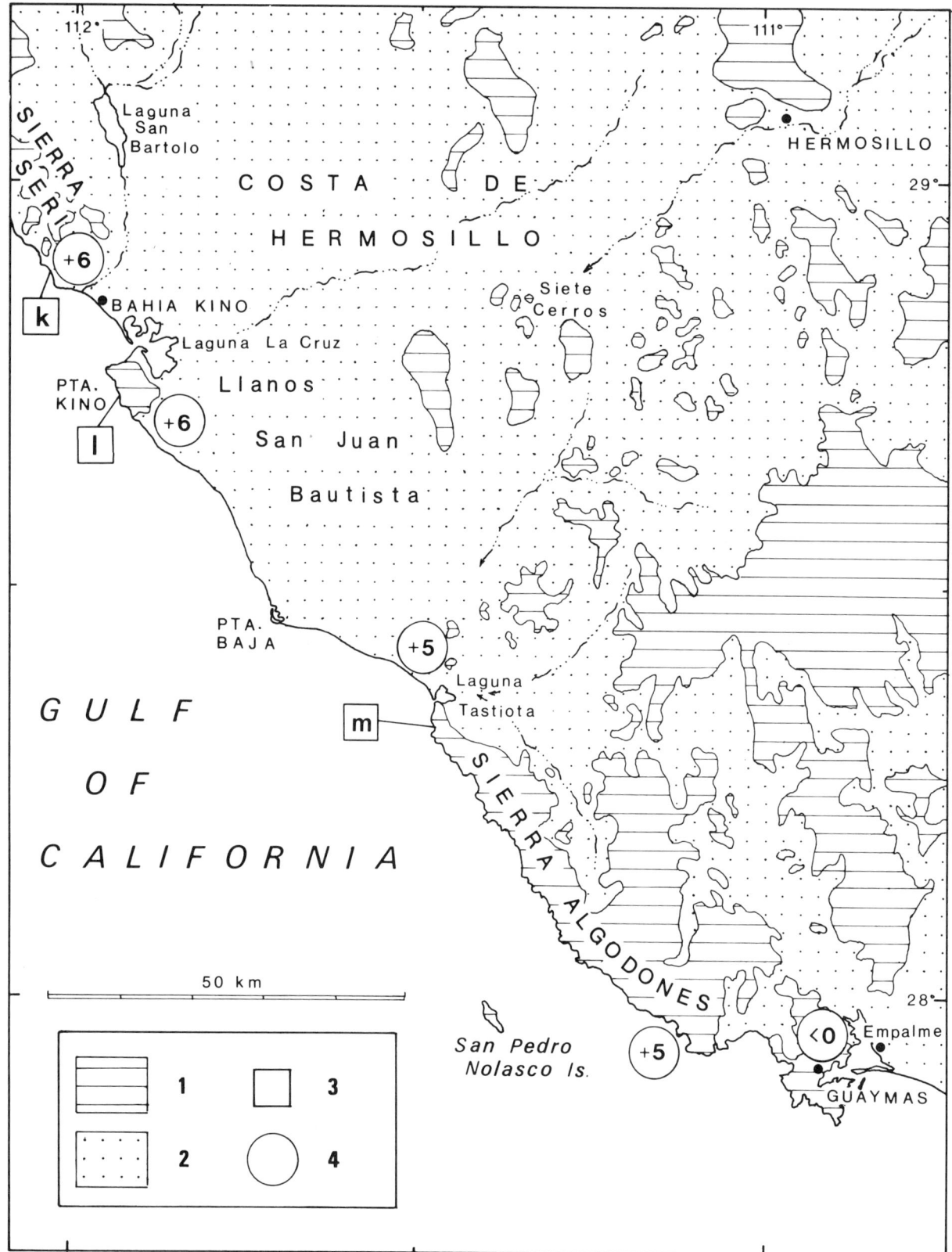

Figure 10. Bahía Kino–Guaymas region: distribution and height of the early late Pleistocene shoreline. 1. Pre-Quaternary units. 2. Quaternary alluvium and eolian sands. 3. Location of the sections of Quaternary coastal deposits (k to m) shown in Figure 9. 4. Maximum elevation (meters above MSL) of the early late Pleistocene (IS 5e) shoreline.

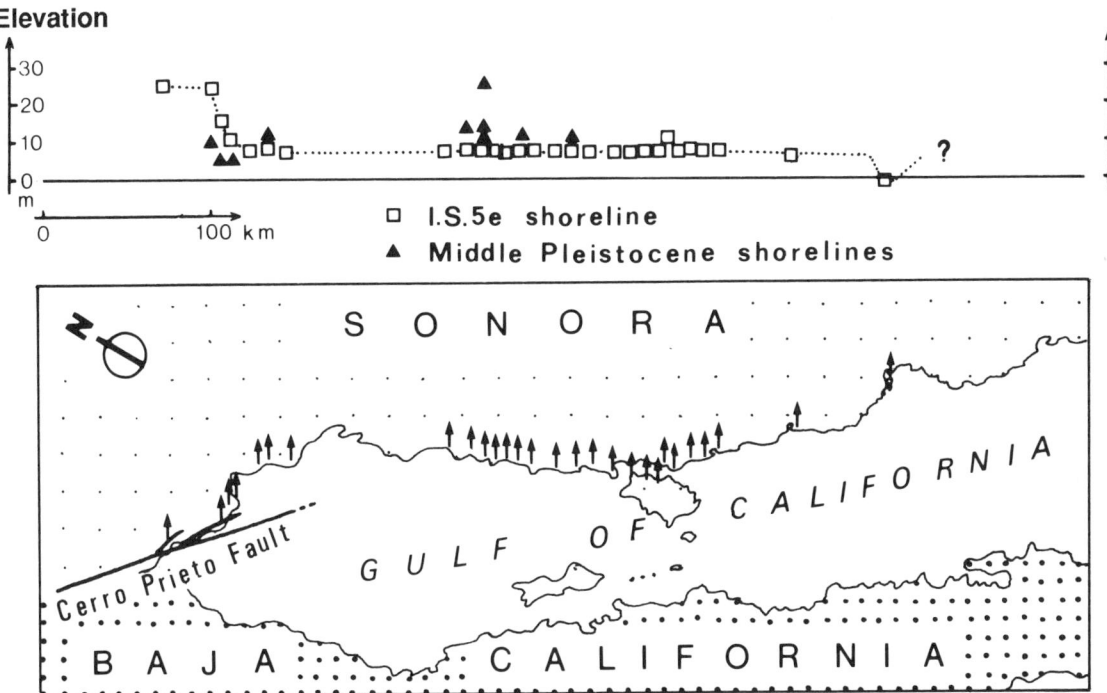

Figure 11. Height distribution of late (IS 5e) and middle Pleistocene shorelines along the Sonoran coast. The even elevation of the IS 5e shoreline along most of this coastal segment indicates that the northeastern Gulf of California did not register vertical motion in the late Quaternary.

IS 5e shoreline is observed, or can be reconstructed, at a nearly constant elevation of $\geq +6$ (± 1) m between Bahia Adair and Guaymas (Fig. 11). Assuming that this $\sim +6$-m elevation is very close to its original "eustatic" position, it may be inferred that during the late Quaternary (i.e., the last 125,000 yr), vertical deformations have been practically nonexistent, or of less than 1 to 2 m (mean uplift rates of 8 to 16 mm/10^3 yr).

The eastern bank of the Rio Colorado mouth is the only area that underwent important vertical displacements (maximum rate of 170 mm/10^3 yr). This displacement is interpreted to be the result of transpressive folding directly related to the Cerro Prieto fault activity. The uplifted area is limited to the southwestern edge of the Mesa de Sonora. On the southwestern extremity of the Rio Colorado delta and in northern Bahia Adair, the IS 5e shoreline is very close to its "eustatic" position.

Another kind of recent deformation, of much more limited amplitude (a few meters?) than along the Mesa de Sonora, has been described in the Canal del Infiernillo, and corresponds to a late Quaternary reactivation of one of the numerous, NW-SE–oriented, late Cenozoic faults that control the structure of the Sonoran coast.

In conclusion, vertical displacements along the northeastern coast of the Gulf of California appear to have been very limited during the entire Quaternary, and particularly in the late Quaternary. The only segment of the coast that had important vertical deformations associated with strike-slip along the Cerro Prieto fault is located on the eastern side of the Rio Colorado delta (Fig. 11).

A comparison of the neotectonic comportment of both sides of the Gulf of California (Ortlieb, 1990) shows that since the Pliocene the Sonoran coast has been significantly less uplifted than the peninsula of Baja California.

ACKNOWLEDGMENTS

This work was prepared within the framework of the GEOCORTEZ program, which associated the Institut Français de Recherche Scientifique pour le Développement en Coopération (ORSTOM, UR 106), and the Instituto de Geología, Universidad Nacional Autónoma de México. The author benefited from the facilities of the Estación Regional del Noroeste del Instituto de Geología UNAM, Hermosillo (Sonora), from 1975 to 1981, and expresses his gratitude to his colleagues and to the staff of this regional office. Many thanks are also due to D. Cordoba and J. Guerrero, successive Directors of the Instituto de Geología, and to his collaborators in the field, particularly V. Malpica-Cruz, A. Castro del Río, S. Celis-Gutiérrez, R. Gonzales-Gonzalez, and B. Luna-Guerin. He also acknowledges the benefit of many discussions and several field trips with B. Colletta.

The U-series analyses were performed by M. Bernat and C. Gaven (Causse) (ORSTOM–CNRS–Université de Nice), and by

O. Carro (GEOTOP, Université du Québec à Montréal). The amino acid analyses were done by G. Miller (INSTAAR, University of Colorado, Boulder) and by R. Mitterer (University of Texas, Austin). The radiocarbon analyses were made in the Laboratoire de Géologie Dynamique (Université Paris VI), the Laboratoire d'Hydrologie et de Géochimie isotopique (Université Paris XI), and at Queens College (New York).

C. Hillaire-Marcel and H. Faure (Laboratoire de Geologie du Quaternaire, CNRS, Marseilles) provided appreciated comments on an earlier version of this work. G. Miller (University of Colorado) and J. F. Schreiber (University of Arizona) kindly reviewed the chapter, and otherwise helped with editorial comments.

REFERENCES CITED

Aharon, P., 1983, 140,000 yr isotope climatic record from raised coral reefs in New Guinea: Nature, v. 304, p. 720–723.

Anderson, C. A., 1950, 1940 E. W. SCRIPPS cruise to the Gulf of California; Part 1, Geology of the islands and neighboring land areas: Geological Society of America Memoir 43, 53 p.

Arnal, R. E., 1961, Limnology, sedimentation, and micro-organisms of the Salton Sea, California: Geological Society of America Bulletin, v. 72, p. 427–478.

Beal, C. H., 1948, Reconnaissance of the geology and oil possibilities of Baja California, Mexico: Geological Society of America Memoir 31, 138 p.

Bernat, M., Gaven, C., and Ortlieb, L., 1980, Datation de dépôts littoraux du dernier interglaciaire (Sangamon) sur la côte orientale du Golfe de Californie, Mexique: Société Géologique de France, Bulletin, ser. 7, t. 22, no. 2, p. 219–224.

Bischoff, J. L., and Henyey, T. L., 1974, Tectonics elements of the central part of the Gulf of California: Geological Society of America Bulletin, v. 85, p. 1893–1904.

Blake, W. P., 1854, Ancient lake in the Colorado desert: American Journal of Science, v. 17, p. 435–438.

Bloom, A. L., Broecker, W. S., Chappell, J.M.A., Matthews, R. K., and Mesolella, K. J., 1974, Quaternary sea level fluctuations on a tectonic coast; New ^{230}Th/^{234}U dates from the Huon Peninsula, New Guinea: Quaternary Research, v. 4, p. 185–205.

Bradley, W. C., and Griggs, G. B., 1976, Form genesis and deformation of central California wavecut platforms: Geological Society of America Bulletin, v. 87, p. 22–40.

Broecker, W. S., 1963, A preliminary evaluation of U-series disequilibrium as a tool for absolute age measurements of marine carbonates: Journal of Geophysical Research, v. 68, p. 2817–2834.

Broecker, W. S., Thurber, D. L., Goddard, J., Ku, T. L., Matthews, R. K., Mesolella, K. J., 1968, Milankovitch hypothesis supported by precise dating of coral reefs and deep-sea sediments: Science, v. 159, p. 297–300.

Celis-Gutiérrez, S., 1975, Estudio microfaunístico y paleoecológico de una sección estratigráfica, entre Punta Onah y Punta Tormenta, Sonora, México [Profession thesis in biology]: Mexico, Facultad de Ciencias, Universidad Nacional Autónoma de México, 47 p.

—— , 1979, Les foraminifères quaternaires des anciennes lignes de rivage de la côte de Sonora et de Basse Californie, Mexique [Speciality doctoral thesis]: Paris, Université P. et M. Curie, 110 p.

—— , 1980, Estudio microfauníistico de una sección del Pleistoceno superior en la región de Puerto Libertad, Sonora: Universidad Nacional Autónoma de México, Instituto de Geología, Revista, v. 4, p. 76–81.

Celis-Gutierrez, S., and Malpica-Cruz, V., 1981, Interpretación paleoambiental y diagenética de sedimentos marinos del Pleistoceno tardío al noroeste de Puerto Libertad, Sonora, México: Geological Society of America Abstracts with Programs, v. 12, p. 48–49.

Chappell, J. M. A., 1974, Geology of coral terraces, Huon Peninsula, New Guinea; A study of Quaternary tectonic movements and sea level changes: Geological Society of America Bulletin, v. 85, p. 553–570.

Chappell, J. M. A., and Shackleton, N. J., 1986, Oxygen isotopes and sea level: Nature, v. 324, p. 137–140.

Chavez, E., 1975, Fauna de invertebrados de Bahía Kino, Sonora [Professional thesis in biology]: México, Facultad de Ciencias, Universidad Nacional Autónoma de México, 52 p.

Colletta, B., and Ortlieb, L., 1979, Neotectonic evolution of the northernmost coastal area of the Gulf of California, Mexico: Geological Society of America Abstracts with Programs, v. 11, p. 403–404.

—— , 1981, La actividad tectónica cuaternaria en le extremidad meridional del sistema de San Andrés, in Memoria del Simposio sobre asentamientos humanos en la Falla de San Andrés, Tijuana, 1979: México, Secretaría de Asentamientos Humanos y Obras públicas, p. 79–90.

—— , 1984, Deformations of middle and late Pleistocene deltaic deposits at the mouth of the Rio Colorado, northwestern Gulf of California, in Malpica-Cruz, V., Celis-Gutiérrez, S., Guerrero-García, J., and Ortlieb, L., eds., Neotectonics and sea level variations in the Gulf of California area; A symposium, Hermosillo, México, Universidad Nacional Autónoma de México, Instituto de Geología, p. 31–53.

Connally, G. G., 1984, Soil stratigraphy and inferred tectonic history of the West Mexican Coastal Plain, in Malpica-Cruz, V., Celis-Gutiérrez, S., Guerrero-García, J., and Ortlieb, L., eds., Neotectonics and sea level variations in the Gulf of California area; A symposium, Hermosillo: Mexico, Universidad Nacional Autónoma de México, Instituto de Geología, p. 55–73.

Curray, J. R., and Emmel, F. J., and Crampton, P. J., 1969, Holocene history of a strand plain lagoonal coast, Nayarit, Mexico, in Lagunas costeras, un simposio, México, 1967: Mexico, Universidad Nacional Autónoma de México, p. 63–100.

Dibblee, T. W., 1954, Geology of the Imperial Valley region, California, in Jahns, R., ed., Geology of Southern California: California Division of Mines Geology Bulletin, v. 170, p. 21–28.

Durham, J. W., 1950, 1940 E.W. SCRIPPS cruise to the Gulf of California: Part 2, Megascopic paleontology and marine stratigraphy: Geological Society of America Memoir 20, 68 p.

Eberly, L. D., and Stanley, T. B., 1978, Cenozoic stratigraphy and geologic history of southwestern Arizona: Geological Society of America Bulletin, v. 89, p. 921–940.

Elders, W. A., Rex, R. W., Meidav, T., Robinson, P. T., and Biehler, S., 1972, Crustal spreading in Southern California: Science, v. 178, p. 15–24.

Fuis, G. F., Mooney, W. D., Healy, J. H., Lutter, W. J., and McMechan, G. A., 1982, Crustal structure of the Imperial Valley region, in The Imperial Valley, California, earthquake of October 15, 1979: U.S. Geological Survey Professional Paper 1247, p. 25–49.

Gastil, R. G., and Krummenacher, D., 1974, Reconnaissance geologic map of coastal Sonora between Puerto Lobos and Bahia Kino: Geological Society of America Map and Charts Series MC-16, scale 1:150,000.

—— , 1977, Reconnaissance geology of coastal Sonora between Puerto Lobos and Bahia Kino: Geological Society of America Bulletin, v. 88, p. 189–198.

Gastil, R. G., and Ortlieb, L., 1981, Geology of coastal Sonora between Puerto Lobos and Bahia Kino, roadlog for a field trip, in Ortlieb, L., and Roldan, J., eds., Geology of northwestern Sonora and southern Arizona, fieldguides and papers: Hermosillo, Universidad Nacional Autónoma de México, Instituto de Geología, p. 52–72.

Gilmore, T. D., 1985, Episodic regional uplift in southeastern California preceding major historic earthquakes in the Imperial Valley (abs.): EOS Transactions of the American Geophysical Union, v. 66, p. 856.

—— , 1986, Historic vertical displacements in the Salton trough and adjacent parts of southeastern California: U.S. Geological Survey Open-File Report 86–380, 177 p.

Gilmore, T. D., and Castle, R. O., 1983, Tectonic preservation of the divide between Salton basin and the Gulf of California: Geology, v. 11, p. 474–477.

Giresse, P., and Davies, O., 1980, High sea levels during the last glaciation; One of the most puzzling problems of sea levels studies: Quaternaria, v. 22, p. 211–236.

González-González, R. M., 1982, Comparación sedimentológica y faunística (*Mollusca, Gastropoda*) de diferentes paleoambientes costeros del Pleistoceno superior de la región Tepopa, Sonora [Professional thesis in biology]: México, Facultad de Ciencias, Universidad Nacional Autónoma de México, 174 p.

Gonzalez-Gonzalez, R. M., Luna-Guerin, B., and Ortlieb, L., 1984, Análisis faunístico (*Bivalvia* and *Gastropoda*) de depósitos litorales del Pleistoceno tardío en diversas localidades del área de Tepopa, Sonora, *in* Malpica-Cruz, V., Celis-Gutiérrez, S., Guerrero-García, J., and Ortlieb, L., eds., Neotectonic and sea level variations in the Gulf of California area; A symposium, Hermosillo: Mexico, Universidad Nacional Autónoma de México, Instituto de Geología, p. 125–137.

Gorsline, D. S., 1967, Sedimentologic studies of the Colorado delta: University of Southern California, Report USC-Geol. 67-1, submitted to Geogr. Branch, Office of Naval Research, 89 p.

Gutiérrez-Estrada, M., and Castro del Río, A., 1984, Terrazas marinas pleistocénicas en el área de Mazatlán, Sinaloa, México, *in* Abstracts Symposium on Neotectonics and Sea Level Variations in the Gulf of California Area, Hermosillo: México, Universidad Nacional Autónoma de México, Instituto de Geología, Abstracts, p. 27.

Harrison, J. C., and Mathur, S. P., 1964, Gravity anomalies in the Gulf of California, *in* van Andel, H., and Shor, G. G., eds., Marine geology of the Gulf of California: American Association of Petroleum Geologists Memoir 3, p. 76–89.

Hertlein, L. G., and Emerson, W. K., 1956, Marine Pleistocene invertebrates from near Puerto Peñasco, Sonora, Mexico: San Diego Society of Natural History Transactions, v. 12, p. 154–176.

Hudnut, K., Beavan, J., and Bilham, R., 1985, Salton sea level data; Active transpression on the southern San Andreas fault [abs.]: EOS Transactions of the American Geophysical Union, v. 66, p. 383.

Ives, R. L., 1951, High sea levels of the Sonora shores: American Journal of Science, v. 249, p. 215–223.

——, 1959, Shell dunes of the Sonoran shore: American Journal of Science, v. 257, p. 449–457.

——, 1964, The Pinacate region, Sonora, Mexico: California Academy of Science Occasional Papers 47, 43 p.

——, 1971, An archeological sterile area in northern Sonora: Kiva, v. 36, p. 1–10.

Johnson, N. M., and 5 others, 1983, Rates of late Cenozoic tectonism in the Vallecito–Fish Creek basin, western Imperial Valley, California: Geology, v. 11, p. 664–667.

Jones, P. L., 1975, Petrology and petrography of beachrock (Pleistocene?), Sonora coast, northern Gulf of California [M.S. thesis]: Tucson, University of Arizona, 47 p.

Kaufman, A., Broecker, W. S., Ku, T. L., and Thurber, D. L., 1971, The status of U-series methods of mollusks dating: Geochimica et Cosmochimica Acta, v. 35, p. 1155–1183.

Kennedy, G. L., Lajoie, K. R., and Wehmiller, J. F., 1982, Aminostratigraphy and faunal correlations of late Quaternary marine terraces, Pacific coast, U.S.A.: Nature, v. 299, p. 545–547.

Kern, J. P., 1977, Origin and history of Upper Pleistocene marine terraces, San Diego: Geological Society of America Bulletin, v. 88, p. 1553–1566.

Kniffen, F. B., 1932, The natural landscape of the Colorado delta: University of California Publications in Geography, v. 5, p. 149–244.

Ku, T. L., 1976, The Uranium-series methods of age determination: Annual Review of Earth and Planetary Sciences, v. 4, p. 347–379.

Lajoie, K. R., and 7 others, 1979, Quaternary shorelines and crustal deformations, San Diego to Santa Barbara, California, *in* Abbott, P. L., ed., Geological excursions in Southern California: San Diego, California, San Diego State University, p. 3–15.

Lancin, M., 1979, Géomorphologie des littoraux du Canal de l'Infiernillo et du Canal Ballenas-Salsipuedes, Golfe de Californie, Mexique [Speciality doctoral thesis]: Paris, Université Paris I, 184 p.

——, 1985, Geomorfología y génesis de las flechas litorales del Canal del Infiernillo, Estado de Sonora: Universidad Nacional Autónoma de México, Instituto de Geología, Revista, v. 6, p. 52–72.

Lecolle, J., 1980, Morphologie, formation et aspects sédimentologiques de la Lagune Tepoca (Sonora), Mexique: Cahiers ORSTOM, sér. Géologie, v. XI-2, p. 205–224.

Lecolle, J., and Ortlieb, L., 1978, Etude préliminaire de l'évolution paléogéographique au Quaternaire supérieur de la Laguna Tepoca, Golfe de Californie, Mexique: *in* Abstracts 10th International Congress of Sedimentology, Jerusalem, Abstracts, v. 1, p. 272–273.

Luna-Guerin, B., 1981, Comparación sedimentológica y faunística (phylum *Mollusca,* clase *Pelecypoda*) en diferentes paleoambientes costeros del Pleistoceno Superior en la región de Tepopa, Sonora [Thesis manuscript]: 202 p.

Maloney, N. J., 1986, Late Holocene uplift of Bat Cave buttes, Salton trough, California: Geological Society of America Abstracts with Programs, v. 18, p. 114.

Malpica, V., Ortlieb, L., and Castro del Rio, A., 1978, Transgresiones cuaternarias de la costa de Sonora: Mexico: Universidad Nacional Autónoma de México, Instituto de Geología, Revista, v. 2, p. 90–97.

McGee, W. J., and Johnson, W. D., 1896, Seriland: National Geographic Magazine, v. 8, p. 125–133.

McLaughlin, R. J., Lajoie, K. R., Sorg, D. H., Morrinson, S. D., and Wolf, J. A., 1983, Tectonic uplift of a middle Wisconsin marine platform near Mendocino triple junction, California: Geology, v. 11, p. 35–39.

Meckel, L. D., 1975, Holocene sand bodies in the Colorado delta area, northern Gulf of California, *in* Broussard, M. L., ed., Deltas, models for exploration: Houston Geological Society, p. 239–265.

Merriam, R. H., 1965, San Jacinto fault in northwestern Sonora, Mexico: Geological Society of America Bulletin, v. 76, p. 1051–1054.

Moore, D. G., 1973, Plate-edge deformation and crustal growth, Gulf of California structural province: Geological Society of America Bulletin, v. 84, p. 1883–1905.

Mörner, N. A., 1971, The position of the ocean sea level during the interstadial at about 30,000 BP; A discussion from a climatic-glaciologic point of view: Canadian Journal of Earth Sciences, v. 8, p. 132–143.

Nichols, M. M., 1965, Composition and environment of recent transitional sediments on the Sonoran coast, Mexico [Ph.D. thesis]: Los Angeles, University of California, 401 p.

Ortlieb, L., 1978, Relative vertical movements along the Gulf of California, Mexico, during the late Quaternary: Geological Society of America Abstracts with Programs, v. 10, p. 466.

——, 1979, Terrasses marines dans le nord-ouest mexicain, étude au long d'une transversale entre la côte Pacifique et le Sonora en passant par la péninsule de Basse California, *in* Suguio, K., Fairchild, T. R., Martin, L., and Flexor, J. M., eds., Proceedings, International Symposium on coastal evolution in the Quaternary, Sao Paulo: Instituto Astronomico e Geofisico, Universidade de São Paulo, p. 453–474.

——, 1980, Neotectonics from marine terraces along the Gulf of California, *in* Mörner, N.-A., ed., Earth rheology, isostasy, and eustasy: New York, John Wiley and Sons, p. 479–504.

——, 1981, Recent investigations on Quaternary geology of the coast of central Sonora, Mexico, *in* Ortlieb, L., and Roldan, J., eds., Geology of northwestern Sonora and southern Arizona, Fieldguides and Papers: Hermosillo, Universidad Nacional Autónoma de México, Instituto de Geología, p. 137–149.

——, 1982a, Geochronology of Pleistocene marine terraces in the Gulf of California region, northwestern Mexico, *in* Abstracts, 11th International Union for Quaternary Research Congress, Moscow, v. 2, p. 229.

——, 1982b, La ligne de rivage du dernier interglaciaire autour de la péninsule de Basse Californie (Mexique); Reconnaissance générale et implications néotectoniques: Cahiers ORSTOM, sér. Géologie, v. 12, p. 103–115.

——, 1984a, Fieldtrip Guidebook prepared for the Symposium on Neotectonics and Sea Level Variations in the Gulf of California Area: Mexico, Universidad Nacional Autónoma de México, Instituto de Geología, 152 p.

——, 1984b, Pleistocene high stands of sea level and vertical movements in the Gulf of California area, in Abstracts Symposium on Neotectonics and Sea Level Variations in the Gulf of California Area, Hermosillo: Mexico, Universidad Nacional Autónoma de México, Instituto de Geología, p. 131–132.

——, 1984c, Radiometric and aminoacid dating of late Pleistocene fossils in the Gulf of California area, Mexico; Available results and problems of interpretation, in Abstracts Symposium on Neotectonics and Sea Level Variations in the Gulf of California Area, Hermosillo: Mexico, Universidad Nacional Autónoma de México, Instituto de Geología, p. 133–134.

——, 1987, Néotectonique et variations du niveau marin au Quaternaire dans la région du Golfe de Californie, Mexique [Doct. Sc. thesis]: Marseilles, Université Aix-Marseille II, 2 volumes, 779 p. + 442 p. [Published in Etudes et Thèses ORSTOM, Paris, 2 vol., 779 p. + 257 + 4 microforms].

——, 1990, Quaternary vertical movements along the coasts of Baja California and Sonora, in Dauphin, J. P., and Simoneit, B.R.T., eds., The Gulf and Peninsular Province of the Californias: Tulsa, American Association of Petroleum Geologists Memoir 47.

Ortlieb, L., and Malpica-Cruz, V., 1978, Reconnaissance des depôts pléistocènes marins autour du Golfe de Californie, Mexique: Cahiers ORSTOM, sér. Géologie, v. 10, p. 177–190.

Ortlieb, L., and Triclot, M. P., 1984, Etude préliminaire sur la structure et la composition géochimique de test de *Dosinia ponderosa* (Gray, 1838) actuels et fossiles du Golfe de Californie, Mexique, in Malpica-Cruz, V., Celis-Gutiérrez, S., Guerrero-García, J., and Ortlieb, L., eds., Neotectonics and Sea Level Variations in the Gulf of California Area; A symposium, Hermosillo: Mexico, Universidad Nacional Autónoma de México, Instituto de Geología, p. 269–294.

Ortlieb, L., and 5 others, 1989, Geodetic and tectonic analyses along an active plate boundary; The central Gulf of California: Tectonics, v. 8, p. 429–441.

Palmer, L. A., 1967, Marine terraces of California, Oregon, and Washington [Ph.D. thesis]: Los Angeles, University of California, 379 p.

Richards, H. G., 1973, A Quaternary elevated beach along the Gulf of California in Sonora, Mexico, in Abstracts, 9th International Union for Quaternary Research Congress: Christchurch, New Zealand, p. 306.

Rose, M. W., 1975, Sedimentology of Estero La Cholla, northwest coast of Sonora, Mexico [M.S. thesis]: Tucson, University of Arizona, 99 p.

Sanchez, O., Ness, G. E., Couch, R. W., and Dauphin, J. P., 1985, Present-day faulting and major structural blocks of the northern Gulf of California [abs.]: EOS Transactions of the American Geophysical Union, v. 66, p. 844.

Sandusky, C. L., 1969, Sedimentology of Estereo Marúa, Sonora, Mexico [M.S. thesis]: Tucson, University of Arizona, 84 p.

Shackleton, N. J., and Opdyke, N. D., 1973, Oxygen isotope and paleomagnetic stratigraphy of equatorial Pacific core V28–238; Oxygen isotope temperatures and ice volumes on a 10^5 and 10^6 year time scale: Quaternary Research, v. 3, p. 39–55.

——, 1976, Oxygen isotope and paleomagnetic stratigraphy of Pacific core V28–239; Late Pliocene to latest Pleistocene, in Cline, R. M., and Hays, J. D., eds., Investigation of late Quaternary paleoceanography and paleoclimatology: Geological Society of America Memoir 145, p. 449–464.

Sharp, R. V., 1982, Tectonic setting of the Imperial Valley region, in The Imperial Valley, California, earthquake of October 1979: U.S. Geological Survey Professional Paper 1254, p. 5–14.

——, 1986, Holocene sedimentation in the Imperial Valley, California: Geological Society of America Abstracts with Programs, v. 18, p. 183.

Sherwin, R. W., 1971, Coastal landforms and vegetation associations of the Straits of Infiernillo region, Sonora, Mexico [M.S. thesis]: Tucson, University of Arizona, 78 p.

Sirkin, L., 1984, Late Pleistocene stratigraphy and environments of the West Mexican Coastal Plain, in Malpica-Cruz, V., Celis-Gutiérrez, S., Guerrero-García, J., and Ortlieb, L., eds., Neotectonics and Sea Level Variations in the Gulf of California Area; A Symposium, Hermosillo: Mexico, Universidad Nacional Autónoma de México, Instituto de Geología, p. 309–318.

Smith, J. T., and 6 others, 1985, Fossil and K-Ar age constraints on upper middle Miocene conglomerate, SW Isla Tiburon: Geological Society of America Abstracts with Programs, v. 17, p. 409.

Stanley, G. M., 1962, Deformation of Pleistocene Lake Cahuilla shoreline, Salton basin, California: Geological Society of America Special Paper 87, p. 165.

Stearns, C. E., 1976, Estimates of the position of sea level between 140,000 and 75,000 years ago: Quaternary Research, v. 6, p. 445–449.

——, 1980, A molluscan revival?; Actes du Colloque Niveaux marine et tectonique quaternaires dans l'aire méditerranéenne: Paris, Centre National de la Recherche Scientifique and Université Paris I, p. 15–26.

Stump, T. E., 1975, Pleistocene molluscan paleoecology and community structure of the Puerto Libertad region, Sonora, Mexico: Palaeogeography, Palaeoclimatology, Palaeoecology, v. 17, p. 177–226.

——, 1981, Some Pleistocene facies and faunas of coastal Sonora and Tiburon Island, Mexico, in Ortlieb, L., and Roldan, J., eds., Geology of northwestern Sonora and southern Arizona Fieldguides and Papers: Hermosillo, Universidad Nacional Autónoma de México, Instituto de Geología, p. 125–131.

Tarbet, L. A., 1941, Imperial Valley, California, in Possible future petroleum provinces of North America: American Association of Petroleum Geologists Bulletin, v. 35, p. 260–263.

Thom, B. G., 1973, The dilemna of high sea levels during the last glaciation: Progress in Geography, v. 5, p. 170–246.

Thomas, R. G., 1963, The late Pleistocene 150-foot fresh water beach line of Salton Sea area: California Academy of Science Bulletin, part 1, p. 62, p. 9–17.

Thompson, R. W., 1968, Tidal flat sedimentation on the Colorado River delta: Geological Society of America Memoir 107, 133 p.

Trinidad-Reyes, R., and Rueda-Gaxiola, J., 1982, La historia sedimentaria y climática de la secuencia cortada por el pozo Extremeño 301 en el delta del Colorado, in Abstracts 6th Convención Geológica Mexicana, Mexico: Sociedad Geológica Mexicana, p. 109–110.

Van den Kamp, P. C., 1973, Holocene continental sedimentation in the Salton basin, California; A reconnaissance: Geological Society of America Bulletin, v. 84, p. 827–848.

Veeh, H. H., and Valentine, J. W., 1967, Radiometric ages of Pleistocene fossils from Cayucos, California: Geological Society of America Bulletin, v. 78, p. 545–550.

Viñas-Gomez, F., 1982, Estudio bioestratigráfico basado en nanoplanctón calcáreo de sedimentos calcáreos en el Mar de Cortez: in Abstracts 6th Convención Geológica Mexicana, Mexico: Sociedad Geológica Mexicana, p. 33.

——, 1984, Calcareous nannoplankton in the well Extremeño 301, in Abstracts Symposium on Neotectonics and Sea Level Variations in the Gulf of California Area, Hermosillo: Mexico, Universidad Nacional Autónoma de México, Instituto de Geología, p. 151.

Walker, T. R., and Thompson, R. W., 1968, Late Quaternary geology of the San Felipe area, Baja California, Mexico: Journal of Geology, v. 76, p. 479–485.

Waters, M. R., 1983, Late Holocene lacustrine chronology and archeology of ancient Lake Cahuilla, California: Quaternary Research, v. 19, p. 373–387.

Wehmiller, J. F., and 8 others, 1977, Correlation and chronology of Pacific coast marine terrace deposits of continental United States by fossil amino-stereochemistry; Technique evaluation, relative ages, kinetic model ages, and geologic implications: U.S. Geological Survey Open-File Report 77–680, 191 p.

MANUSCRIPT ACCEPTED BY THE SOCIETY APRIL 18, 1990

Printed in U.S.A.

Coal in Sonora, Mexico

Luis Obregon-Andria and Francisco Arriaga-Arredondo
Comision Federal de Electricidad, Estudios Carboniferos, Piedras Negras, Coahuila, Mexico

ABSTRACT

Coal deposits in the state of Sonora are distributed mainly in two regions. One is the Cabullona region, in the northeast, where the coal is contained in the Cintura Formation of Early Cretaceous age; its physical and chemical characteristics place it in the hypo-bituminous rank. The other region is in La Barranca, east-central Sonora, where the coal is contained in the Santa Clara Formation of Triassic-Jurassic age; there the coal ranks as anthracite. This region has been divided into the San Marcial, San Javier, Santa Clara, and San Enrique areas.

The Cabullona region is currently in the regional exploration stage, whereas in the La Barranca region, although there has been previous exploitation, exploration has not been completed. Therefore, the coal potential of Sonora has not yet been fully determined.

INTRODUCTION

Coal is one of the nonrenewable natural resources of Sonora that occurs in the Cabullona area and La Barranca region (Fig. 1).

The first documentation of the occurrence of coal in Sonora was in 1866 when Remond published a report on his geological studies in the Los Bronces area, in the Barranca region. Since this report, the region has been the object of sporadic geological exploration, along with some drilling activity and a few mining operations. From 1880 to 1911 the Sud-Pacific Company recovered coal from the San Marcial and Santa Clara fields. Later, from 1942 to 1955, these deposits were worked by the Compañia de Carbón Sonora, S.A. from Hermosillo.

More recent reports are those by Toron and Esteve (1946), Wilson and Rocha (1946), and Pesquera and Carbonel (1960). Wilson and Rocha (1946) studied the Santa Clara and San Javier coal fields for the Comite Directivo par las Investigaciones de los Recursos Minerales de México. That same year, Toron and Esteve (1946) did a study for the Banco de México. Presquera and Carbonel continued this work in the San Marcial coal field, the results of which were published by the Consejo de Recursos Minerales (C.R.M.). The same C.R.M. restarted exploration in 1980. The Comisión Federal de Electricidad did regional exploration in the La Barranca area from 1982 to 1984, and is currently studying the two areas that are the subject of this chapter.

There have been some geological studies in the Cabullona region since the early 1990s (Ransome, 1904; Taliaferro, 1933; Imlay, 1939; Rangin, 1977), but the publications related to coal are only the ones by Bustillos (1963) and Yza and others (1984). Salas (1976), Ojeda (1978), Verdugo and Arriaga (1984), and Pérez-Segura (1985) briefly discuss both regions.

The object of this chapter is to integrate all the existing geological information on coal, and to characterize and classify chemically and petrographically the coal in the two regions.

For this purpose, sampling was done in some outcrops in Cabullona, whereas in the La Barranca area, samples were taken from mine dumps. Both petrographic (reflectance measures and maceral analysis) and chemical analyses were performed on the Cabullona samples, whereas the La Barranca samples were studied only petrographically. Chemical analyses for the La Barranca samples were taken from Toron and Esteve (1946). Coal rank was determined on the basis of physical and chemical characteristics.

CABULLONA REGION

The Cabullona region lies within the Naco and Agua Prieta municipalities; it is geographically limited by the international border to the north and by the town of Fronteras to the south (Fig. 2).

The physiography is composed of low mountains, such as the Sierras Anibacachi, La Ceniza, Cabullona, El Caloso, and San José. The region lies within the northern portion of the Sierra Madre Occidental geomorphic province, in the parallel ranges

Obregon-Andria, L., and Arriaga-Arredondo, F., 1991, Coal in Sonora, Mexico, *in* Pérez-Segura, E., and Jacques-Ayala, C., eds., Studies of Sonoran geology: Geological Society of America Special Paper 254.

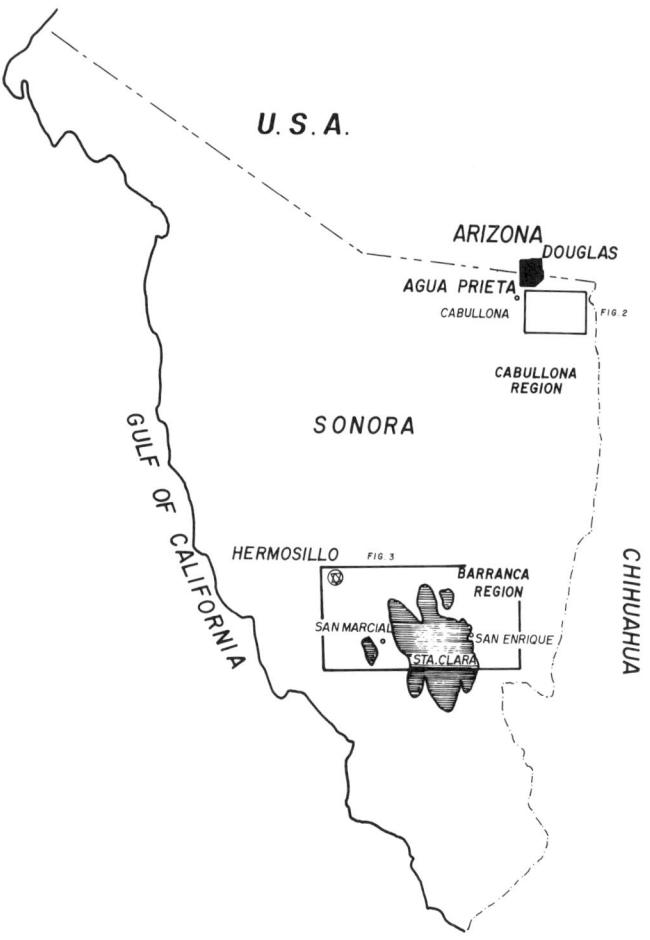

Figure 1. Location of the Cabullona and La Barranca regions, Sonora, Mexico.

and valleys subprovince (Alvarez, 1969). The average trend of the ranges is N10°W, and these are separated by large alluvial valleys. The ranges have abrupt escarpments, which are probably the result of faulting.

Cintura Formation (middle Albian)

The Cintura Formation is the uppermost unit of the Bisbee Group, and was first described by Ransome (1904) in the Bisbee area, southeastern Arizona. He gave the name "Cintura Formation" to a sequence of reddish feldspathic sandstone, siltstone, and shale beds 550 to 600 m thick.

The Cintura Formation is well exposed in the Sierra de San José; in the Cerro El Caloso, it is reduced in thickness, and it is missing in the Sierra de Anibacachi, probably because of erosion (Taliaferro, 1933). This discordance and erosion of the Cintura, according to Rangin (1977) testifies to a tectonic movement after the Bisbee Group was deposited, and before the Cabullona Group.

The thickness of the Cintura is approximately 800 m; the lower contact, with the Mural Formation, is transitional, and the upper one, with the Snake Ridge Formation, is discordant. The Cintura has been assigned a middle Albian age by Araujo and others (1984), using fossils found in the lower part of the unit.

The lower part of the Cintura was deposited in a shallow-marine environment. Toward the top it changes to fluvial–deltaic plains facies (Araujo and others, 1984).

The Cintura Formation was studied and sampled in the San Marcos coal field, located north of the Sierra El Caloso. The lower third of the unit consists of intercalations of coarse to fine-grained sandstone, siltstone, and shale. Color varies from purple to tan and greenish gray. The middle and upper thirds consist of greenish gray sandstone, shale, carbonaceous shale, and coal beds. The lower part of the unit is thin bedded, whereas the middle part has thin to thick beds. Although the development of cycles is not always complete, there is an appreciable lithologic gradation from coarse sandstone to siltstone to shale and coal beds.

Sampling and analysis

Geological reconnaissance and sample of the coal deposits in the San Marcos coal field was done. To locate the samples a section was measured, taking as a reference an easy-to-locate inclined shaft 15 m deep. The section was measured using a compass, tape, and distancemeter (Fig. 2). Fifteen carbonaceous beds ranging in thickness from 0.15 to 3.4 m were sampled. Of these, only 8 correspond to coal beds, and the rest to carbonaceous shale.

To find the chemical properties of coal, a proximate analysis that could determine the presence of certain coal components such as moisture, volatile matter, and ash was done (fixed carbon is estimated by subtraction). These analyses were complemented with the determination of sulfur, calorific value, and density.

A petrographic study was done to determine coal evolution, and to assist in the interpretation of depositional environment. The study consisted of determination of reflectance of vitrinite and maceral analysis. The natural coke samples were only analyzed for vitrinite reflectance.

Characterization and classification

Table 1 shows the results of the physical-chemical and petrographic analyses of the San Marcos coal field, where the following can be defined:

1. Volatile-matter values show a wide variation because samples were collected from outcrops that were weathered and consequently oxidized. The same is true for calorific values, which decrease and thus are not trustworthy as classification parameters.

2. Two samples corresponding to the same bed are classified as natural coke, based on their optical properties.

3. The rank of coal corresponds to hypo-bituminous, with vitrinite reflectance varying from 0.60 to 0.87, except for the samples mentioned in 2.

4. The type of coal corresponds to vitric, for it has a vitrinite percentage greater than 65. This type of coal is interpreted to be similar to that formed in other basins of the northern hemisphere.

5. The contents of ash lower than 30 percent and "mixtures" with ash values between 30 and 80 percent place the studied samples in the "coal facies."

Reserves

The Cabullona region is in the regional exploration stage. Most work has been restricted to the San Marcos field, although there are other locations with traces of coal that have not yet been prospected, such as the Sierra de San José and the Encino coal field, where three coal beds have been reported in the Cintura Formation. The coal beds have been described only in outcrops, so their continuity at depth is not known. Based on the geological

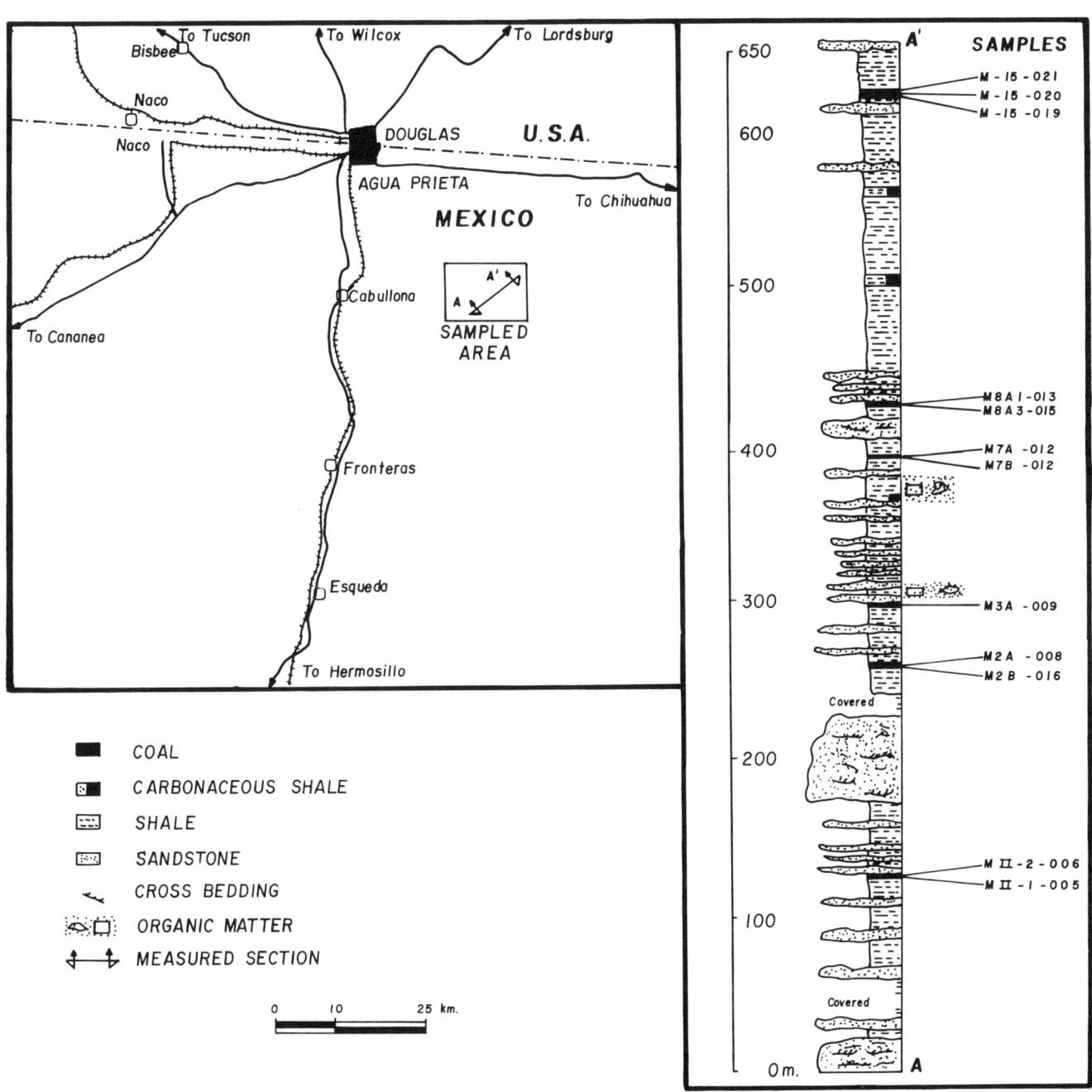

Figure 2. Location of the measured and sampled section of the Cintura Formation in the Cabullona area.

TABLE I PHYSICAL-CHEMICAL AND PETROGRAPHIC ANALYSIS OF THE SAN MARCOS COALFIELD (CABULLONA REGION)

| SAMPLES | THICKNESS (m.) | PROXIMATE ANALYSIS (1) | | | COMPLEMENTARY ANALYSIS | | | PETROGRAPHIC ANALYSIS | | | | MACERAL ANALYSIS | | | TRANSFORMED DATA | | CLASSIFICATION |
| | | | | | | | | REFLECTANCE | | | | | | | | | SOLID FOSSIL FUELS |
		MOISTURE	VOLATILE MATTER	FIXED CARBON	ASH	SULPHUR	DENSITY	CALORIFIC VALUE Btu/lb	MEASURES	RANDOM	STANDAR DEVIATION	VITRINITE	EXINITE	INERTINITE	VOLATILE MATTER (2)	CALORIFIC VALUE (3) Btu/lb	
MⅡ-1-005	0.15	5.05	24.41	27.94	42.60	2.48	1.88	6238	100	0.68	0.06	78.05	4.88	17.07	46.63	11430	HYPO BITUMINOUS, VITRIC, MIXTE
MⅡ-2-006	0.32	7.65	36.91	32.04	23.40	2.34	1.71	7589	100	0.66	0.05	69.77	5.09	25.14	53.53	10040	HYPO BITUMINOUS, VITRIC, COAL
M-2A-008	0.69	1.02	11.28	53.46	34.24	0.20	–	7171	67	2.34	0.17	–	–	–	17.42	11370	NATURAL COKE
M-3A-009	0.69	7.23	21.97	23.39	47.41	0.21	2.16	3094	100	0.87	0.08	97.32	0.89	1.79	48.43	6316	HYPO BITUMINOUS, VITRIC, MIXTE
M-7A-012	1.20	12.12	35.52	35.79	16.57	1.11	1.71	7761	100	0.60	0.05	86.93	6.00	7.07	49.81	9401	HYPO BITUMINOUS, VITRIC, COAL
M-8A1-013	0.25	7.19	23.91	23.79	45.11	0.23	2.11	4327	50	0.63	0.05	75.71	0.00	24.29	50.13	8418	HYPO BITUMINOUS, VITRIC, MIXTE
M-8A3-015	3.40	8.40	25.08	30.23	36.29	0.34	2.01	4905	75	0.63	0.05	78.95	14.03	7.02	45.34	8040	HYPO BITUMINOUS, VITRIC, MIXTE
M-2B-016	1.30	4.28	16.38	57.37	21.97	0.48	1.77	9511	100	1.67	0.09	–	–	–	22.21	12371	NATURAL COKE
M-7B-017	0.35	10.31	29.67	31.43	28.59	0.35	1.90	6037	100	0.67	0.07	93.83	0.00	6.17	48.56	8712	HYPO BITUMINOUS, VITRIC, COAL
M-15A-019	0.30	10.56	26.51	36.79	26.14	0.38	1.72	7092	81	0.73	0.07	93.44	0.00	6.56	41.88	9862	HYPO BITUMINOUS, VITRIC, COAL
M-15A-020	0.80	11.85	28.39	37.57	22.19	0.32	1.76	7201	100	0.87	0.06	82.90	0.00	17.10	43.04	9455	HYPO BITUMINOUS, VITRIC, COAL
M-15A-021	0.20	8.84	26.27	25.26	39.63	0.33	1.87	5706	100	0.78	0.07	76.79	20.98	2.23	50.98	9554	HYPO BITUMINOUS, VITRIC, MIXTE

(1) As received (in percent)
(2) Dry-ash free
(3) Moist-ash free

OBREGON AND ARRIAGA (1987)

work done in the San Marcos coal field, only hypothetical resources can be estimated. According to Verdugo and others (1984), the reserves amount to about 68 million tons.

LA BARRANCA REGION

La Barranca region has been divided in the San Marcial, San Javier, Santa Clara, and San Enrique coal fields. These are located in the Sierra Madre Occidental geomorphic province, within the Parallel Ranges and Valleys subprovince (Fig. 3). The San Marcial coal field is included in the Sonoran Desert Province of Alvarez (1969).

The stratigraphy of the region includes strata ranging in age from Paleozoic to Recent, comprising sedimentary, volcano-sedimentary, metamorphic, and igneous rocks. The coal occurs in rocks of Late Triassic–Early Jurassic age. These were originally studied by Dumble (1900), who defined it as Division Barranca, after which King (1939) assigned the name of Barranca Formation, and Alencaster (1961) raised it to group status, dividing it into three formations, defined from bottom to top as Arrayanes, Santa Clara, and Coyotes. The most important unit for coal prospecting is the Santa Clara Formaion, which contains anthracite.

The Arrayanes Formation consists of quartzitic sandstone layers with interbedded thin layers of shale and conglomerate. The type locality is 3 km northeast of Rancho La Barranca in the Arroyo Arrayanes. It overlies the Paleozoic basement on a surface of angular unconformity and grades upward into the Santa Clara Formation.

The approximate thickness in the Sierra de San Javier is 700 m. It forms some of the most remarkable topographic features in the area. It is assigned a post-Paleozoic and pre-Carnian age because the layers that overlie it are of Carnian age (Alencaster, 1961).

The Santa Clara Formation is the middle part of the Barranca Group; its type locality is in Estación Santa Clara. It consists of coarse to fine-grained sandstone, siltstone, shale, carbonaceous shale, and coal beds forming fining-upward cycles. Coal is not present in all the cycles. These sequences were deposited by meandering streams in a low-gradient coastal plain with abundant swamps (Cojan and Potter, this volume; Stewart and Roldán, this volume). The total thickness of the formation is approximately 370 m; it has been assigned a Carnian age based on marine invertebrate fossils (Alencaster, 1961). The lower and upper contacts are transitional with the Arrayanes and Coyotes Formations, respectively.

The thickness and number of coal beds are not well defined. In the San Marcial coal field, Pesquera and others 1960) delineated four beds with thicknesses that vary from 0.20 to 3 m. Wilson and Rocha (1946) distinguished seven beds for the Santa Clara coal field, where there is a 2.8-m bed in the Santa Clara coal mine. In the San Enrique coal fields, SEB and SEB1 drill holes cut ten beds, although it is not clear if these are different deposits or are repeated by faults.

The Coyotes Formation, with an approximate thickness of 650 m, consists of light gray, quartz-rich sandstone with interbedded shale and conglomerate. The conglomerates are made up of black and white chert gravel. The upper contact is an angular unconformity with the andesites and latites of the Cretaceous Tarahumara Formation. The Coyotes is assigned a post-Carnian and pre-Cretaceous age, based on its stratigraphic position.

The region has been affected by several tectonic events. The Barranca Group displays faults and several phases of igneous intrusion. The occurrence of intrusive bodies in the form of dikes and sills caused organic matter to evolve into anthracite, natural coke, and graphite. Vasallo (1985) interprets three deformational events to explain the evolution from coal into graphite: (1) the Nevadan event (Late Jurassic); (2) the Oregonian event (Albian–Cenomanian); and (3) the emplacement of the granitic batholith and/or younger plutons (Late Cretaceous–early Tertiary).

Sampling and analysis (Fig. 3)

Core samples from the San Marcial coal field (SMG1 drill hole) were obtained from the Consejo de Recursos Minerales. Other selected samples were collected from the dump of the Tiro El Salto, abandoned at present. Samples from the San Javier coal field were collected from the dump of the Palo Pinto mine. The Santa Clara coal field was sampled from the following mine dumps: Tiro Santa Clara; Santa Clara's inclined shaft; La Z mine shaft; and the Santa Clara, El Voladero, La Estación, La Calera, El Refugio, and El Yaqui mines. All these are caved in, or otherwise inaccessible for sampling beds in situ. In the San Enrique coal field, systematic sampling was done in outcrops of the Salto La Higuera and Ruinas del Rancho localities. Also, samples from cores SEB and SEB1 were kindly supplied by the Consejo de Recursos Minerales.

Results

Outcrop and dump samples were found to be too weathered, and therefore inadequate for chemical studies (oxidation is expressed as a variation of volatile matter, slight increase of moisture, and a decrease of calorific value). They were analyzed only petrographically. We used the chemical values from Pesquera and Carbonel (1960) for the San Marcial coal field, and from Torón and Esteve (1946) for the Santa Clara, San Javier, and San Enrique fields.

Physical-chemical (proximate analysis, sulfur, calorific value, and density) and petrographic (reflectance of vitrinite) analyses were performed on the core samples. Table 2 shows the results of the analyses of the Santa Clara, San Javier, and San Marcial samples. Interpretation of these data led to the following conclusions:

1. Volatile-matter content allows classification of all the coals as anthracite, with the exception of those from the El Voladero and El Tren mines, which fall into the semi-anthracite rank.

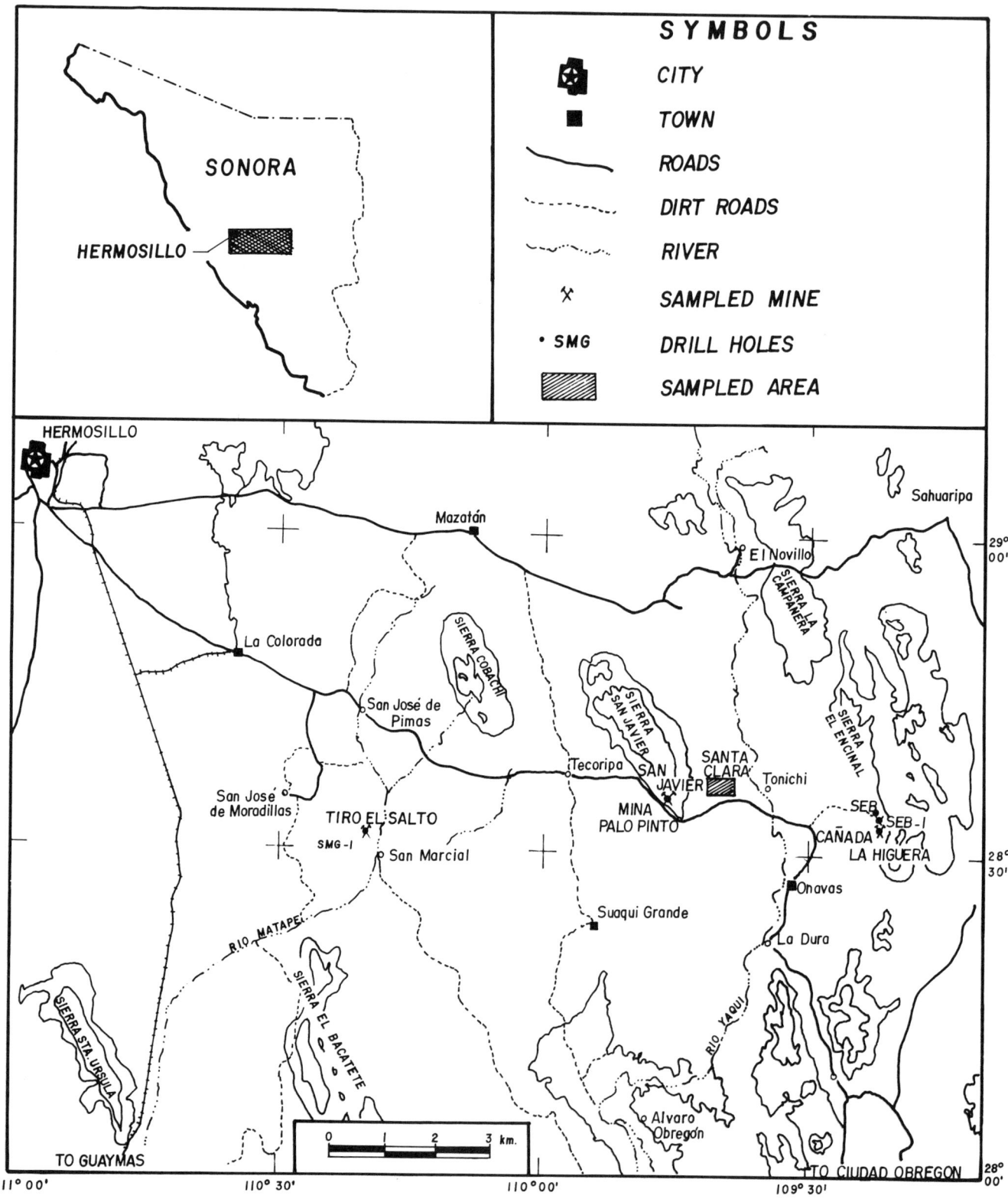

Figure 3. Location of samples from the La Barranca region.

TABLE II PHYSICAL CHEMICAL AND PETROGRAPHIC ANALYSIS

SAMPLES	PROXIMATE ANALYSIS [1]			ULTIMATE ANALYSIS [2]					CALORIFIC VALUE Btu/lb	FUSION ASH TEMP°C			DENSITY	REFLECTANCE			DATA USED TO CLASSIFY			ASTM COAL CLASSIFICATION
	MOISTURE 110°C	VOLATILE MATTER	FIXED CARBON	ASH	CARBON	HIDROGEN	OXIGEN	NITROGEN	SULPHUR		INITIAL DEFORMATION TEMPERATURE	SOFTENIG TEMPERATURE	FLUID TEMPERATURE		MEASURES	RANDOM	STANDAR. DEVIATION	VOLATILE MATTER [3]	CALORIFIC VALUE Btu/lb [2]	REFLECTANCE	
SANTA CLARA (LA BARRANCA)																					
TIRO SANTA CLARA	11.02	4.25	78.46	6.27	96.00	0.77	2.31	0.59	0.33	11836	1178	1212	1340	1.95	100	4.63	0.58	5.16	14421	4.63	ANTHRACITE
SANTA CLARA MINE	11.37	6.01	76.97	5.65	96.61	0.55	1.72	0.58	0.24	11906	1199	1232	1338	1.96	100	4.23	0.39	7.24	14455	4.23	ANTHRACITE
SANTA CLARA INCLINED SHAFT	10.26	3.21	80.30	6.23	94.90	0.88	3.39	0.59	0.24	11806	1160	1204	1354	1.96	100	4.40	0.29	3.84	14257	4.40	ANTHRACITE
EL VOLADERO MINE	7.56	7.63	80.38	4.43	92.72	0.67	5.72	0.45	0.44	11908	1192	1228	1349	1.95	100	4.56	0.45	8.67	13681	4.56	SEMIANTHRACITE
ESTACION MINE	8.18	4.45	75.51	11.86	95.56	0.93	2.61	0.50	0.40	11453	1185	1220	1340	1.96	100	4.73	0.50	5.56	13518	4.73	ANTHRACITE
LA Z MINE SHAFT	3.94	6.58	78.76	10.72	93.38	0.48	5.28	0.44	0.42	11543	1191	1224	1360	1.95	100	4.85	0.53	7.71	13714	4.85	ANTHRACITE
LA CALERA MINE	6.72	5.10	80.04	8.14	95.37	0.88	2.89	0.45	0.41	12133	1270	1298	1390	1.89	100	4.70	0.52	6.00	13773	4.70	ANTHRACITE
EL TREN MINE	6.80	6.88	74.99	11.33	93.91	0.48	3.82	0.53	1.26	11264	1277	1377	1449	1.91	100	5.34	0.82	8.42	13944	5.34	SEMIANTHRACITE
EL TREN MINE	1.82	7.88	76.05	14.25	93.39	0.54	4.54	0.48	1.05	11439	1280	1383	1458	2.10	100	4.64	0.43	9.34	13842	4.64	NATURAL COKE [4]
EL REFUGIO MINE	4.34	5.74	76.84	13.08	93.61	1.11	4.19	0.36	0.73	11602	1393	1504	1560	1.96	100	4.84	0.54	6.95	14207	4.84	ANTHRACITE
EL YAQUI MINE	3.00	5.20	73.00	18.80	95.78	0.98	1.70	0.77	0.77	11323	1110	1170	1266	1.99	100	4.54	0.37	6.65	14715	4.54	NATURAL COKE [4]
SAN JAVIER																					
PALO PINTE MINE	5.62	6.35	78.09	9.94	94.27	0.73	4.19	0.36	0.45	11323	1230	1343	1415	1.96	100	5.00	0.74	7.52	14715	5.00	ANTHRACITE
SAN MARCIAL																					
CORE DRILL SMG-I DEPTH 76.50-78.50	0.50	5.08	65.83	28.59	–	–	–	–	1.04	9648	–	–	–	–	100	3.57	0.54	7.16	13950	3.57	ANTHRACITE
CORE DRILL SMG-I DEPTH 90.00-91.40	1.39	4.06	56.31	38.24	–	–	–	–	0.12	8128	–	–	–	2.10	–	–	–	6.72	13833	–	ANTHRACITE
TIRO EL SALTO NIVEL 40	3.45	4.28	80.35	11.92	95.89	1.03	2.17	0.23	0.68	12056	1148	–	–	1.90	50	3.54	0.49	5.06	13825	3.54	ANTHRACITE
TIRO EL SALTO NIVEL 105	1.88	3.87	77.51	16.74	96.03	0.89	2.18	0.19	0.71	11343	1216	–	–	1.90	50	4.86	0.54	4.75	13834	4.86	ANTHRACITE

(1) As received (in percent)
(2) Moist-ash free
(3) Dry-ash free
(4) Classified petrographically

TORON AND ESTEVE (1946) MODIFIED BY OBREGON AND ARRIAGA (1987)

TABLE III PHYSICAL-CHEMICAL AND PETROGRAPHIC ANALYSIS OF THE SAN ENRIQUE COALFIELD (LA BARRANCA)

SAMPLES	PROXIMATE ANALYSIS (1)				COMPLEMENTARY ANALYSIS			PETROGRAPHIC ANALYSIS			DATE USED TO CLASSIFY			ASTM COAL CLASSIFICATION
	MOISTURE	VOLATILE MATTER	FIXED CARBON	ASH	SULPHUR	DENSITY	CALORIFIC VALUE Btu/lb	MEASURES	RANDOM	STANDARD DEVIATION	VOLATILE MATTER (2)	CALORIFIC VALUE Btu/lb (3)	REFLECTANCE	
OUTCROP GALERIA SALTO LA HIGUERA	7.10	5.50	73.76	13.64	0.67	1.89	11293	100	4.37	0.65	6.94	14434	4.37	ANTHRACITE
OUTCROP RUINAS DEL RANCHO	8.39	8.60	70.64	12.37	0.65	1.97	10553	100	4.88	0.79	10.85	13532	4.88	SEMIANTHRACITE
CORE DRILL SEB DEPTH 76.00 – 77.10 m.	2.40	4.14	76.91	16.55	0.66	1.75	10701	–	–	–	5.11	13015	–	ANTHRACITE
CORE DRILL SEB DEPTH 77.30 – 78.60 m.	1.54	6.54	63.57	28.35	0.66	1.89	8145	–	–	–	9.33	11714	–	SEMIANTHRACITE
CORE DRILL SEB DEPTH 87.50 – 89.35 m.	1.87	6.94	70.79	20.40	0.66	1.83	1027	100	4.46	0.73	8.93	13168	4.46	SEMIANTHRACITE
CORE DRILL SEB DEPTH 93.60 – 93.80 m.	1.90	2.77	67.34	27.99	0.66	1.73	8658	–	–	–	3.95	12387	–	ANTHRACITE
CORE DRILL SEB DEPTH 139.15 – 141.90 m.	1.67	5.43	74.25	18.65	0.29	1.83	10094	–	–	–	6.81	12632	–	ANTHRACITE
CORE DRILL SEB DEPTH 152.35 – 153.43 m.	1.12	7.64	69.00	22.24	0.66	–	9757	–	–	–	9.97	12825	–	SEMIANTHRACITE
CORE DRILL SEB DEPTH 153.75 – 155.80 m.	1.80	3.78	79.24	15.18	0.40	1.90	10627	–	–	–	4.55	12700	–	ANTHRACITE
CORE DRILL SEB DEPTH 176.00 – 177.06 m.	1.74	5.16	56.01	37.09	0.54	1.95	6931	–	–	–	8.43	11539	–	SEMIANTHRACITE
CORE DRILL SEB DEPTH 192.15 – 192.45 m.	1.92	5.63	70.84	21.61	0.66	–	9351	–	–	–	7.36	12177	–	ANTHRACITE
CORE DRILL SEB DEPTH 224.00 – 226.50 m.	1.43	5.28	77.85	15.44	0.59	1.85	10744	100	4.72	0.65	6.35	12880	4.72	ANTHRACITE
CORE DRILL SEB-1 DEPTH 63.13 – 66.35 m.	2.44	4.53	60.62	32.41	0.66	1.93	8179	100	4.01	0.43	6.95	12562	4.01	ANTHRACITE
CORE DRILL SEB-1 DEPTH 72.00 – 73.50 m.	3.29	3.47	51.65	41.59	0.16	2.16	6033	–	–	–	6.29	10945	–	ANTHRACITE
CORE DRILL SEB-1 DEPTH 79.00 – 82.50 m.	1.76	11.40	73.08	13.76	0.66	1.87	10720	–	–	–	13.49	12574	–	SEMIANTHRACITE
CORE DRILL SEB-1 DEPTH 85.56 – 86.90 m.	2.47	6.64	65.62	25.27	3.60	2.11	9721	100	3.90	0.54	9.19	13287	3.90	SEMIANTHRACITE
CORE DRILL SEB-1 DEPTH 144.20 – 147.52 m.	2.05	7.58	63.37	27.00	0.66	–	8690	105	4.20	0.48	10.68	12245	4.20	SEMIANTHRACITE
CORE DRILL SEB-1 DEPTH 152.05 – 153.74 m.	2.12	7.50	74.28	16.10	0.66	–	10119	–	–	–	9.17	12231	–	SEMIANTHRACITE
CORE DRILL SEB-1 DEPTH 164.50 – 164.80 m.	1.80	3.26	80.55	14.39	0.66	1.96	11271	100	3.90	0.73	3.89	13332	3.90	ANTHRACITE
CORE DRILL SEB-1 DEPTH 187.00 – 187.25 m.	1.48	7.16	76.73	14.63	0.66	–	10288	–	–	–	8.53	12200	–	SEMIANTHRACITE
CORE DRILL SEB-1 DEPTH 255.17 – 260.00 m.	1.85	4.80	56.42	36.93	2.88	2.18	7189	–	–	–	7.84	11840	–	ANTHRACITE
CORE DRILL SEB-1 DEPTH 260.93 – 262.20 m.	1.23	3.61	70.38	24.78	0.66	–	9412	–	–	–	4.88	12832	–	ANTHRACITE

(1) As received (in percent)
(2) Dry-ash free
(3) Moist-ash free

OBREGON AND ARRIAGA (1987)

2. The quality of the studied samples is very good, having low contents of ash and high calorific values; only the core samples from the San Marcial coal field have high ash contents.

3. The content of hydrogen is less than one percent in almost all of the samples, defining them as meta-anthracites.

4. The reflectance of vitrinite classifies these coals, according to the Solid Fossils Fuel Classification System (Alper, 1981), as meta-anthracite; that is, reflectance values are higher than 4, which corresponds to the classification obtained with hydrogen values.

5. The natural coke samples were classified on the basis of their optical properties, because the chemistry showed no difference with the anthracite values.

Table 3 shows the results of the chemical, physical, and petrographic analysis of the San Enrique coal field from which the following observations are possible:

1. The sampling was done in 1980, and the drilling campaign between 1976 and 1979; therefore, volatile-matter values present variations possibly due to weathering of the core, which leads to classification of some samples as anthracite and the others as semi-anthracite.

2. Ash-content values of the mine samples vary considerably. These values range from 12.37 percent from the Ruinas del Rancho outcrop to 41.59 percent from SEB1 core sample, which has a depth of 72.00 and 73.50 m.

3. The results of vitrinite reflectance place the coal in the meta-anthracite, rank with the exception of two samples (reflectance 3.90) that fall into the meso-anthracite category.

Reserves

In the La Barranca region the economic potential is unknown because exploration has been restricted to the areas where coal beds are exposed, although several authors have reported that the La Barranca Group is distributed in an area of 30,000 km^2. In this region the mines are at present caved-in and are inaccessible for a complete evaluation. No drilling has been done in the Santa Clara and San Javier coal fields. Therefore, the existing evaluation for the La Barranca does not represent the potential of the region. The amount of reserves shown here was taken from Pesquera and Carbonel (1960) for the San Marcial coal field, and from Toron and Esteve (1946) for the Santa Clara, San Javier, and San Enrique coal fields (Table 4).

CONCLUSIONS

Coal deposits in the state of Sonora are geographically distributed in two regions: Cabullona in the northeastern part, and La Barranca in the east-central part.

Coal from the Cabullona region is found in the Cintura Formation of Early Cretaceous age, consisting of eight horizons with thicknesses that vary from 0.15 to 3.40 m. Of these, seven correspond to hypo-bituminous coal with an average vitrinite reflectance of 0.71, and one is a coke because it has been locally metamorphosed.

Coal from the La Barranca region is located in the Santa Clara Formation of the Barranca Group of Late Triassic–Early Jurassic age. Four areas have been studied and mined: the San Marcial, the San Javier, the Santa Clara, and the San Enrique coal fields. Coal rank varies between semi-anthracite and anthracite, with few horizons of natural coke because of contact methamorphism.

TABLE IV. COAL RESERVES FROM LA BARRANCA REGION

COAL FIELD	CATEGORY OF RESERVES		
	MEASURED	INFERRED	POSSIBLE
SAN MARCIAL*	4'000,000	9'000,000	18'000,000
STA. CLARA**	4'936,297	2'156,255	1'501,295
SAN JAVIER**	3'402,821	3'920,756	3'001,778
SAN ENRIQUE**	754,427	1'062,176	1'608,638
TOTAL	13'093,545	16'139,187	24'111,711

*Pesquera and Carbonel (1960).
**Toron and Esteve (1946).

Coal from the Cabullona region, which has the characteristics of hypo-bituminous rank, has an average calorific value of 9,776 BTU/lb. On the other hand, coal from the La Barranca region, which rans as anthracite, shows an average calorific value of 13,374 BUT/lb, and the low ash content indicates that it does not require processing for industrial use, falling in the category of steam coal (Fig. 4).

Taking into consideration that the currently known areas

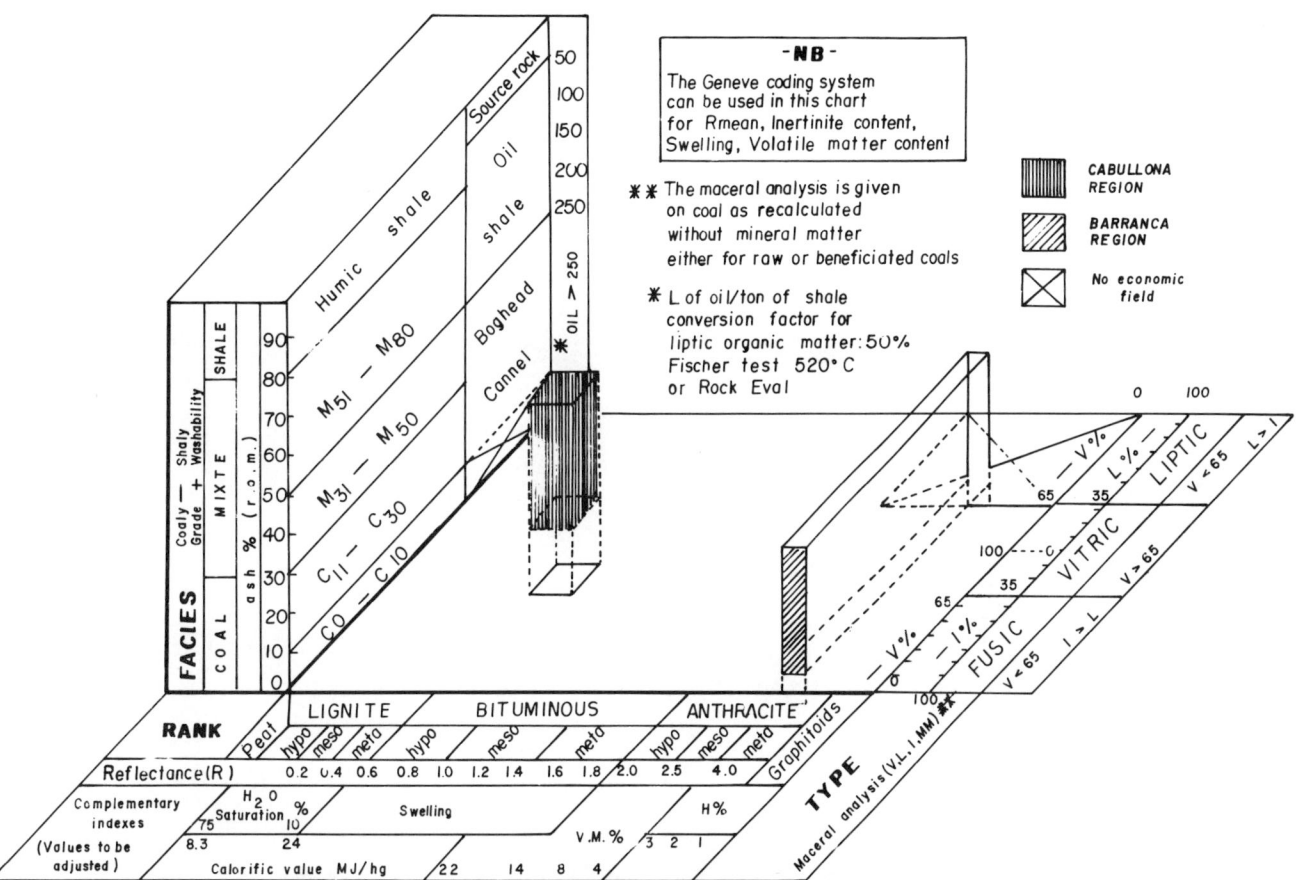

Figure 4. Coal classification of Sonora.

with presence of coal represent only a minimum portion in the La Barranca, as well as in the Cabullona regions, exploration and evaluation of the total area can considerably increase the coal reserves.

REFERENCES CITED

Alencaster, G., 1961, Paleontología del Triásico Superior de Sonora: Univ. Nac. Autón. de México, Inst. Geol., Paleontología Mexicana No. 11, p. 18–83.

——, 1981, Pour une classification synthétique universalle des combustibles solides: Bull. Cen. Rech. Explor. Prod. Elf-Aquitaine 5, p. 171–290.

Alvarez, M., Jr., 1969, Apuntes de Geologí de México, Facultad de Ingeniería de la UNAM.

Araujo, J., and others, 1984, Estudio Estratigráfico-Sedimentológico del Jurásico-Cretácico -Prospecto Cucurpe, Sonora: Instituto Mexicana del Petróleo Proyecto C-1151 (unpublished report).

Bustillos, A. G., 1963, Depósitos de carbón en la Sierra de Cabullona, Municipio de Agua Prieta, Son. Consejo de Recursos Minerales, Residencia Hermosillo, Hermosillo, Sonora (unpublished report).

Dumble, E. T., 1900, Triassic coal and coke in Sonora, Mexico: Geological Society of America Bulletin, v. 11, p. 10–14.

Imlay, W. R., 1939, Paleogeographic studies in northeastern Sonora, Mexico: Geological Society of America Bulletin, v. 50, p. 1723–1744.

King, E. R., 1939, Geological reconnaissance in northeastern Sierra Madre Occidental of Mexico: Geological Society of America Bulletin, v. 50, p. 1625–1722.

Ojeda, R. J., 1978, Main coal regions of Mexico: Geological Society of America Special Paper 17, p. 73–74.

Pérez-Segura, E., 1985, Carta Metalogenética de Sonora: Gobierno del Estado de Sonora -Univerisdad de Sonora, Publication No. 5, 64 p.

Pesquera-V, R., and Carbonel-C, M., 1960, Geología y exploración de los depósitos de carbón de la región de San Marcial, Estado de Sonora: Consejo de Recursos Naturales No. Renovables, México, Boletín no. 59, 41 p.

Rangin, C., 1977, Tectónicas sobrepuestas en Sonora septentrional: Univ. Nac. Autón. de México, Instituto de Geología, Boletín, v. 1, no. 1, p. 44–47.

Ransome, F. L., 1904, The geology and ore deposits of the Bisbee Quadrangle, Arizona, U.S.A.: U.S. Geological Survey Professional Paper, 168 p.

Remond, A., 1866, Notice of geological explorations in northern Mexico: Proceedings National Academy of Sciences, v. 3, p. 244–257.

Salas, G. P., 1976, Reservas y exploración por carbón en México: Geomimet, no. 83, p. 35–55.

Taliaferro, N. L., 1933, An occurrence of Upper Cretaceous sediments in Sonora, México: Journal of Geology, v. 41, p. 12–37.

Toron, L., and Esteve, A., 1946, La cuenca carbonífera del Yaqui: Banco de México, Investigaciones Industriales, 200 p.

Vasallo, L., 1985, Sobre la evolución geológica de la parte central del Estado de Sonora, México, y su relación con los depósitos de grafito: III Congreso Latinoamericano de Paleontología, México, Memoria, p. 87–100.

Verdugo, F., and Arriaga, F., 1984, Exploración de carbón en México: Symposium Latinoamericano del Carbón, Memorias, p. 881–922.

Wilson, I. F., and Rocha-V, S. 1946, Los yacimientos de carbón de la región de Santa Clara, municipio de San Javier, estado de Sonora. Com. Dir. Inv. Rec. Min. México. Bol. 9, 108 p.

Yza-D., R., Alcántara, J., and Silva-M., N., 1984 Exploración carbonífera en el Estado de Sonora: Sociedad Geológica Mexicana, Boletín, Tomo XLV, p. 17–40.

Manuscript Accepted by the Society April 18, 1990